WITHDRAWN
UTSA LIBRARIES

RENEWALS 458-4574

DATE DUE			
APR 13			

GAYLORD / PRINTED IN U.S.A.

Optical Transmission Systems Engineering

For a listing of recent titles in the *Artech House Optoelectronics Library,* turn to the back of this book.

Optical Transmission Systems Engineering

Milorad Cvijetic

Artech House, Inc.
Boston • London
www.artechhouse.com

Library of Congress Cataloging-in-Publication Data
A catalog record for this book is available from the U.S. Library of Congress.

British Library Cataloguing in Publication Data
Cvijetic, Milorad
  Optical transmission systems engineering—(Artech House optoelectronics library)
  1. Optical communications
  I. Title
  621.3'827

  ISBN 1-58053-636-0

Cover design by Igor Valdman

© 2004 ARTECH HOUSE, INC.
685 Canton Street
Norwood, MA 02062

All rights reserved. Printed and bound in the United States of America. No part of this book may be reproduced or utilized in any form or by any means, electronic or mechanical, including photocopying, recording, or by any information storage and retrieval system, without permission in writing from the publisher.
  All terms mentioned in this book that are known to be trademarks or service marks have been appropriately capitalized. Artech House cannot attest to the accuracy of this information. Use of a term in this book should not be regarded as affecting the validity of any trademark or service mark.

International Standard Book Number: 1-58053-636-0

10 9 8 7 6 5 4 3 2 1

Contents

Preface — ix
Acknowledgments — xi

CHAPTER 1
Introduction to Optical Bandwidth and Lightwave Paths — 1

1.1	Optical Transmission and Networking	1
1.2	Optical Transmission System Definition	3
1.3	Organization of the Book	11
1.4	Summary	12
	References	13

CHAPTER 2
Optical Components as Constituents of Lightwave Paths — 15

2.1	Semiconductor Light Sources		17
	2.1.1	Light Emitting Diodes	17
	2.1.2	Semiconductor Lasers	18
	2.1.3	Wavelength Selectable Lasers	22
2.2	Optical Modulators		24
	2.2.1	Direct Optical Modulation	24
	2.2.2	External Optical Modulation	25
2.3	Optical Fibers		28
	2.3.1	Single-Mode Optical Fibers	31
	2.3.2	Optical Fiber Manufacturing and Cabling	35
2.4	Optical Amplifiers		39
2.5	Photodiodes		45
2.6	Key Optical Components		47
	2.6.1	Optical Couplers, Isolators, Variable Optical Attenuators, and Optical Circulators	48
	2.6.2	Optical Switches	52
	2.6.3	Optical Filters	53
	2.6.4	Optical Multiplexers and Demultiplexers	59
2.7	Summary		61
	References		61

CHAPTER 3
Optical Signal, Noise, and Impairments Parameters — 65

- 3.1 Optical Signal Parameters — 65
  - 3.1.1 Output Signal Power — 66
  - 3.1.2 The Extinction Ratio — 67
  - 3.1.3 Optical Amplifier Gain — 67
  - 3.1.4 Photodiode Responsivity — 69
- 3.2 Noise Parameters — 70
  - 3.2.1 Mode Partition Noise — 73
  - 3.2.2 Laser Intensity and Phase Noise — 74
  - 3.2.3 Modal Noise — 75
  - 3.2.4 Quantum Shot Noise — 76
  - 3.2.5 Dark Current Noise — 79
  - 3.2.6 Thermal Noise — 79
  - 3.2.7 Spontaneous Emission Noise — 81
  - 3.2.8 Noise Beat Components — 83
  - 3.2.9 Crosstalk Noise Components — 84
- 3.3 Signal Impairments — 86
  - 3.3.1 Fiber Attenuation — 86
  - 3.3.2 Insertion Losses — 87
  - 3.3.3 Frequency Chirp — 87
  - 3.3.4 Chromatic Dispersion — 89
  - 3.3.5 Polarization Mode Dispersion — 103
  - 3.3.6 Self-Phase Modulation — 108
  - 3.3.7 Cross-Phase Modulation — 115
  - 3.3.8 Four-Wave Mixing — 116
  - 3.3.9 Stimulated Raman Scattering — 119
  - 3.3.10 Stimulated Brillouin Scattering — 122
- 3.4 Summary — 125
- References — 125

CHAPTER 4
Assessment of the Optical Transmission Limitations and Penalties — 127

- 4.1 Attenuation Impact — 128
- 4.2 Noise Impact — 129
- 4.3 Modal Dispersion Impact — 132
- 4.4 Polarization Mode Dispersion Impact — 136
- 4.5 Impact of Nonlinear Effects — 140
- 4.6 Summary — 147
- References — 147

CHAPTER 5
Optical Transmission Systems Engineering — 149

- 5.1 Transmission Quality Definition — 149
- 5.2 Receiver Sensitivity Handling — 154
  - 5.2.1 Receiver Sensitivity Defined by Shot Noise and Thermal Noise — 155
  - 5.2.2 Receiver Sensitivity Defined by Optical Preamplifier — 158

		5.2.3	Optical Signal-to-Noise Ratio	159
5.3	Power Penalty Handling			160
	5.3.1	Power Penalty Due to Extinction Ratio		162
	5.3.2	Power Penalty Due to Intensity Noise		163
	5.3.3	Power Penalty Due to Timing Jitter		166
	5.3.4	Power Penalty Due to Signal Crosstalk		167
	5.3.5	Comparative Review of Power Penalties		169
	5.3.6	Handling of Accumulation Effects		172
5.4	Systems Engineering and Margin Allocation			175
	5.4.1	Systems Engineering of Power-Budget Limited Point-to-Point Lightwave Systems		177
	5.4.2	Systems Engineering of Bandwidth-Limited Point-to-Point Lightwave Systems		179
	5.4.3	Systems Engineering for High-Speed Optical Transmission Systems		182
	5.4.4	Optical Performance Monitoring		187
	5.4.5	Computer-Based Modeling and Systems Engineering		189
5.5	Summary			193
	References			194

CHAPTER 6
Optical Transmission Enabling Technologies and Trade-offs — 195

6.1	Enabling Technologies		195
	6.1.1	Optical Amplifiers	196
	6.1.2	Advanced Dispersion Compensation	204
	6.1.3	Advanced Modulation Schemes	220
	6.1.4	Advanced Detection Schemes	230
	6.1.5	Forward Error Correction	233
	6.1.6	Wavelength Conversion and Optical 3R	237
6.2	Transmission System Engineering Trade-offs		243
	6.2.1	Optical Fiber Type Selection	244
	6.2.2	Spectral Efficiency	245
	6.2.3	Chromatic Dispersion Management	247
	6.2.4	Optical Power Level	248
	6.2.5	Optical Path Length	249
6.3	Summary		250
	References		251

CHAPTER 7
Optical Transmission Systems Engineering Toolbox — 253

7.1	Physical Quantities, Units, and Constants Used in This Book	253
7.2	Electromagnetic Field and the Wave Equation	254
7.3	The Propagation Equation for Single-Mode Optical Fiber	257
7.4	Frequency and Wavelength of the Optical Signal	260
7.5	Stimulated Emission of Light	260
7.6	Semiconductors as Basic Materials for Lasers and Photodiodes	262
7.7	Laser Rate Equations	267

7.8 Modulation of an Optical Signal 269
7.9 Optical Receiver Transfer Function and Signal Equalization 269
7.10 Summary 271
References 271

List of Acronyms 273

About the Author 277

Index 279

Preface

Optical transmission systems deliver information between two distinct physical locations through optical fiber, while achieving a specified system performance. System performance is dependent on a number of parameters related to the transmission signal and conditions and effects that might occur during signal generation, propagation, amplification, and detection. Optical transmission systems engineering can be understood as a process in which these parameters are combined in the most favorable way possible with respect to a defined performance goal.

This book is aimed at enabling the reader to attain a broader perspective of the parameters involved in the transmission of optical signals, gain insight into the systems engineering process, and discuss potential trade-offs between different system parameters and transmission system optimization. It is structured to provide straightforward guidance to readers looking to capture systems engineering fundamentals and gain practical knowledge that can be easily applied. As such, it provides an understanding of parameters and processes involved in overall engineering considerations and establishes conservative (worst case) scenarios that serve as a reference and reality check for any other calculations. In addition, it builds up the knowledge and skills necessary for using numerical modeling and computer-based calculations.

The theme of the book is simple: it consists of the stated goal (optimized optical system engineering), the key system parameters that make the system operate properly, and the impairments that degrade the system characteristics. These three elements will be put together to form a seamless functional relationship. The reader will be guided from the basics (described in an encyclopedic manner), through relevant parameters and impairments, to systems engineering and trade-offs. The process of optical transmission system engineering is explained in detail with several examples related to real-world applications. There are a number of tables throughout the book that contain the practical data related to system engineering parameters.

The principal objective of this book is to serve as a handbook to systems engineers dealing with optical transmission line design and planning in both vendor and carrier communities. This includes engineers, product managers, and network planners. Developers and researchers interested in entering the field of optical networking will also find this book beneficial since it contains topics related to advanced optical transmission technologies and solutions. The secondary objective of the book is to act a reference manual to an audience of graduate and undergraduate students of electrical engineering, as well as the attendants of the short courses organized by leading technical conferences related to optical communication topics.

Finally, the book can serve as a guide for technical managers and marketers who wish to become more familiar with the large spectrum of issues and solutions related to optical transmission systems.

The subject of the book is not limited to any particular geographical region, or any specific transmission scenario. The background knowledge necessary to study this book and fully understand the topics is that of a typical senior-year undergraduate engineering student.

Acknowledgments

I would like to express my appreciation to the scientists and engineers whose contributions from open literature were the basic material for this book. I am very grateful to numerous colleagues from both industry and academia for the useful discussions we have had in the past. I also extend my personal thanks to my friends from NEC Corporation for their helpful suggestions and comments. Finally, I would like to express my deep gratitude to my wife Rada and my daughters Neda and Marija for their unconditional moral support and understanding.

CHAPTER 1
Introduction to Optical Bandwidth and Lightwave Paths

This chapter introduces all relevant parameters and defines terms associated with optical transmission systems engineering. It should help the reader to establish a clear picture about both the subject and the goal of the engineering process. In addition, this chapter should help the reader to recognize the place and the role of optical components and modules placed along the lightwave path.

A number of parameters related either to optical signal or to different impairments will have an impact on the transmission quality. Optical transmission systems engineering can be understood as a process in which these parameters are combined in the most favorable way with respect to a defined goal. In most cases the defined goal is identified through the required bit error rate of the optical bandwidth channel over specified transmission distance, which is associated with the lightwave path.

Optical bandwidth channels and lightwave paths, which are introduced in this chapter, are the main objects of optical systems transmission engineering. In addition, the structure and basic principles of digital signal transmission over optical fibers and the role of the key optical elements are also described in this chapter.

1.1 Optical Transmission and Networking

We live in time officially proclaimed as "the information era," which has been characterized by the insatiable demand for information capacity and distance-independent connectivity. Internet data traffic has become the driving force behind this demand, requiring a networks infrastructure of huge information capacity, as well as equipment and methods that can tailor the bandwidth and deliver it to an end user. Optical networking technology has become closely related to Internet technology with the same ultimate goal: to satisfy these never-ending demands for bandwidth and connectivity in a sophisticated and automated manner.

Optical transmission links have been built all around the globe. High-capacity submarine optical transmission links are being built to connect continents and provide protected transmission of data, while multiple terrestrial physical optical fiber–based networks have been built both for transmission and distribution of different bandwidth channels. The optical fiber connection has been coming all the way down to the curb, building, home, and the desk.

Both theoretical and operational aspects of optical networking and optical transmission have been widely analyzed in open literature, such as [1–5]. The operational aspect has also been captured under recommendations from different national and international standards institutions [6–12]. There are also a number of publications that provide more detailed treatment of specific subjects related to optical transmission and networking [13–18].

In terms of ownership, networks and transmission systems can belong either to private enterprises or be owned by telecommunication carriers. Ownership can be related either to networking equipment and infrastructure associated with a specified network topology or to a logical entity known as the optical virtual private network that resides within the physical network topology.

The optical networking structure can be represented by three concentric circles, as shown in Figure 1.1. The central part of this structure is a long-haul core network interconnecting big cities or major communication hubs. The connections between big cities on different continents have been made by using submarine optical transmission systems. The core network is a generic name, but very often it can be recognized as either a wide area network (WAN) if it belongs to an enterprise or as an interchange carrier (IXC) network if operated by telecommunication carriers.

The second part of the optical network structure is the edge optical network, which is deployed within a smaller geographical area (usually in a metropolitan area or smaller geographic region). The edge network is often recognized as a metropolitan area network (MAN) if it belongs to an enterprise or as a local exchange carrier (LEC) network if operated by telecommunication carriers. Finally, the access network is a peripheral part of the optical network related to the last-mile access and bandwidth distribution to the individual end users (corporate, government, medical, entertainment, scientific, and private). Access networks examples include the enterprise local area networks (LAN) and a distribution network that connects the central office location of a carrier with individual users.

The physical network topology that best supports traffic demand is generally different for different parts of the optical networking structure presented in Figure 1.1. It could vary from mesh (deployed in the core and edge networks), to ring

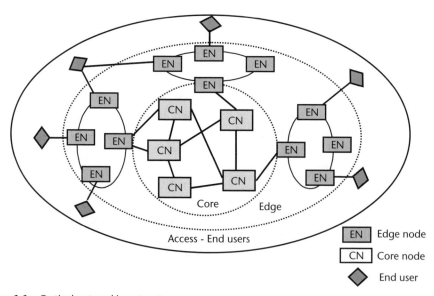

Figure 1.1 Optical networking structure.

(deployed in all portions of a global network), or to star topology (deployed mostly in an access network). From the optical transmission engineering perspective, the optical network configuration is just a mean to support an end-to-end connection via the lightwave path. Therefore, optical transmission engineering is related to the physical layer of an optical network, and takes into account the optical signal propagation length, characteristics of deployed optical elements (fibers, lasers, amplifiers, and so forth), the impact of the networking topology and bandwidth distribution, the characteristics of the signal that need to be delivered from the source to destination, and the quality of services (QoS) requirements.

In this book we will introduce the fundamentals of optical transmission system engineering by assuming that a digital electrical signal carries information that enters and exits the optical transmission system. The term "digital" is related to the waveform at the input of the optical transmission system. We assume that it is an electrical signal with two discrete levels, which occupy a specified time slot. The lower level is recognized as the logical space, or zero (0) bit, while the upper level is recognized as the logical mark, or one (1) bit. The bits are aligned along the time coordinate and form a data stream, which is characterized with the corresponding bit rate. The bit rate is expressed in bits per second, which means that it measures a number of bits, either 0 or 1, occurring during a second-long time interval. In practice, that number is very high, and bit rate is expressed in units such as kilobit per second (Kbps), megabit per second (Mbps), and gigabit per second (Gbps).

The ultimate goal of proper optical system engineering is to deliver the information bandwidth from one physical location to the other in the most economical way, while achieving required QoS. This can be done by establishing the most favorable relationship between a number of variables that characterize the signal and different impairments (such as noise, nonlinear effects) within a specific transmission scenario. The scope of the optical transmission system engineering is to understand how to minimize the effect of different impairments, learn how to allocate system margin to cope with remaining destructive effects, and how to make trade-offs between different design parameters to achieve the goal mentioned above.

1.2 Optical Transmission System Definition

The simplest optical transmission system is a point-to-point connection that utilizes a single optical wavelength, which propagates through an optical fiber. The upgrade to this topology is deployment of the wavelength division multiplex (WDM) technology, where multiple optical wavelengths are combined to travel over the same physical route.

The WDM has proven to be a cost-effective means of increasing the bandwidth of installed fiber plant. While the technology originally only served to increase the size of the fiber spans, it gradually became the foundation for optical networks. In an optical networking scenario, different signals are transported over arbitrary distances, while the wavelength routing can take a place at specified locations. The WDM technology is sometimes named with different prefixes—dense-WDM (DWDM), course-WDM (CWDM), or ultra-dense-WDM (UDWDM)—used to reflect a specific multiplexing technique used.

The general scheme of an optical transmission system is shown in Figure 1.2. We should assume that the signal transmission is generally bidirectional (the unidirectional character of Figure 1.2 is just for illustration purposes). Several optical channels, associated with specified information bandwidth, have been combined by WDM technology and sent to the optical fiber line. The aggregated signal is then transported over some distance before it is demultiplexed and converted back to an electrical level by a photodetection process. The optical signal transmission path can include a number of optical amplifiers, optical cross-connects, and optical add-drop multiplexers. The optical signal on its way from the source to the destination can be processed by various optical components, such as optical filters, optical attenuators, and optical isolators (please also refer to Figure 2.1).

There are several types of optical transmission systems, as presented in Chapter 5. If using the transmission length as a criterion, all systems can be classified as (1) very short reach (VSR) with lengths measured in hundreds of meters, (2) short reach (SR) with lengths measured in kilometers, (3) long reach (LR) with length measured in tens and hundreds of kilometers, and (4) ultra-long reach (ULR) with lengths measured in thousands of kilometers. If using the bit rate as a criterion, optical transmission systems can be (1) low-speed with bit rates measured in tens of megabits per second, (2) medium-speed with bit rates measured in hundreds of megabits per second, and (3) high-speed with bit rates measured in gigabits per second. Finally, from the application perspective, transmission systems are either power budget limited (or loss limited), or transmission speed (or bandwidth) limited.

The set of parameters that are related either to the enabling technologies and components or to the transmission and networking issues can be attached to Figure 1.2. The ultimate goal of optical signal transmission is usually defined as achieving the specified bit error rate (BER) between two end users, or between two separate intermediate points. Optical transmission system needs to be properly engineered in order to provide stable and reliable operation during its lifetime, which includes the management (setup, fine tuning, trade-offs) of key engineering parameters.

The arrangement of relevant engineering parameters is shown in Figure 1.3. As we can see, most of them are either optical power related or optical wavelength related. They can also be constant or time dependent, as illustrated in Figure 1.3.

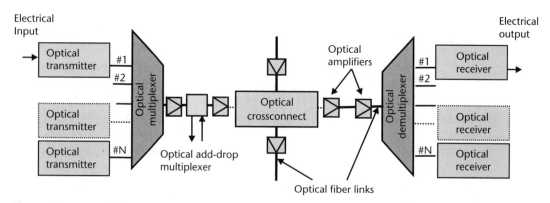

Figure 1.2 Network elements in an optical transmission system.

1.2 Optical Transmission System Definition

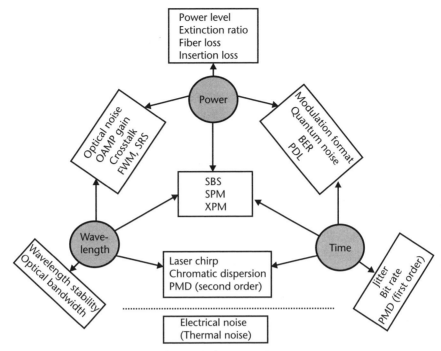

Figure 1.3 Major parameters related to optical transmission.

The optical power–related system parameters are optical power per channel, optical fiber attenuation, the extinction ratio, and optical component losses. The optical wavelength–related parameters are the optical wavelength itself (through the wavelength stability) and optical channel bandwidth. The third group of parameters, time-dependant parameters, includes optical channel bit rate, first-order polarization mode dispersion (PMD), and signal timing jitter.

The next group of parameters is optical power related but time-dependent, and it includes optical modulation format, quantum noise, polarization-dependent losses (PDL), and the BER. There is yet another parameter group that contains parameters dependent on both optical power and optical wavelength, which include optical amplifier (OAMP) gain, optical noise, crosstalk between optical channels, and some nonlinear effects such as four-wave mixing (FWM) and stimulated Raman scattering (SRS). Parameters that are dependent on wavelength, but are also functions of time, are chromatic dispersion, laser frequency chirp, and second-order PMD. Finally, there are some parameters that are dependent on both optical power and optical wavelength, and are also functions of time. Among these are stimulated Brillouin scattering (SBS), self-phase modulation (SPM), and cross-phase modulation (XPM). Please notice from Figure 1.3 that any detection scenario should include the impact of thermal noise, which is an electrical parameter associated with the optical receiver design.

Parameters presented in Figure 1.3 do not have the same significance in terms of the impact to the transmission system performance. We can say that the following parameters have a predominant impact in the most practical scenarios:

- Optical power, which defines the signal level at any specific point, such as output from transmitters or amplifiers, or input to an optical receiver. This power is effectively attenuated during propagation through optical fibers and other optical elements.
- Pulse dispersion, which is an effect that leads to the pulse spreading in optical fibers, thus contributing to the performance degradation. This effect occurs for several reasons, and they are related to physical properties of the optical fiber and the light source.
- Noise generated during photodetection process. This noise comes from three major sources: statistical and uncontrolled movement of electrons in resistors; the quantum nature of the incoming light, which makes the photodetection a statistical process related to timing of incoming photons of the optical signal; and the detection of light that is not associated with the signal and presets the optical noise.

The impact of other parameters is closely related to each specific scenario, as pointed out in Chapter 5. Table 5.1 offers a summary review of the importance of parameters presented in Figure 1.3.

Optical transport systems engineering involves accounting for all effects that can alter an optical signal on its way from the source (laser) to destination (photodiode), and then to the threshold decision point. Different impairments will degrade and compromise the integrity of the signal before it arrives to the decision point to be recovered from corruptive additives (noise, crosstalk, and interference). The transmission quality is measured by the received signal-to-noise ratio (SNR), which is defined as the ratio of the signal level to the noise level at the threshold point. The other parameter used to measure signal quality is the BER. The BER parameter is interrelated with SNR and defines the probability that a digital signal space (or 0 bit) will be mistaken for a digital signal mark (or 1 bit), and vice-versa. Evaluating the BER requires determination of the received signal level at the threshold point, calculation of the noise power, and quantification of the impacts of various impairments.

Optical fiber is the central point of an optical signal transmission. It offers wider available bandwidth, lower signal attenuation, and smaller signal distortion than other wired physical media. The total available optical bandwidth is illustrated in Figure 1.4, which shows a typical attenuation curve of a single-mode optical fiber used today. The total bandwidth is approximately 400 nm, or around 50 THz, if it is related to the wavelength region with fiber attenuation lower than 0.5 dB/km. The resultant information capacity can be calculated by multiplying the available optical bandwidth with the spectral efficiency, which is expressed in bits of information per hertz of bandwidth. The total information capacity can be more than 50 Tbps by using advanced modulation techniques that improve spectral efficiency over 1 bit per hertz.

The usable optical bandwidth is commonly split into several wavelength bands. The bands around the minimum attenuation region, usually referred to as C and L bands, have been considered as the most suitable ones for high channel count DWDM transmission, and have been already widely used for transmission purposes. The C band occupies wavelengths approximately from 1,530 to 1,560 nm, while the L band includes wavelengths between 1,580 and 1,610 nm. The S band

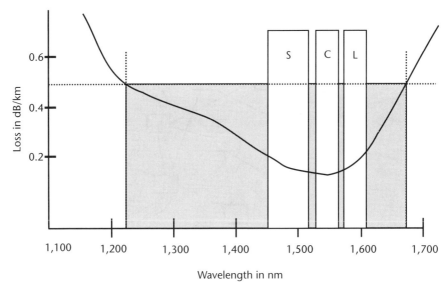

Figure 1.4 Optical fiber bandwidth.

covers shorter wavelengths usually above 1,460 nm, where optical fiber attenuation is slightly higher than in the C and L bands.

The wavelength region around 1,300 nm is less favorable for optical signal transmission since signal attenuation is higher than attenuation associated with wavelengths from S, C, and L bands. On the other hand, the bandwidth around 1,300 nm is quite usable for some specific purposes, such as transmission of cable television (CATV) signals. In addition, the CWDM technique can be easily employed in this band [9].

Optical transmission systems engineering is related to the lightwave path, which is defined as the trace that the optical signal passes between the source and destination without experiencing any opto-electrical-opto (O-E-O) conversion. Two examples of the lightwave transmission paths are presented in Figure 1.5. In general, the lightwave paths differ in lengths and in the information capacity that is carried along. For example, the lightwave path in a submarine transmission system can be several thousand kilometers long and carry information capacity measured by terabits per second. On the other hand, a typical lightwave path within a metropolitan area is measured by tens of kilometers with the information capacity measured in gigabits per second. The lightwave path length is one of the most important variables, since most of the impairments have an accumulating effect.

Several optical signals from different lightwave paths can be joined in a composite signal and continue to travel together throughout the network. In such a case, an equalization of the optical signal parameters from different lightwave paths might be necessary in order to provide an equal transmission quality for all channels. Such equalization can include signal level adjustment, adaptive dispersion compensation, and optical filtering.

It is not always possible to establish a transparent lightwave bandwidth path between two end users since signal degradation becomes vary intense and the BER goes down below an acceptable level. Therefore, some digital regeneration method

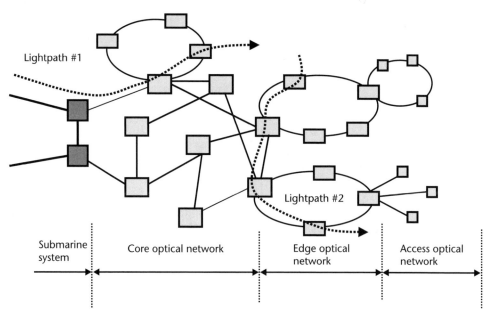

Figure 1.5 Definition of the lightwave signal path.

should be applied to recover the signal content. A full regeneration is known as 3R (reamplification, retiming, and reshaping) and can be done either at the optical or electrical signal level. Optical regeneration will not disrupt the lightwave path, which means that optical regenerators can be considered as an integral part of the lightwave path. However, signal cleaning and regeneration is often done electronically by O-E-O conversion since it can be a more efficient solution. The lightwave path is effectively disrupted, which means that any optoelectronic regeneration creates two independent lightwave pats that should be engineered separately.

The lightwave path, as an optical connection between two distinct locations, is realized by assigning a dedicated optical channel or dedicated optical band between these locations. It can have a permanent character if the lightwave path is set up during the network deployment and there is no change whatsoever. It can also have a temporary character if the lightwave path can be set up and torn down on request. This type of lightwave path is often related to so-called wavelength services offered by network providers.

The lightwave paths are the foundation for lower bandwidth services between two end users. The lightwave path can be considered as a bandwidth wrapper for lower speed transmission channels, which form virtual circuit services. In such cases, the time division multiplexing (TDM) technique is applied to aggregate the bandwidth of virtual circuits before it is wrapped in the lightwave path.

There are two forms of multiplexing of virtual circuits, known as fixed multiplexing and statistical multiplexing. In fixed multiplexing, the aggregate bandwidth is divided among individual virtual circuits in such a way that each circuit receives a guaranteed amount of the bandwidth, often referred to as a *bandwidth pipe*. If there is an unused portion of the bandwidth within any specific bandwidth pipe, it is wasted since a dynamic exchange of contents of any pair of bandwidth pipes is not possible. Therefore, it can happen that some bandwidth pipes are almost empty at any given moment, while others can be quite full.

Statistical multiplexing can minimize the waste of bandwidth, since it can be used more efficiently. This is done by breaking up the data content in each bandwidth pipe into data packets, which can be handled independently. By doing this, so-called packet switching is enabled. If there is an overload in any specific bandwidth pipe, some packets can be redirected to other unloaded pipes. In addition, statistical multiplexing permits a finer granulation of the aggregated bandwidth among individual virtual circuits.

The fixed multiplexing of virtual circuits is defined by the Synchronous Optical Network (SONET) standard applied in North America [6] and the Synchronous Digital Hierarchy (SDH) standard applied worldwide [10]. Both standards define a synchronous frame structure, bandwidth partitioning, multiplexing patterns, and supervisory functions for transport of different kinds of digital signals. Originally, only digital voice signals with TDM were packed into a SONET/SDH frame, but it gradually became a means for framing and transport of different kind of data channels [11, 12]. These data channels are usually arranged according to specified standards [13–18].

A review of the most-used digital bandwidth channels is presented in Table 1.1. There are two separate columns in this table, related to TDM/synchronous channels and data/asynchronous bandwidth channels. The bandwidth channels are lined up based on the bit rate. We can conditionally divide all channels to low-speed bandwidth channels (the first four rows) and high-speed ones (the last four rows), although this division is quite relative.

The basic building block of the TDM hierarchy is 64 Kbps, corresponding to one digitalized voice channel. The next level is obtained by multiplexing either 24 channels (DS-1 format in North America) or 30 channels (E-1 format outside North America). The additional higher levels in asynchronous TDM hierarchy were built up just by multiplexing several lower level signals. It has been common to take four lower level streams and combine then into a higher-level aggregate signal. In such a way, a third level corresponding to DS-3 (44.736 Mbps) in North America, E-3 (34.368 Mbps) in Europe, or DS-3J (32.064 Mbps) in Japan has been obtained.

Table 1.1 Bit Rates of Different Bandwidth Channels

TDM/Synchronous Bandwidth Channels	Bit Rate	Data/Asynchronous Bandwidth Channels	Bit Rate
DS-1	1.544 Mbps	10-BaseT Ethernet	10 Mbps
E-1	2.048 Mbps	100-BaseT Ethernet	100 Mbps
OC-1	51.84 Mbps	FDDI	100 Mbps
OC-3=STM-1	155.52 Mbps	ESCON	200 Mbps
		Fiber Channel-I	200 Gbps
OC-12=STM-4	602.08 Mbs	Fiber Channel-II	400 Gbps
		Fiber Channel-III	800 Gbps
OC-48=STM-16	2.488 Gbps	Gb Ethernet	1 Gbps
OC-192=STM-64	9.953 Gbps	10Gb Ethernet	10 Gbps
OC-768=STM-256	39.813 Gbps	40Gb Ethernet	40 Gbps

The lack of a unified approach becomes an obstacle for high-speed connection around the globe, due to interoperability issues. The introduction of SONET/SDH standards was the first step towards a global networking. Although the corresponding building blocks, which are optical carrier one (OC-1) in SONET and synchronous transport module one (STM-1) in SDH, are different in bit rates, they belong to the same synchronous hierarchy.

Optical transmission systems widely deployed today operate at bit rates equal to 2.488 and 9.953 Gbps. These bit rates are commonly referred to as 2.5 and 10 Gbps. There is the expectation that high-speed transmission systems operating at 40 Gbps, which corresponds to OC-768/STM-256, will be introduced in the near future.

Another ITU standard based on fixed multiplexing has been specially designed for optical transport networks and bit error rates up to 40 Gbps [8]. It defines optical channel (OCh) in a flexible way that provides more efficient data accommodation and better operation, administration, and maintenance (OA&M) capabilities. In addition, it defines forward error correction (FEC) as an optional feature.

Data channels presented on the right side of Table 1.1 do not follow any strict hierarchy. Some of them, such as Enterprise Serial Connection (ESCON) or Fiber Channel, are designed for interconnecting computers with other computers and peripherals (such as large memories and data storages.) They have been widely used in storage area networks (SAN) for data transfer related to data storage, backup operations, and financial transactions. The bit rates presented in Table 1.1 are related to data payload, or useful data. However, these bit rates are increased when some line coding is applied before the data is sent. For example, the ESCON data channel uses so-called (8B, 10B) line code, which increases the line bit rate to 250 Mbps since each 8 bits of the information signal are replaced by 10 bits by adding 2 redundant bits in accordance with the established coding rule, while fiber distributed data interface (FDDI) utilizes (4B, 5B) line code, which increases the line bit rate to 125 Mbps.

It is important to notice that the Fiber Channel standard defines three data-payload bit rates of 200, 400, and 800 Mbps [17]. These bit rates eventually become equal to 256.6, 531.2, and 1,062.6 Mbps, respectively, after (8B, 10B) line code is applied, which inserts 2 redundant bits after each 8 data bits to improve the transmission robustness and monitoring capabilities.

Both ESCON and Fiber Channel, if deployed individually, use low-speed and low-cost optical components. However, these channels can be also multiplexed in order to improve the transport efficiency. Multiplexing could be done by using the WDM technique with optical channels loaded by data bit rates presented in Table 1.1. It could be also done by placing them into the SONET/SDH frame in accordance with new standard specifications [11].

The FDDI and the Ethernet data channels have been originally designed for data bandwidth sharing in the LAN environment. The FDDI is based on the token ring access method [18], while the Ethernet standard is based on the well-known medium access scheme, Carrier Sense Medium Access/Collision Detection (CSMA/CD). The Ethernet technology has been recently enhanced by resilient packet ring (RPR) technology [19]. The size of shared bandwidth channel, as defined by the Ethernet standards, went from 10 Mbps a while ago to 10 Gbps today—it is widely anticipated that a 40-Gbps bit rate will be realized by Ethernet standards in the near future. The enhancement of both bandwidth channel size and the bandwidth sharing capability has been accompanied by an effort to define Ethernet as a transport standard, which would be suitable not just for LAN and MAN environments, but for WAN applications as well. It is also worth mentioning that the Gigabit Ethernet bit rate goes up to 1.25 Gbps after line coding is applied. Therefore, from the transmission perspective, the line bit rate is a parameter that should be taken into account.

An optical channel is associated with a specific optical wavelength. As such, it is data content agnostic and does not belong to any digital hierarchy. However, an optical channel can be considered as a building block of the overall optical bandwidth that can be carried along the lightwave path. Several optical channels can be grouped together to form an optical subband. The amount of bandwidth carried by an optical subband depends on the number of wavelengths inside the subband and the bit rates of signals carried by each individual wavelength. It is important from a routing and provisioning perspective that a wavelength subband be treated the same way as a single wavelength channel is treated. The next level above the optical subband is the optical band (refer to Figure 1.4). Each individual wavelength band from Figure 1.4 can be considered as an aggregation of optical channels and wavelength subbands.

The optical bandwidth building blocks have been arranged by using WDM, which is in essence a well-known frequency division multiplex (FDM) technique. All optical bandwidth channels that are transported along a lightwave path can be considered as optical wrappers of the digital signal streams from Table 1.1. There is an attempt well underway to adopt the generalized multiprotocol label switching (GMPLS) scheme, which would facilitate the process of lightwave path management (creation, tearing down, switching, and administration) [20].

1.3 Organization of the Book

The chapters that follow are intended to provide a straightforward guide to optical transmission system engineering. The sequence of topics is arranged to take the reader from a description of optical components, through identification of key parameters that define the signal quality, to the analysis of transmission penalties and trade-offs involved in systems engineering. The reader will be guided to recognize the key steps and most relevant parameters in the systems engineering process.

Chapter 2 describes the characteristics of the optical elements that might be deployed along a lightwave path, thus having an impact on the optical signal quality. Chapter 3 deals specifically with the parameters that define the bit error rate at the receiving side.

Chapter 4 is related to the assessment of the limitations and penalties associated with bandwidth transmission along the lightwave path. The reader should be able to understand the importance and impact of each specific parameter to the system performance.

The basic principles of optical transmission systems engineering are introduced in Chapter 5. This is accompanied by a discussion of several practical scenarios that should help the reader to establish a clear picture that connects the following three entities: the engineering goal, the transmission obstacles, and the transmission enabling elements. This chapter also contains descriptions of simulation tools and computer-aided methods that can be used in the systems engineering process.

Chapter 6 describes the advanced technologies and methods that enhance the overall transmission performance and discusses the trade-offs related to the systems engineering process. It also explains how to use benefits brought by the application of advanced methods and schemes to offset the impacts of impairments.

Finally, Chapter 7 serves as a toolbox for optical transmission systems engineering purposes. It contains an overview of units, definitions, and mathematical formulas related to the physics of key components, and to the telecommunication theory. This chapter should help the reader to better understand the material presented in the book, as it provides a more comprehensive picture with respect to optical signal transmission.

There will be a number of acronyms used in this book. In most cases they will be accompanied by the words explaining their meaning, but in some cases that explanation might be missing if there is an impression that the reader has become familiar with the meaning of the acronym. If not, the reader is encouraged to look at the list of acronyms given at the end of the book.

There is a list of references provided at the end of each chapter. Some of them were used as source material in preparation of this book, while some others contain more detailed information and can be used if the reader would like to gain some additional knowledge with respect to a specific topic.

This book contains a number of examples and calculations that should help the reader to establish practical reference points with respect to the impact of impairments, and to signal quality. There are also several tables that contain typical data of optical parameters associated with different optical components. All numerical inputs that are presented in tables, and which were used in calculations, are taken either from product data sheets that are publicly available, or from catalogs and engineering sourcebooks, such as [21].

Since optical fiber communication technology is still evolving at a relatively fast rate, it is logical to expect that there will be new approaches of how to deal with optical signal generation, amplification, multiplexing, detection, and signal processing. New materials may be also introduced in the future to further improve the features of the optical components. However, we may not see a "quantum leap" in achievements, such as the one between the pioneering results achieved soon after Kao and Hockham published their revolutionary concept a few decades ago [22] and the heroic experiment [23] published at the very beginning of this century. What we may see are some events telling us that the photon has just been following the way the electron has passed a while ago. Although these changes in technology are widely anticipated, they should not have a major impact on the fundamentals presented in the book, since the book is structured on enduring concepts that should help to understand the impact of new technologies.

1.4 Summary

The reader should be able to associate the transmission system engineering with the lightwave path and understand the content of the information channel associated with each specific path. The reader should also understand that there are a number of parameters involved in the transmission engineering process. However, some of them may have a minor impact to the transmission characteristics in any given scenario, and so may be neglected. This will be discussed in more detail in Chapter 5. On the other hand, the engineering goal can be achieved by following several different ways, while the system characteristics can be optimized the by the fine-tuning of some parameters. The engineering process involves not just calculation and fine-tuning, but also the selection and trade-offs with respect to both the system parameters and the components within the system (refer to Section 6.2).

References

[1] Gower, J., *Optical Communication Systems*, 2nd ed., Upper Saddle River, NJ: Prentice Hall, 1993.

[2] Agrawal, G. P., *Fiber Optic Communication Systems*, 3rd ed., New York: Wiley, 2002.

[3] Ramaswami, R., and K. N. Sivarajan, *Optical Networks, a Practical Perspective*, 2nd ed., San Francisco, CA: Morgan Kaufmann Publishers, 2002.

[4] Kazovski, L., S. Benedetto, and A. Willner, *Optical Fiber Communication Systems*, Norwood, MA: Artech House, 1996.

[5] Cvijetic, M., *Coherent and Nonlinear Lightwave Communications*, Norwood, MA: Artech House, 1996.

[6] Bellcore, "GR. SONET Transport Systems: Common Generic Criteria," GR-253-CORE, Bellcore, 1995.

[7] ITU-T Rec. G.681, "Functional Characteristics of Interoffice and Long-Haul Line Systems Using Optical Amplifiers, Including Optical Multiplexing," ITU-T (10/96), 1996.

[8] ITU-T Rec. G.709/Y1331, "Interfaces for the Optical Transport Network (OTN)," ITU-T (02/01), 2001.

[9] ITU-T Rec. G.694.2, "Spectral Grids for WDM Applications: CWDM Wavelength Grid," ITU-T (06/02), 2002.

[10] ITU-T Rec. G.704, "Synchronous Frame Structures Used at 1544, 6312, 2048, 8448 and 44 736 kbit/s Hierarchical Levels," ITU-T (10/98), 1998.

[11] ITU-T Rec. G.7041/Y.1303, "Generic Framing Procedure (GFP)," ITU-T (12/01), 2001.

[12] ITU-T Rec. G.7042/Y.1305, "Link Capacity Adjustment Scheme (LCAS) for Virtual Concatenated Signals," ITU-T (11/01), 2001.

[13] Kartalopoulos, S. V., *Understanding SONET/SDH and ATM*, Piscataway, NJ: IEEE Press, 1999.

[14] Loshin, P., F. Kastenholz, and P. Loshin, *Essential Ethernet Standards: RFCs and Protocols Made Practical*, New York: Wiley, 1999.

[15] Huitema, C., *Routing in the Internet*, 2nd ed., Upper Saddle River, NJ: Prentice Hall, 1999.

[16] Calta, S. A., et al., "Enterprise System Connection (ESCON) Architecture—System Overview," *IBM Journal of Research and Development*, Vol. 36, 1992, pp. 535–551.

[17] Sachs, M. W., and A. Varma, "Fiber Channel and Related Standards," *IEEE Commun. Magazine*, Vol. 34, 1996, pp. 40–49.

[18] Ross, F. E., "FDDI—A Tutorial," *IEEE Commun. Magazine*, Vol. 24, 1986, pp. 10–17.

[19] IEEE 802.17 Resilient Packet Ring (RPR) Architecture, IEEE Standard Forum, Draft 2.0, IEEE 802.17 (12/02), 2002.

[20] IETF extension of RFC 3031, Generalized Multiprotocol Label Switching (GMPLS) Architecture, Working Draft, 09/02.

[21] *Lightwave Optical Engineering Sourcebook, 2003 Worldwide Directory*, Lightwave 2003 Edition, Nashua, NH: PennWell, 2003.

[22] Kao, K. C., and G. A. Hockman, "Dielectric Fiber Surface Waveguides for Optical Frequencies," *Proc. IEE*, Vol. 133, 1966, pp. 1151–1158.

[23] Fukuchi, K., et al., "10.92 Tb/s (273x40 Gbps) Triple Band Ultra Dense WDM Optical-Repeated Transmission Experiment," *Optical Fiber Conference OFC 2001*, Anaheim, CA, PD 26.

CHAPTER 2
Optical Components as Constituents of Lightwave Paths

This chapter describes the characteristics and engineering parameters of the optical elements that might be deployed along a lightwave path, thus having an impact on the optical signal quality. The description is done without more rigorous mathematical treatment and serves to better understand the building blocks of the optical transmission system.

The end-to-end signal transmission from Figure 1.2 includes both electrical and optical signal paths. The conversion from electrical to optical level is done in the optical transmitter, while conversion from the optical level to an electrical signal takes place in the optical receiver. The key elements on the optical signal path are shown in Figure 2.1, and they are:

- *Semiconductor light sources* convert an electrical signal to optical radiation. This is achieved by the bias current flow through a semiconductor p-n junction, which stimulates recombination of electrons and holes. The optical signal that has been generated is a flow of photons produced by recombination of electrons and holes. If the current flowing through the p-n junction is higher than a certain threshold, so-called *stimulated emission of radiation* can be achieved in special semiconductor structures. In these structures, known as semiconductor lasers, the recombination occurs in an organized way, with strong correlation in phase, frequency, and direction of radiated photons that form the output optical signal. Semiconductor lasers could either be directly modulated by an electrical signal or just biased by a dc voltage, and they operate in combination with an external optical modulator. Each laser generates a specified optical wavelength or optical carrier, but some spectral linewidth is associated with the generated optical signal as well. These lasers are known as single-mode lasers (SML), characterized by a distinguished single longitudinal mode in the optical spectrum. If a set of separated longitudinal modes can be recognized under the optical spectrum envelope, the lasers are called multimode lasers (MML) (refer to Figure 2.4).
- *Optical modulators* operate in combination with semiconductor lasers. This scheme is applied in high-speed transmission systems since it is more effective to combat the impact of laser frequency chirp and the influence of chromatic dispersion in optical fibers (refer to Sections 3.3.3 and 3.3.4). Optical modulators accept a continuous wave (CW) optical signal from the laser and change its parameters (usually amplitude and phase) by an applied

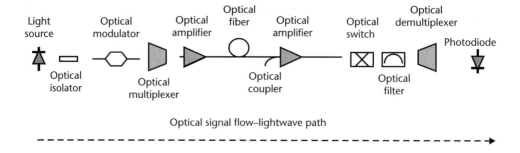

Figure 2.1 Optical components and their likely position along the lightwave path.

modulation voltage. There are two types of optical modulators used today: Mach-Zehnder (MZ) modulators and electro-absorption (EA) modulators.

- *Optical fibers* transport an optical signal to its destination. The combination of low signal loss and extremely wide transmission bandwidth allows high-speed optical signals to be transmitted over long distances before regeneration becomes necessary. There are two groups of optical fibers. The first group, multimode optical fibers, transfers light through multiple spatial or transversal modes. Each mode, defined through a specified combination of electrical and magnetic field components, occupies a different cross-sectional area of the optical fiber core and takes a slightly distinguished path along the optical fiber. The difference in mode path lengths causes a difference in arrival times at the receiving point. This phenomenon is known as multimode dispersion and causes signal distortion and limitations in transmission bandwidth. The second group of optical fibers effectively eliminates multimode dispersion by limiting the number of modes to just one through a much smaller core diameter. These fibers, called single-mode optical fibers, do, however, introduce another signal impairment known as chromatic dispersion. Chromatic dispersion is caused by a difference in velocities among different wavelength components within the same pulse. There are several methods to minimize chromatic dispersion at a specified wavelength, involving either the introduction of new single-mode optical fibers or the utilization of different dispersion compensation schemes.

- *Optical amplifiers* amplify weak incoming optical signals through the process of stimulated emission, without any conversion back to the electrical level. Optical amplifiers should be capable of amplifying at several optical wavelengths simultaneously. There are different types of optical amplifiers currently in use: semiconductor optical amplifiers (SOA), erbium doped fiber amplifiers (EDFA), and Raman amplifiers. Amplifier parameters are: the total gain, the gain flatness over amplification bandwidth, output optical power, signal bandwidth, and noise power. The noise generated in an optical amplifier occurs due to a spontaneous emission process that is not correlated with the signal. All amplifiers degrade the SNR of the output signal because of amplified spontaneous emission (ASE), which adds itself to the signal during the amplification process. The SNR degradation is measured by the noise figure.

- *Photodiodes* convert an incoming optical signal back to the electrical level through a process opposite to one that takes place in lasers. Photodiodes can be classified into PIN or avalanche photodiodes (APD). The process within the PIN photodiodes is characterized by quantum efficiency, which is the probability that each photon will generate an electron-hole pair. In APDs each primary electron-hole pair is accelerated in a strong electric field, which can cause the generation of several secondary electron-hole pairs through the effect of impact ionization. This process is random in nature and avalanche-like.
- *Optical components* have to be used to either enhance the signal quality by preventing detrimental effects or perform a specific function related to the optical signal processing. The first group of optical components includes optical isolators, optical circulators, and optical filters, while the second group includes optical couplers, optical switches, and optical multiplexers. Although there is a specific set of parameters for each set of optical components, which should be taken into account during the system engineering process, some parameters, such as insertion loss and sensitivity to optical signal polarization state, are common for all of them.

In this chapter we will describe operational principles of the optical elements placed along the lightwave path in order to better understand the contribution they have on lightwave path creation. A quantitative assessment of the parameters associated with these components will be done in Chapter 3.

2.1 Semiconductor Light Sources

The vast majority of light sources used in optical communications systems are based on semiconductor chips. As already mentioned, the light generation process occurs in certain semiconductors when they are under direct bias voltage due to recombination of electrons with holes at the n-p junction (refer to Sections 7.5 and 7.6). The recombination process can be either spontaneous or stimulated [1, 2]. The nature of the recombination process determines the type of light sources. There are two major types of the semiconductor optical sources: light emitting diodes and semiconductor lasers.

2.1.1 Light Emitting Diodes

The light emitting diode (LED) is a monolithically integrated semiconductor device with a sandwich-like p-n junction and terminal contacts, as illustrated in Figure 2.2. The light is generated by spontaneous emission in the junction region, but propagates outside the junction region and out of the semiconductor's structure. This simplest structure is known as the surface radiation LED, which is shown in Figure 2.2(a). If there is at least one more semiconductor layer, as in Figure 2.2(b), the active region can be structured as a waveguide, since the relationship $n_1 > n_2$, and $n_1 > n_3$ is valid for the refractive indexes belonging to the junction and surrounding layers. In such a case, the radiation is lateral through the edges of the semiconductor structure. The output power of the LED is relatively low, usually up to 0.1 mW, for bias current exceeding 100 mA.

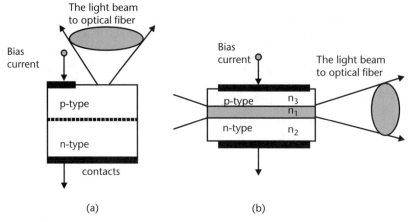

Figure 2.2 Light emitting diodes: (a) surface emission, and (b) edge emission.

The spectral emission patterns, expressed through the dependence of the output power on the wavelength, can be approximated by Gaussian curve [2], with the full-width half maximum (FWHM) ranging from 30 nm at central wavelengths around 850 nm, to more than 60 nm if operated in the wavelength region above 1,300 nm.

The large diffusion capacitance in the active region limits the frequency response of LED. The following relation connects the output optical power P_{out} with the frequency f of the modulation current

$$P_{out}(f) = P_0 \left[1 + \left(2\pi f \tau_{eff}\right)^2\right]^{-1/2} \qquad (2.1)$$

where P_0 is the output optical power for dc current, while τ_{eff} is the effective carrier lifetime, usually ranging from 5 to 10 ns.

The features mentioned above limit the application of LED to lower speeds (up to 250 Mbps) and lower distances (up to several tens of kilometers).

2.1.2 Semiconductor Lasers

Semiconductor lasers also convert an electrical signal to the optical radiation. The bias current flows through the laser p-n junction and stimulates the recombination of electrons and holes, which leads to the generation of photons (refer to Section 7.6). If the current is higher than a certain threshold, recombination occurs in an organized way, with strong correlation in phase, frequency, and direction of radiated photons that form the output optical signal. This case corresponds to so-called stimulated emission of radiation. The basic structure of a semiconductor laser is illustrated in Figure 2.3(a). If reflection coatings, acting as mirrors, cover the semiconductor facets, generated light will be captured within resonant cavity and will make multiple paths between the mirrors. When the reflection coefficient of one of the facets is lower than 100%, a portion of the light will come out, while the rest will continue the back-and-forth oscillation. This device that produces the light amplification of stimulated emission of radiation (laser) is well known as the Fabry-Perot (FP) type. The output edge

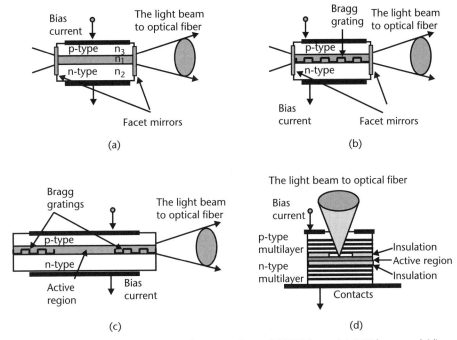

Figure 2.3 Semiconductor lasers: (a) Fabry-Perot laser, (b) DFB laser, (c) DBR laser, and (d) VCSEL.

light radiation presented in Figure 2.3(a) is coherent in nature, which makes both the spectral width and beam divergence angle much smaller than those in LED. At the same time, the output power is much higher, since the light is amplified by multiple reflections and walks between the mirrors. There are a large number of references dealing with semiconductor optical sources that can be used to find more details about any specific semiconductor source type [3–15].

It is important to notice that there is a bias current threshold where the stimulated radiation process really starts, or where the amplification of the light within cavity picks up. It is easily recognized at the functional curve presenting output light versus bias current (refer to Figure 2.6). The radiation pattern of semiconductor lasers can contain several distinct wavelength peaks placed under the spectral envelope, as shown in Figure 2.4(a). These peaks are known as longitudinal modes,

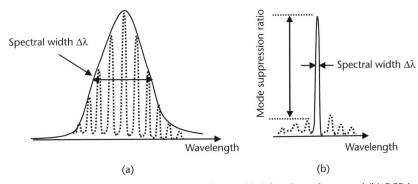

Figure 2.4 Output spectrum of semiconductor lasers: (a) Fabry-Perot laser, and (b) DFB laser.

and they are typical for any structure based on the Fabry-Perot resonator. The existence of several longitudinal modes is the reason why these lasers are well recognized as MMLs.

Further improvement in the output light coherence has been achieved through additional structural elements that effectively select just one mode from the Fabry-Perot spectrum and suppress the remaining longitudinal modes. This also contributes to the improvement of the laser modulation speed, thus making them more suitable for high-speed applications. The improvement is usually done by inserting the grating-type element either in the resonant cavity or outside of the cavity. If the selective grating is placed within the cavity, lasers are referred to as the distributed feedback (DFB) ones, and have a structure shown in Figure 2.3(b). The lasers with an outside Bragg grating, or gratings placed instead of the facet mirrors, are referred to as distributed Bragg reflector (DBR) lasers, as shown in Figure 2.3(c).

There is another type of semiconductor laser known as vertical cavity surface emitting laser (VCSEL), which is a monolithic multilayer semiconductor device, as shown in Figure 2.3(d). This type is usually based on In-GaAs-P layers that act as Bragg reflectors, thus enabling the positive feedback and stimulated emission. The layers are very thin, with a size of about one-fourth wavelength. The VCSEL devices emit a regular circular light beam, as opposed to the other semiconductor lasers mentioned above. This feature is much more convenient in terms of launching the light into an optical fiber.

The VCSEL lasers are very promising candidates for different applications. At present time they can sustain modulation bit rates up to 15 Gbps, with output powers up to 3 to 4 mW. A rapid progress in development and an enhancement of their features are anticipated. All three laser types mentioned above (DFB, DBR, VCSEL) are known as single-mode lasers, which are characterized by a distinguished single longitudinal mode in the optical spectrum, as shown in Figure 2.4(b).

The spectral curve of the SML is a result of the transition between discrete energy levels that reside within conduction and valence bands, respectively (refer to Section 7.6). This spectral curve can be represented by the so-called Lorentzian line, given as [2, 3]

$$g(\nu) = \frac{\Delta \nu}{2\pi\left[(\nu - \nu_0)^2 + (\Delta \nu / 2)^2\right]} \qquad (2.2)$$

where ν_0 is the central optical frequency, and $\Delta \nu$ represents the spectral linewidth. The function $g(\nu)$ is normalized so that the total area under the curve equals 1.

The spectral linewidth, defined as the FWHM of the Lorentzian spectral line, can be expressed as

$$\Delta \nu = \frac{n_{sp} G \left[1 + \alpha_{chirp}^2\right]}{4\pi P} \qquad (2.3)$$

where n_{sp} is the spontaneous emission factor, G is the net-rate of stimulated emission, α_{chirp} is the amplitude-phase coupling parameter often known as the chirp-factor, and P presents the output power of the radiated optical signal. The reader is advised to consult Chapter 7 to find more details with respect to characteristics of radiated optical signals. It is important to notice that the linewidth Δν decreases with the increase of optical power. In addition, it can be reduced by decreasing chirp factor values and the rate of spontaneous emission. Parameter α_{chirp} can be reduced by using multiquantum-well (MQW) laser design [5], while the rate of spontaneous emission can be reduced by increasing the resonant cavity length. Although with some special design the linewidth of DFB single-mode lasers can go down to several hundred kilohertz, the linewidth of most DFB lasers is in the range from 5 to 10 MHz for an output power level of approximately 10 mW.

The frequency response of the semiconductor laser is largely defined by the relaxation frequency, since the modulation response starts to decrease rapidly when modulation frequency exceed the relaxation frequency value. The 3-dB modulation bandwidth, defined as the frequency where modulation output is reduced by two times (or 3 dB) with respect to output corresponding to the CW case, is given as [6]

$$f_M = \frac{f_R \sqrt{3}}{2\pi} \approx \left(\frac{3GP_b}{4\pi^2 \tau_p}\right)^{1/2} \qquad (2.4)$$

where f_R is the relaxation frequency, G is the net-rate of stimulated emission, P_b is the output power for bias current, and τ_p is the photon lifetime related to the excited energy level. It is worth noting the square root dependence of the frequency response on the output power. The modulation bandwidth of the semiconductor lasers can go above 25 GHz [7]. In addition to direct modulation, semiconductor lasers can be biased only by a dc voltage to produce a continuous wave radiation and to operate in combination with an external optical modulator.

As a summary, the following laser parameters are important from the system engineering perspective: the output power level, the operational wavelength, the spectral linewidth, and the laser frequency chirp (which is discussed in Section 3.3.3). As for the laser operation, it is important to maintain the stability of both the optical signal amplitude and the optical signal wavelength [12–16]. That stability is dependent on the material type, aging, bias current value, and the temperature of the laser chip. Designers have already achieved the temperature stabilization within a fraction of a degree K by using thermoelectric coolers. As for the aging, there is usually an automatic control of the output optical power and signal wavelength, but some system margin should be also allocated to account for a gradual degradation of laser parameters.

Typical values of optical source parameters that are important from the system engineering perspective are summarized in Table 2.1. The numbers shown are

Table 2.1 Typical Values of Semiconductor Sources Parameters

Parameter	LED	FP Lasers	DFB Lasers	VSCEL
Output power	Up to 150 μW	Up to 10 mW	Up to 20 mW	Up to 4 mW
Spectral width (FWHM)	30–60 nm	1–3 nm	0.00004–0.0004 nm (5–50 MHz)	0.1–1 nm
Modulation bandwidth	Up to 250 MHz	Up to 3 GHz	Up to 30 GHz	Up to 15 GHz

typical values extracted from product-related literature, such as [16], or data sheets of different manufacturers. It is important to mention that the output powers for both FP and DFB lasers can be much higher if they are designed to serve as power pumps in optical amplifier schemes. In such cases the power can be as high as 400 to 500 mW for FP lasers, and up to 300 mW for DFB lasers [16].

2.1.3 Wavelength Selectable Lasers

It is highly desirable from the application perspective to have lasers that can be frequency (or wavelength) manageable in an organized manner. In such a way, the same physical device can be used for different optical channels within the DWDM system, just by changing the wavelength of the output optical signal. This feature is quite welcome from operation and maintenance perspective. In addition, some other applications, such as optical packet switching, are feasible if the change in the output wavelength occurs very quickly.

The simplest way to change the operating wavelength of a semiconductor laser is by changing its injection current. If the injection current changes, an effect identical to the adiabatic chirp described in Section 3.3.3 takes place, and the output wavelength is also changed. The wavelength change with current is typically around 0.02 nm/mA. Next, the operating wavelength can be tuned by changing the laser operating temperature with a rate of about 0.08 nm/K—it is not advisable to tune the wavelength by a larger temperature change. Any temperature change greater than 20 degrees will cause a drop in the output signal level and will have a negative impact on the reliability and lifetime expectations (refer to Figure 2.6).

There are several practical methods that are used today to design a semiconductor laser with tunable wavelength range. The first of them is related to the deployment of an external tunable filter between the mirrors. The external mirror is used to extend the resonant cavity and to easily accommodate the placement of the tunable filter. By changing the optical filter spectrum, either through the refractive index tuning or by temperature tuning, the output wavelength of the laser can be effectively changed. It can also be done if some diffraction grating, such as that in Figure 2.5(a), is placed at the end of the external resonator. The Bragg reflection condition is changed by changing the grating position, which means that different wavelengths can be selected to comply with the lasing regime.

The same principle, based on an adjustable Bragg grating, can be used in different DBR constructions. In this case, there is no mechanical movement involved in order to change the position of the Bragg mirror since the Bragg condition is changed by current injection. To make this principle workable, a DBR laser is made as a multisection device, such as one shown in Figure 2.5(c). The number of sections can vary from two to four and more [17]. In a two-section device there are two pairs of electrodes to carry injection currents, one pair for the regular active region, and the other for controlling Bragg reflection conditions through the refractive index change. The two-section structure prevents the output optical power from falling down, but it does not provide wider continuous tunable range.

In the three-section structure, presented in Figure 2.5(c), there is a third section inserted between the active region and Bragg grating section. This section is called the phase-shift section and serves to change the phase of the light entering into the Bragg

2.1 Semiconductor Light Sources

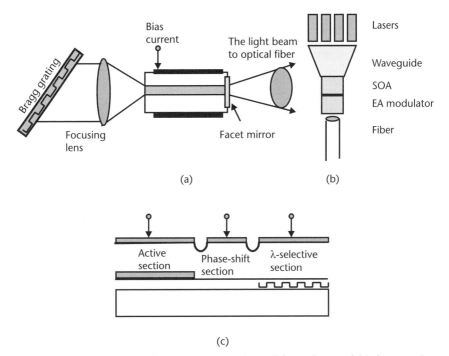

Figure 2.5 Tunable lasers: (a) external cavity laser, (b) multilaser chip, and (c) three-section tunable laser.

grating section. The phase shift, achieved through independent current bias, helps to spread the continuous tunable range (about 9 nm). Any structure involving four or more sections can be considered as a cascade of two or more Fabry-Perot filters, each of them with different free spectral range (FSR). The free spectral range of a cascade is equal to sum of individual FSRs (Section 2.6.3). In a cascaded multisectional structure, the Bragg mirror function occurs not just once but a number of times, which effectively provides continuous tunable range of more than 30 nm [17].

Wavelength selectivity can also be achieved through combining the outputs of several semiconductor lasers into one output, as illustrated in Figure 2.5(b). Such a multilaser structure on the same wafer can be combined with other elements, such as semiconductor optical amplifiers and EA modulators [18]. The current bias is switched between individual laser structures, so that just one of them radiates at a time. The bias switching process can be combined with small wavelength tuning within each of the lasers to achieve a finer tuning granularity.

The current versions of wavelength selectable lasers are mainly based on schemes described above. The wavelength range covered by a single laser device varies from several nanometers to more than 100 nm, while the output power can be up to 20 mW [16]. The wavelength selectable lasers operate in combination with external modulators, and very often they are equipped with a wavelength locker that helps to preserve the values of the output wavelength after the tuning takes place. A broadly tunable light source can also be produced by using optical fiber loops with integrated optical filters that are the integral part of the loop. The fiber loop serves as an active medium that amplifies the light, while the filter helps to select a specified wavelength at the output side [19].

2.2 Optical Modulators

2.2.1 Direct Optical Modulation

The optical signal generated by the light source still needs to be modulated by an information data signal. There are two broadly used modulation schemes: direct modulation and external modulation of the optical carrier. The direct modulation process is initiated by a digital modulation current that is added on the top of the dc bias current. The level of the dc bias current is usually chosen in such a way that both logical levels of the modulation current, which correspond to 0 and 1 bits, are above the laser threshold current. In some cases the level corresponding to 0 bits might be lower that the threshold value, which is more appropriate for lower speed operations (up to several hundred megabits per second). As for modulation of light emitting diodes, the modulator design is more flexible, and the main goal is to achieve the highest output power for 1 bits under specified conditions.

It is important to know that the modulation curve, which represents changes of the output light signal with the direct current through laser diode, is not quite stable and changes with aging and temperature, as shown in Figure 2.6(a). This curve is sometimes called the L-I curve. Both effects, the aging and temperature changes, cause L-I curve degradation and a decrease in the output power. The degradation of the modulation curve can have a severe impact on total system characteristics since it degrades the SNR at the receiving side. Therefore, a permanent monitoring of the output power, which is accompanied by the feedback control of the temperature and bias current, is needed.

Direct modulation can be efficiently used for modulation signal bit rates that scale up to 1 Gbps. The transmission distance is determined by the point at which the signal degradation, which is characterized by the BER, exceeds the specified level. The signal degradation is mainly caused by chromatic dispersion, which is an effect that scales proportionally to the laser frequency chirp and dispersion properties of the optical fiber.

The frequency chirp, which is described in Section 3.3.3, arises mainly due to variation of the modulation current between two logical levels. This variation of the

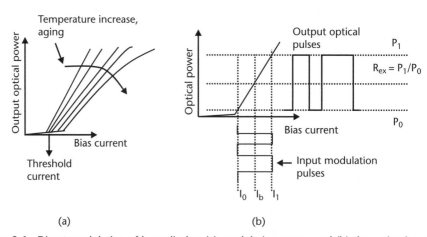

Figure 2.6 Direct modulation of laser diodes: (a) modulation curve, and (b) the extinction ratio.

current is accompanied by the variation of carriers' density within the laser cavity, which brings the changes to the refractive index in the active region. The variations of the refractive index will change the phase of the generated photons and cause the variations in the output frequency. It is desirable, therefore, to minimize the laser frequency chirp by limiting the variations of the modulation current.

The extinction ratio is another parameter, besides the frequency chirp, that is important from the systems engineering perspective. This parameter is the ratio between output optical powers corresponding to bits 1 and 0, respectively, as shown in Figure 2.6(b). It is desirable that the extinction ratio is as high as possible in order to achieve high SNR, which can be achieved by allowing broader variations in the modulation current. However, this requirement is contrary to the need to minimize variations in modulation current in order to suppress the impact of the frequency chirp.

2.2.2 External Optical Modulation

The other option for modulating an optical signal is through an external process that takes place after the light is generated. In this case the laser diode produces a CW signal, which serves as an input to the external modulator. Switching between two logical levels occurs within the modulator under the impact of a digital modulation voltage. This process is known as intensity modulation, since it produces an intensity modulated optical signal at the modulator output. Some other modulation types, such as phase modulation, can also be performed by an external modulator structure (refer to Section 7.8 to find more details about different modulation types). The external modulation process is more complex than the direct modulation one, but it can provide a significant practical advantage. Such an advantage, which is measured by an increase in both the modulation bite rate and transmission distance, is enabled by the smaller frequency chirp.

There are two types of external modulators commonly used in practice: Mach-Zehnder (MZ) and electroabsorption (EA). The operation principle of these two types is illustrated in Figure 2.7. It is based on the well-known electro-optic effect that occurs in some materials when the refractive index is changed under the impact of an external electric field [20]. Some of materials that are the most suitable for electro-optical effect employment are Lithium Niobate (LiNbO$_3$), Indium Phosphate (InP), Gallium Arsenide (GaAs), and some polymer materials. The refractive index change occurs relatively quickly and in accordance with the changes in the electric field applied, which enables a high-speed operation.

The Mach-Zehnder modulator is a planar waveguide structure deposited on the substrate, as shown in Figure 2.7(a). There are two pairs of electrodes applied to the waveguide, the first one for dc bias voltage, and the second one for the high-speed ac voltage that represents the modulation data signal. The electrodes for ac and dc voltages are deposited along the interferometer arms. There are various combinations with respect how to apply these voltages, in terms of numbers and the physical positions of the electrodes. The physical layout is extremely important since the way an electrical field is applied has a significant impact to the modulation characteristics of the MZ modulator.

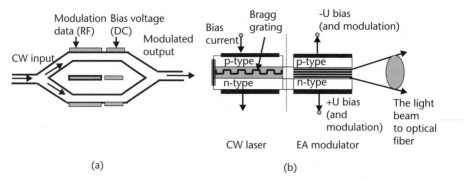

Figure 2.7 External optical modulators: (a) Mach-Zehnder modulator, and (b) electroabsorption modulator.

The continuous lightwave coming from the laser is equally split between two arms of the modulator. These portions are eventually combined again at the modulator output. The modulation principle of the MZ modulator is as follows: any applied voltage to the modulator arms increases the refractive index of the material and slows down the light speed, thus effectively retarding the phase of the optical signal. The phases of two incoming portions will determine the nature of the eventual recombination at the modulator's output. If the refractive indexes at any given moment are equal, two light streams arrive in phase, and their amplitudes are added to each other. This situation at the detection point corresponds to the 1 bit, since a high pulse level has been produced at the modulator's output. On the other hand, the output signal level will be lower if there is a difference in refractive indexes between two interferometer arms. The extreme case occurs if the difference between the phases equals π (or 180°), since two signal streams interfere destructively by canceling each other [21]. It is easy to understand that this case corresponds to 0 bits after photodetection takes place. Therefore, we can say that by applying the modulation voltage to the Mach-Zehnder interferometer arms, the CW light is effectively chopped and a modulation by digital data stream is performed.

The phase shift between two optical streams is proportional to the difference in voltages applied to two waveguide arms. The voltage difference is usually measured with respect to the voltage value V_π that is needed to shift the phase by 180°, and switch the output level from space to mark. The modulation curve of the MZ modulator, which is shown in Figure 2.8, can be expressed by the following relation:

$$P_{out} = P_{in} \cos^2\left(\frac{\pi V}{2 V_\pi}\right) \qquad (2.5)$$

where P_{in} and P_{out} are the incoming and outgoing signals from the modulator, respectively, while V is the total voltage applied (which is a sum of bias voltage and modulation data signal).

Either positive or negative slope of the modulation curve can be selected by choosing a proper value of the dc bias voltage, and that effectively determines the properties of the modulated signal and the value of the induced frequency chirp (refer to Section 3.3.3). As we already mentioned, a 180-degree phase shift between interferometer arms produces the total mutual cancellation of the contributing arms signals. That would produce an indefinite value of the extinction ratio (which is the ratio between levels associated with 1 and 0 bits). However, the situation is quite different in practice since the mutual cancellation is not perfect and the extinction ratio takes some defined value, which is usually in the range around 20 dB.

The EA modulator shown in Figure 2.8(b) is a semiconductor-based planar waveguide that consists of multiple p-type and n-type layers, which form multiple quantum wells (MWQs) [22, 23]. The multiple p/n-type layer design serves to support the quantum Stark Effect more effectively [20]. Since the layered structure of these devices has some similarity with the laser structures, the design of both the lasers and EA modulators can be done on the same substrate, as indicated in Figure 2.8(b). However, in such a case, the laser and modulator must be electrically isolated from each other. Although the EA modulator can be a single packaged device that is connected to the laser by a fiber pigtail, integration with the laser is a much better practical solution and has a wider application in high-speed communication systems.

The EA modulators operate just the opposite to semiconductor lasers since they are reverse biased, which resembles a photodiode regime. The EA modulator is practically transparent to the incoming continuous wave optical signal if there in no bias voltage applied to the modulator waveguide structure. That transparency is not quite ideal since some incoming photons will eventually produce electron-hole pairs and cause some signal attenuation.

The situation changes when some bias voltage is applied. The applied voltage can separate electron-hole pairs and generate a photocurrent, which effectively increases the waveguide attenuation and signal losses. Therefore, we can say that the output signal from the EA modulator is the highest when there is no voltage applied and decreases as the bias voltage increases. It is worth noting that the modulated optical signal at the EA output is opposite in phase to the modulation digital stream applied through the electrodes. This is because 0 level in voltage generates 1 level in optical signal, and vice-versa. A typical modulation curve of the EA modulator is shown in Figure 2.8(b).

The modulation speed of EA modulators is comparable with the speed of MZ modulators, but the extinction ratio is generally smaller, typically around 10 to 13 dB. The ability to integrate semiconductor lasers with EA modulators on the

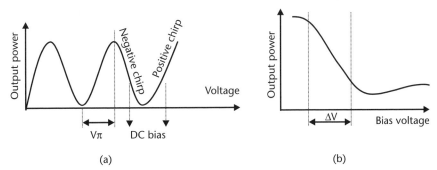

Figure 2.8 Modulation curves of (a) a Mach-Zehnder modulator, and (b) an electroabsorption modulator.

same substrate is a big competitive advantage since it reduces the insertion losses, simplifies packaging by making everything almost like a regular laser package, and reduces the total cost of the device. In addition to integration with the lasers, EA modulators could be integrated with other semiconductor chips, such as semiconductor optical amplifiers and multichip wavelength selectable lasers [18]. Besides the features mentioned above, it is important to note that the EA modulator is more resilient to changes in polarization of the incoming light signal than the MZ modulator. However, the output optical power from the EA modulator is generally lower than the power at the output of the MZ modulator.

As a summary, we can outline the optical modulator parameters that are important from the systems engineering perspective. They are:

- The insertion loss of the modulator that determines the output power;
- The frequency chirp characterized by chirp factor α_{chirp}, as defend in Section 3.3.3;
- The extinction, or contrast ratio, characterized be the ratio of optical powers corresponding to 1 and 0 bits, respectively.
- The modulation speed that is characterized by the frequency response in general, and by the cutoff modulation frequency in particular.

Typical values of optical parameters related to MZ and EA modulators are shown in Table 2.2. These values, referring to modulators as standalone devices, are extracted from the sourcebook [16], and can serve for reference purposes.

2.3 Optical Fibers

Optical fibers serve as the foundation of an optical transmission system since they transport optical signals from source to destination. The combination of low attenuation and extremely wide transmission bandwidth allows high-speed optical signals to be transmitted over long distances before regeneration becomes necessary. A low-loss optical fiber is manufactured from several different materials. The base row material is ultra pure silica, which is mixed with different additives, or dopants, in order to adjust the refractive index of the optical waveguide, and to influence the propagation characteristics of the fiber. The optical fiber consists of two key waveguide layers protected by buffer coating, as shown in Figure 2.9. The inner layer is known as the fiber core, and serves as a medium to guide the optical signal

Table 2.2 Typical Values of Optical Modulator Parameters

Parameter	EA	MZ
Insertion loss	7–15 dB	4–7 dB
Extinction ratio	10–13 dB	10–50 dB
Modulation bandwidth	Up to 75 GHz	Up to 85 GHz
Chirp factor	−0.2 to 0.8	−1.5 to 1.5

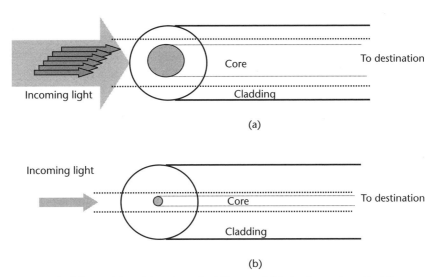

Figure 2.9 Optical fibers: (a) multimode optical fiber, and (b) single-mode optical fiber.

[24–27]. The majority of the optical signal power is concentrated within the core area, although some portion can spread outside the fiber core to the surrounding layer, which is known as the optical fiber cladding.

There is a difference in refractive indexes between the core and cladding, which is achieved by a mix of dopants usually added to the optical fiber core. The fiber core and the fiber cladding together form a waveguide structure that effectively supports light propagation over longer distances. This waveguide structure is covered by a buffer coating for protection purposes. The coating is plastic based and can absorb mechanical stresses and isolate the waveguide structure from impurities. The optical fiber can be physically treated as a wire that should be cabled before it is ready for an application. During the cabling process, the structure from Figure 2.9 is additionally protected with layers that provide both strength and insulation.

There are two groups of optical fibers shown in Figure 2.9. The first group, multimode optical fibers, transfers light through a collection of spatial transversal modes. Each mode, defined through a specified combination of electrical and magnetic field components, occupies a different cross-sectional area of the optical fiber core and takes a slightly distinguished path along the optical fiber. The cross-sectional distribution of the optical power of four fundamental modes is shown in Figure 2.10. The modes are presented according to linear polarization (LP) approximation, which is widely used in the literature for analysis of optical fiber characteristics [24–29]. The darker area means a high concentration of the optical energy, while brighter areas mean a low energy concentration. As we can see, there is a fundamental mode known as the LP_{01} mode that occupies the central part of the optical fiber, which has an energy maximum at the axis of the optical fiber core. The radial distribution of the power of the LP_{01} mode can be approximated by Gaussian curve, as illustrated in the upper part of Figure 2.10.

The number of modes that can effectively propagate though an optical fiber is determined by the V-parameter, defined as

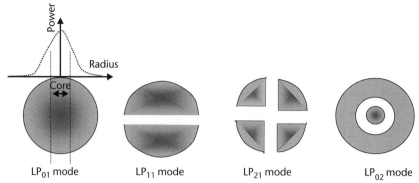

Figure 2.10 Fundamental modes in an optical fiber.

$$V = \frac{2\pi a}{\lambda}\sqrt{n_{co}^2 - n_{cl}^2} \quad (2.6)$$

where a is the fiber core radius, λ is the carrier wavelength of the optical signal, and n_{co} and n_{cl} are the refractive indexes related to the fiber axes and the cladding region, respectively. For a typical multimode optical fiber, the core radius is 50 µm, while the difference between refractive indexes n_{co} and n_{cl} usually stays below 1%. The number of modes supported can vary from a few to several hundreds, and even several thousands.

Each mode propagating through the fiber is characterized by its own propagation constant β (refer to Sections 7.2 and 7.3 for the basics of electromagnetic wave propagation). The dependence of the electric and magnetic fields on the axial coordinate z is expressed through the factor $\exp(-j\beta z)$. Parameter β, which determines the speed at which the optical pulse energy travels through an optical fiber, should satisfy the condition $2\pi n_{cl}/\lambda < \beta < 2\pi n_{co}/\lambda$ for every mode supported by the optical fiber waveguide structure. It is useful to know the functional dependence of the mode propagation constant on the optical signal wavelength in order to evaluate transmission characteristics of the optical fiber. The normalized propagation constant b is defined for that purpose, and it is often used instead of the propagation constant β. The normalized propagation constant b is defined as [24, 25]

$$b = \frac{\beta^2 - (2\pi n_{cl}/\lambda)^2}{(2\pi n_{co}/\lambda)^2 - (2\pi n_{cl}/\lambda)^2} \quad (2.7)$$

Another useful relationship connects the normalized propagation constant b with the V parameter defined by (2.6); this is [24]

$$b(V) \approx (1.1428 - 0.9960/V)^2 \quad (2.8)$$

The approximation given by (2.8) can be effectively used if parameter V ranges from 1.5 to 2.5 [26]. As we will see shortly, this is the range of parameter V that is most interesting from an application perspective.

The difference in mode path lengths in multimode optical fibers produces a difference in arrival times at the receiving point. This phenomenon is known as multimode dispersion and causes signal distortion and limitations in transmission bandwidth.

2.3.1 Single-Mode Optical Fibers

The second group of optical fibers effectively eliminates multimode dispersion by limiting the number of propagating modes to a fundamental one. The single mode that remains is the LP_{01} mode, which has the energy distribution illustrated in Figure 2.10. The single-mode operation is defined through the parameter V, which should be lower than the cutoff value $V_c = 2.405$. The required value of parameter V is maintained mainly through a smaller core diameter and smaller relative difference $\Delta = (n_{co} - n_{cl}) / n_{co}$ between indexes n_{co} and n_{cl}. The typical value of core radius is around $5\,\mu$m for a standard single-mode fiber, while the Δ parameter resides in the range between 0.2% and 0.3%.

Single-mode optical fibers do, however, introduce another signal impairment known as chromatic dispersion. Chromatic dispersion is caused by a difference in velocities among different spectral components within the same mode. There are two components that contribute to the total value of the chromatic dispersion. The first one, commonly known as material dispersion, is related to the fact that the refractive index of the fiber material is not constant but depends on the signal wavelength. This functional dependence can be approximated by the Sellmeier equation:

$$n(\lambda) = \left[1 + \sum_i^M \frac{B_i \lambda^2}{\lambda^2 - \lambda_i^2}\right]^{1/2} \tag{2.9}$$

Parameters B_i and λ_i can be determined empirically for a specified material by using an interpolation procedure based on the measured results, usually for $M = 3$ [27]. These coefficients strongly depend on the dopant concentration implanted in the base silica. Therefore, different wavelength components within the same temporal pulse shape will travel at different speeds since the refractive index is a function of wavelength. This effect is known as material dispersion.

The second component of the chromatic dispersion, known as waveguide dispersion, is related to the physical design of the optical fiber. We already mentioned that the relative difference $\Delta = (n_{co} - n_{cl}) / n_{cl}$ between indexes n_{co} and n_{cl} is fairly small, which means that optical energy is not strictly confined within the fiber core but goes out of the core region and propagates through the fiber cladding at a speed that is different than the speed in the fiber core [28].

Therefore, changing the power distribution across the cross-sectional area can change the overall picture related to the chromatic dispersion. The power is narrowly distributed for the larger parameter Δ, which means that a large portion will be confined within the core area, and vice-versa. In addition, it is important to note that the power distribution is also a function of wavelength. The distribution of the mode power and the total value of waveguide dispersion can be manipulated by multiple cladding layers, which are different in thickness and the refractive index values, as shown in Figure 2.11. This is commonly done in optical fibers designed

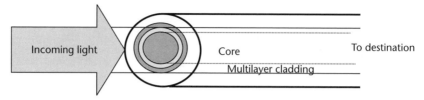

Figure 2.11 Optical fiber with multilayer cladding.

for special application purposes, such as chromatic dispersion suppression or compensation. The waveguide dispersion can be applied against the material dispersion in some wavelength regions, and that is the physical background that explains the changes in total value of chromatic dispersion (refer to Section 3.3.4 to learn more about the chromatic dispersion effect).

It is also worth noting that the chromatic dispersion is an increasing function with respect to optical signal wavelength. There is a wavelength region where chromatic dispersion changes the sign from negative to positive, thus effectively going through the zero dispersion point. The wavelength region where the chromatic dispersion changes the sign is known as the zero dispersion region, and that is one of key the parameters that can be found in the manufacturer's product data sheet. Another important parameter is the slope of the chromatic dispersion curve. Both of these parameters will be discussed in more detail in Sections 3.3.4 and 6.1.2.

Several examples of the refractive index profiles related to different designs of single-mode optical fibers are shown in Figure 2.12 (refer to Section 3.3.4). Classical single-mode fiber (SMF) has a simple step-index profile with two distinguished values corresponding to n_{co} and n_{cl}, as shown in Figure 2.12(a). More complex refractive index profiles have been introduced to serve different purposes, such as:

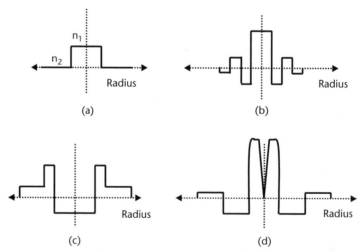

Figure 2.12 Refractive index profiles of single-mode fibers used today: (a) standard SMF, (b) NZDSF with reduced dispersion slope, (c) NZDSF with large effective area, and (d) dispersion compensating fiber.

- To reduce the slope of the chromatic dispersion curve, which might be useful for link dispersion management in some special cases of multichannel optical signal transmission [please see the profile from Figure 2.12(b), which is related to nonzero dispersion shifted fiber (NZDSF) with reduced dispersion slope; the reader is also advised to refer to Figure 3.12 and Table 3.2].
- To increase the effective area of an SMF, which helps to reduce the effect of nonlinearities [the profile from Figure 2.12(c)].
- To introduce large negative waveguide dispersion (which becomes the predominant factor in the total chromatic dispersion), which can be used to compensate for a positive chromatic dispersion accumulated along the transmission line. Such fibers are known as dispersion compensating fibers, and have refractive index profiles similar to that shown in Figure 2.12(d).

Low signal attenuation in optical fibers is the most important feature that enables long distance transmission capability. The attenuation, or signal loss, is relatively small in comparison to losses in other transmission media, such as different metal cables or free space. A typical attenuation curve of a silica-based optical fiber is shown in Figure 2.13. The shape of the curve from Figure 2.13 is determined by a contribution of absorption, scattering, and radiation effects related to the optical signal energy.

Absorption is the dominant factor contributing to the total attenuation. The following physical mechanisms contribute to the total absorption effect:

- *Intrinsic absorption* is caused by the atoms of basic optical fiber material. This absorption is a principal factor that defines the attenuation of an optical signal by setting the lower limit for any particular material. The intrinsic absorption is attributed to the perfectly pure base material that does not have any density variations or material inhomogeneities. It is associated with the interaction of incoming photons with electronic absorption bands whenever the energy of an incoming photon is higher than the electronic band gap of the amorphous glass material. The absorption occurs when photons excite electrons to a higher energy level through the energy transfer. The energy of the photon gradually decreases with an increase of photon wavelength,

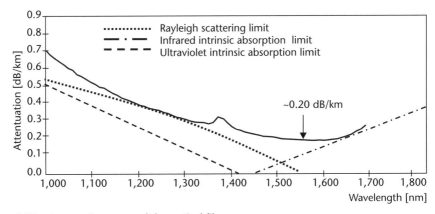

Figure 2.13 Attenuation curve of the optical fiber.

which means that a smaller number of photons will be capable of exciting electrons to higher energy levels. Therefore, this type of absorption, commonly associated with the ultraviolet wavelength region, will gradually decrease with wavelength, as shown in Figure 2.13. For longer optical wavelengths, however, the intrinsic absorption becomes associated with vibration frequencies of the chemical bonds between the atoms of the basic fiber material. The absorption that occurs at longer wavelengths from the infrared region is a result of the transfer of energy from the signal electromagnetic field to the number of bonds in the base fiber material. The absorption loss increases with further increase in optical wavelengths due to increased energy transfer from photons to chemical bonds.

- *Extrinsic absorption* is caused by the presence of impurity ions in the base fiber material. Such ions are the positive ions of some metals (iron, chromium, cobalt, copper), and negative OH ions. The metal ions are present in the original materials in quantities usually around 0.1 ppb (particles per billion). The absorption by these ions occurs due to energy transfer from photons to the electrons associated with the ion subshell. The presence of OH ions is a direct result of the use of oxy-hydrogen flame that is applied to convert chlorides to the implanted dopants (refer to Figure 2.15). A few OH particles per billion would cause an attenuation peak of about 20 dB/km. Such a peak can be observed in the older generation of optical fibers at wavelengths around 950 and 1,400 nm, since it was very difficult at that time to keep the level of OH ions bellow several ppb. The two attenuation peaks separate three wavelength valleys, where the signal attenuation is much lower as compared to surrounding wavelength regions. These three wavelength regions are known as transmission windows. Meanwhile, significant progress has been made in reducing the content of OH ions to less than 1 ppb, which practically eliminates the absorption peaks [29].
- The *imperfections* in fiber material slightly increase the total attenuation effect. Imperfections such as high-density clusters, missing molecules, or even oxygen defects in the glass occur due to defects in the atomic configuration of the base fiber material.

Scattering losses are the second dominant factor contributing to the signal attenuation in an optical fiber. This kind of loss is caused by microscopic variations in the fiber material density, which originates from the fiber manufacturing process. The fiber structure of randomly connected glass molecules contains the regions where the molecular density varies from an average value. Any variation in material density will cause variations of the local value of the refractive index, which will occur over distances smaller than optical signal wavelength. These variations will cause a phenomenon better known as Rayleigh scattering [30].

The scattering loss decreases in proportion to the fourth power of the signal wavelength. That is the reason why the scattering loss, which is a dominant loss mechanism below wavelengths of 1,000 nm, becomes almost negligible in the third transmission window at the wavelengths around 1,550 nm.

The third mechanism that contributes to the total signal attenuation is associated with the radiation of the signal energy outside of the fiber waveguide region and occurs when optical fiber is bent. The fiber bending is usually a consequence of the cabling process. Macrobends occur if the curve radius is measured by centimeters. It is not recommended to allow fibers to undergo any banding where the bending radius is smaller than couple of centimeters since the attenuation increase can be anywhere from 0.01 dB to almost 3 dB. The value of 3 dB occurs when the bending radius is approximately 1 cm.

Microbendings occur if the curve of bending is measured by microns, which usually happens as a result of extrusion of a compressible jacket over the optical fiber. In addition, a number of microbends might appear if an external force is applied to the jacketed fiber. The microbending loss can be evaluated by measuring the total attenuation of an optical fiber before and after the cabling process. Otherwise, it can be characterized by using a statistical evaluation [24]. It is reasonable to assume that high-quality optical cables insert microbending losses that are lower than 0.1 dB/km.

The signal attenuation in optical fibers is not the only cause of the optical signal loss along an optical fiber transmission line. There are also optical fiber splices and fiber connectors that insert additional attenuation. Fiber splices can be either permanent (fused), or removable. A typical mean attenuation inserted by a fused optical splice is somewhere between 0.05 and 0.1 dB, while removable mechanical splices insert a loss comparable or slightly more than 0.1 dB. On the other hand, the optical connectors are designed to be removable, thus allowing many repeated connections and disconnections. The insertion loss for high-quality single-mode optical connectors should not be higher than 0.25 dB. Optical connectors should be very carefully designed since reflection of the incoming optical signal from the surface of receiving part of the connector can occur and cause an additional noise (refer to Section 3.2). Such a design can include angled fiber-end surfaces or some index-matching fluid applied at the fiber surfaces to minimize the refractive index change when the optical signal crosses from one fiber to the other.

The number of optical splices and connectors depends on the length of the optical transmission line. It is up to the transmission line engineer to select where to put a splice and where to put a connector, but the general rule is to use optical splices wherever it is possible. The number of optical fiber joints (splices and connectors) should be taken into account during the optical transmission systems engineering. The engineering is greatly simplified if the total connection losses are distributed over the overall transmission line and added to the fiber attenuation. In such a case, the transmission line is characterized by a per-kilometer attenuation value.

2.3.2 Optical Fiber Manufacturing and Cabling

An optical fiber is made by drawing down a cylindrical preform, as shown in Figure 2.14. The preform is structured the same way the fiber is, with the fiber core and fiber cladding distinguished from each other [31]. The base of the preform is heated in a specially shaped furnace at a temperature usually higher than 2,000K. The fiber drawing can begin when the preform starts melting. During the drawing process, the fiber diameter is continuously monitored and controlled by an automatic servo process, as shown in Figure 2.14. The control is done through adjusting

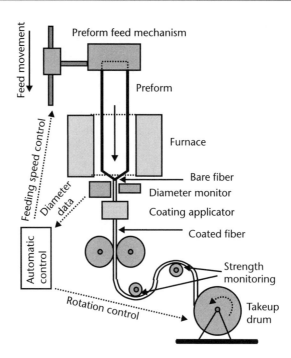

Figure 2.14 Optical fiber drawing process.

the winding rate of the drum, in such a way that it permits just a small diameter variation (usually as small as 0.1%). The protective coating is also inserted during the drawing process to provide mechanical protection and isolation of the fiber waveguide structure. Finally, the protected fiber is drawn to the receiving rotation spool. The diameter of the manufactured fiber structure can vary from several hundreds of microns to close to 1 mm, which depends on the thickness of the protective coating. It usually takes several hours to transfer preform rod, which is usually 1-m long and 2-cm in diameter, into an optical fiber with a length of about 5 km.

The optical fiber preform should have a cylindrical shape to prevent occurrence of polarization mode dispersion (PMD) (refer to Section 3.3.5). There are several established methods for preform preparation, such as vapor axial deposition (VAD), outer vapor deposition (OVD), modified chemical vapor deposition (MCVD), the sol-gel method, and the plasma process [32, 33]. The first three of them have been widely used for massive production of optical fibers. The main purpose of the preform manufacturing is to provide a homogenous radial distribution of dopants that change the refractive index profile. All dopants, such as germanium (Ge), boron (B), phosphorus (P), and fluoride (F), are added in vapor phase together with silicon, which serves as a base material. The chemicals that are added are not pure chemical elements, but rather a mix of chlorines and oxides that are eventually merged with the oxygen.

The chemical interaction takes place under the high temperature torch and leads to the creation of numerous layers. The chemical interaction and deposition of the layers takes place in an organized and controlled manner. The refractive index of a specific layer is controlled by changing the content of dopants that undergo the chemical interaction. An even and radial character of deposition is achieved by the rotation of the tube, and by the movement of the torch.

The MCVD process is characterized by a deposition that is done on the inner surface of the rotating silica tube, while a sliding torch is used to maintain a very high temperature, as illustrated in Figure 2.15. The deposited material will eventually form the core of an optical preform that has a specified refractive index distribution [32]. The temperature of the multiburner torch is raised after the deposition of all layers, and that causes the tube to collapse around the deposited structure to form a solid rode, also known as the fiber preform. The tube used for deposition serves as the cladding of the preform.

A slightly different approach is taken during fiber preform production by using the VAD process [33], which is characterized by a frontal deposition of chemicals and a vertical growth of the preform rod. However, the end result of any manufacturing process is the fiber preform as a macroscopic version of the optical fiber structure.

The manufactured optical fibers should be incorporated in some type of cable structures before any practical application takes place. The cabling is necessary in order to protect fibers from possible damages during transportation and the installation process. In addition, the cabling structure provides a more stable environment for optical fibers during their lifetime. Optical cable structure is generally different for different applications, and it may vary from a simple light-duty structure including just a plastic jacket around the fiber, to very robust mechanical structures that contains some strengthening elements [34].

The light-duty optical cables can be designed with either tight jacket or loose polyethylene tube, as shown in Figure 2.16. A tight jacket, with a thickness of up to 1 mm, is applied directly to the optical fiber primary coating. The tight jacket puts some internal pressure to the optical fiber structure, which generally leads to an increase in microbending losses. The loose-tube design prevents tight mechanical

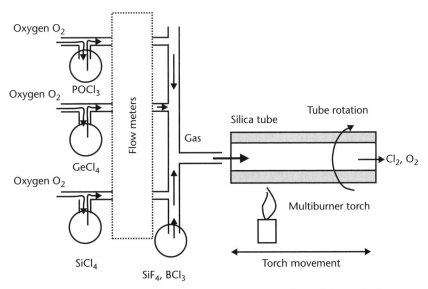

Figure 2.15 Illustration of the fiber preform production process by MCVD method.

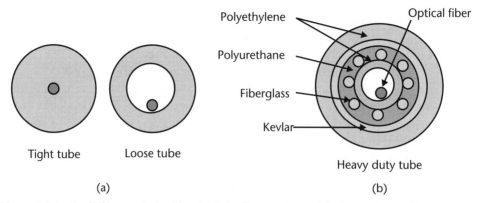

Figure 2.16 Single-fiber optical cables: (a) light-duty cables, and (b) heavy-duty cables.

contacts between fiber and protective jacket, which nearly eliminates microbending losses. Several elementary cable structures can be composed together and wrapped inside an additional tube, which effectively produces a multifiber cable.

A more robust optical fiber structure is usually needed for different outdoor applications. Such cables can be buried directly in the ground, pulled underground or through ducts between two buildings, buried under water, or installed at the outdoor poles. Although each application requires a specific cable design, there are requirements that they should be installed by the same type of equipment and installation technique that is used for installation of other conventional cables.

One of the most important mechanical properties of heavy-duty optical cables is the maximum allowable axial load. Optical fibers are not strong enough to be load-bearing elements, as opposed to copper wires in conventional cables. In addition, optical fibers cannot sustain any serious elongation without irreparable damage. The endurable elongation value before the fiber breaks is usually about 0.5%. However, the elongation that might happen during the cable manufacturing and installation should be limited to up to 0.1% in order to prevent any potential damage.

Either steel or fiberglass wires can be used to reinforce the strength of optical cables, as shown in Figure 2.17. The fiberglass rods embedded in polyurethane are preferred for some applications to reduce the overall optical cable weight and to avoid the effect of electromagnetic induction that might occur if steel wires were

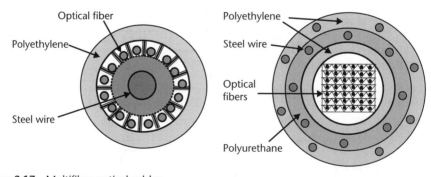

Figure 2.17 Multifiber optical cables.

used. A nonmetallic protection structure might also include a high-tensile strength Kevlar jacket, as well as the polyethylene jacket.

Any construction of heavy-duty optical cables should provide room for fiber to move when the cable is bent or stretched. Fiber ribbons with fibers placed between two polyester tapes are sometimes used to pack a large number of optical fibers within a single optical cable, as shown in the right part of Figure 2.17. Although the number of fibers per ribbon can vary, it usually contains 12 fibers, which are placed between the tapes. The number of ribbons stacked together can also vary, but the best mechanical stability is achieved by a rectangular array shape.

Optical cables should have an outer sheath to protect optical fibers from the impact loads. This is because an optical fiber has a low tolerance to absorb the energy caused by mechanical impacts. The outer sheath that provides protection from transversal forces should also be resistant to corrosion. The outer sheath is usually made from polyethylene, although a metal sleeve can be sometimes used. In addition, the construction of optical cables should prevent water intrusion into the cable structure, which means that special filler should be put in any empty space all along the cable.

Optical fibers are manufactured with some nominal length (usually several kilometers long), which means that longer optical fiber links are formed by joining several nominal lengths together. In addition, there is the need to link an optical fiber with optical transmitters, receivers, and amplifiers in order to establish a lightwave path. Optical fusion splices are commonly used to join optical fiber nominal lengths, while optical connectors are used to link the fiber with transmitters, receivers, and optical amplifiers. Optical fiber ends must be properly prepared, cleaned, and aligned in order to minimize the insertion losses [35].

2.4 Optical Amplifiers

There is a need to restore the optical signal strength from time to time, and also to preserve the SNR; this is because level of an optical signal that travels along a wavelength path gradually decreases due to inserted losses. The restoration can be done either through retransmission, or through an optical amplification process. Optical signal retransmission means that the signal undergoes an O-E-O conversion by using a photodiode, processing electronics, and another optical source. This sequence of functions provides an optoelectronic regeneration process. The signal retransmission is generally rather cumbersome, bit-rate specific, and expensive, especially if applied to multichannel optical transmission systems. It is desirable, therefore, to keep the signal at the optical level as long as possible and use pure optical amplifiers to compensate for optical signal loss accumulated along the lightwave path.

From an application perspective, optical amplifiers are more flexible since they have a large signal bandwidth and relatively high and adjustable gain coefficient. There are three major applications of optical amplifiers along the wavelength path:

1. Booster amplifiers are optical amplifiers placed at the optical transmitter side to enhance the power level of the output optical signal. The purpose of this amplification is to either boost the output power and bring it to the level

that could not be achieved by the laser transmitter alone, or to compensate for loses in the optical elements placed between the laser and optical fibers. Such elements are, for example, optical couplers, optical multiplexers, and external optical modulators.

2. In-line optical amplifiers are placed along the lightwave path to compensate for losses incurred during signal propagation. The chain of in-line optical amplifiers can be quite long, sometimes acquiring more than 100 amplifiers, as in optical submarine transmission systems.
3. Optical preamplifiers are used within optical receivers to increase the optical signal level before optoelectronic conversion takes place in the photodiode. By doing this, the receiver sensitivity is greatly enhanced.

Pure optical amplifiers can be constructed by using the physical principle of stimulated light radiation, the same principle used for light generation in lasers. However, while lasers need some resonant cavity with mirrors to support the lasing regime, optical amplifiers should be designed to prevent the lasing effect. Such a design should be accompanied with a strong pumping to populate upper energy levels and to create a strong population inversion of the carriers. It can be done by either decreasing facets reflectivity, or by excluding mirrors from the design scheme. There are two main groups of optical amplifiers that are based on design principles mentioned above [36–40]. The first one contains semiconducor optical amplifiers (SOAs), which are the devices similar to semiconductor lasers [36, 37]. The name laser, however, cannot really be applied because either direct injection current or facet reflectivity are intentionally kept low, so device operates below the threshold. Multiple reflections still occur at the facets of the Fabry-Perot laser resonator, even if the bias current is below its threshold value, but the feedback is not high enough to cause the lasing regime. Such optical amplifiers are called Fabry-Perot semiconductor optical amplifiers. The incoming optical signal is introduced through an input facet. Another type of semiconductor optical amplifiers is known as traveling wave (TW), where an optical signal does not undergo multiple reflections, but just passes through the cavity [38]. In this case the facet reflectivity is very low (below 10^{-4}), but the bias current is relatively high.

Semiconductor optical amplifiers are good candidates for some applications where larger signal bandwidth (more than 20 nm) and moderate gain (up to 15 dB) are needed. They are very compact devices that can be easily integrated with other semiconductor structures (such as optical modulators and photodetectors). On the other hand, they suffer from polarization sensitivity, relatively high noise, and signal crosstalk.

The second group of optical amplifiers is related to fiber-based amplifiers, where optical fibers serve as the active medium. The optical fibers can either be regular ones or specialty fibers designed for amplification purposes. The specialty optical fibers are silica-based fibers doped with elements such as erbium or praseodymium in order to increase the efficiency of stimulated emission. The population inversion is done by a strong pump optical signal, while stimulated emission is triggered by an incoming optical signal. The phase and frequency of radiated photons are the same as those of incoming photons. The affective amplification of the optical signal occurs since newly generated photons stay within the fiber waveguide structure and start propagation in the same direction as the incoming signal. The best-known amplifiers based on specialty optical fibers are erbium doped fiber amplifiers (EDFA), which are widely used for different purposes. On the other hand, the Raman amplifier is the best example of an amplifier that utilizes regular optical fibers for signal amplification.

The next paragraphs will introduce the basics of EDFAs; details about the design of Raman amplifiers can be found in Section 6.1.1. The most important part of the EDFA is the optical fiber that is doped with erbium ions (Er^{3+} ions) [41–43]. Some additional dopants, such as fluoride or aluminum, are also used to optimize the optical fiber amplifier gain profile with respect to the specified wavelength band.

A general application scheme of EDFAs, which is related to in-line amplification, is shown in Figure 2.18. The design shown in this figure is well known as a two-stage design, where each of the two stages is designed differently. The first stage is optimized to provide both high gain and low noise, while the second stage serves to boost the output optical power. The intermediate stage is commonly used for optical signal conditioning, which may include dynamic gain equalization of different optical channels, dispersion compensation, and an optical signal add/drop function. In either case, there is an insertion signal loss that should be compensated for. The optical pumping is introduced through wavelength selective couplers. It often occurs that a CW semiconductor laser radiating at 980 nm serves as the first pump, while the second-stage pump laser radiates at 1,480 nm. The total noise figure of the two-stage amplifier design is calculated by including contributions of both stages (refer to Section 3.1.3).

The EDFA is produced by incorporating erbium ions into the glass matrix of the core of silica optical fiber. This process leads to a classic three-level lasing system, which is illustrated in Figure 2.19. There are several energy levels that could eventually be used for the electron transitions, but the three levels presented in Figure 2.19(a) are the most suitable to support the amplification process. The population inversion of electrons at upper energy levels is created by optical pumping. Such pumping enhances the energy of electrons, so they are lifted from the ground energy level ($^4I_{15/2}$) to some higher energy levels. For example, the upper energy level ($^4I_{9/2}$) can be populated by using the optical pump radiating at 800 nm, the lower level ($^4I_{11/2}$) can be populated by using the pump radiating at 980 nm, and the optical pump radiating at 1,480 nm can be used to lift electrons from the ground level to the next higher energy level ($^4I_{13/2}$).

The electrons lifted to a level higher than the ($^4I_{13/2}$) level will soon slide down to the ($^4I_{13/2}$) level through the process of nonradiative decay. This occurs because both

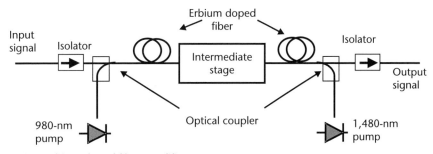

Figure 2.18 Erbium doped fiber amplifier.

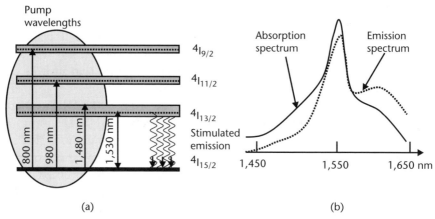

Figure 2.19 EDFA: (a) energy diagram, and (b) the gain spectrum.

upper energy levels are not the stable ones. Nonradiative decay occurs within one microsecond time period calculated from the moment when they populated the upper energy state. On the other hand, the $^4I_{13/2}$ level is known as the metastable state, which means that electrons populating this energy level have a relatively long lifetime, at this energy level. The electron lifetime at the $^4I_{13/2}$ level is around 11 ms [41, 42]. Such a relatively long lifetime of electrons helps to establish a reservoir of energy at the $^4I_{13/2}$ level, which can be used for optical signal amplification. Technically speaking, is better to use the term "energy zone" instead of "energy level" since the discrete energy levels associated with isolated erbium ions have been split into energy zones through a process known as the Stark effect [20].

Optical signal amplification is the process of stimulated radiation of the light when electrons drop from the metastable level to the ground energy level. The radiated photons follow the frequency, phase, and the direction of the incident optical signal, which effectively leads to an amplification of the input optical signal. The difference between individual energy levels within the metastable zone and levels belonging to the ground zone helps to create photons with the emission spectrum covering wavelengths from approximately 1,530 to 1,560 nm. The radiation spectrum can be shifted, or effectively broadened, if some other special codopants are added to the silica glass. The radiation spectrum is widened if there are codoping ions since codopant atoms create an electric field that enhances the Stark effect [43]. A wide variety of options have been tried to optimize characteristics of the EDFAs with respect to the spectrum broadening, gain flattening, or gain wavelength shifting. The highest figure of merit was achieved by adding the cocktail of aluminum and fluoride ions mixed together with erbium ions [44].

The stimulated emission of the light is accompanied by spontaneous emission, when optical radiation occurs randomly without any established pattern. Spontaneous emission is the source of the noise in optical amplifiers, and that effect will be examined in Section 3.2.7. The spontaneous emission is accumulated along the transmission line. In addition, subsequent in-line amplifiers increase the level of the spontaneous emission, so it arrives at the receiving side as the amplified spontaneous emission (ASE) noise.

The high-power semiconductor lasers are mainly used for pumping purposes since they can radiate in the wavelength region from 800 to 1,480 nm, which is needed to excite the electrons from the ground level to metastable one. The 980-nm semiconductor pump lasers are widely employed in the first amplifier stage, while pump lasers radiating at 1,480 nm are used in the second stage (refer to Figure 2.18).

When the 980-nm pump lasers are used, the process of stimulated emission is achieved through several steps. First, the electron is excited from the ground level ($^4I_{15/2}$) to level ($^4I_{11/2}$). The second step takes place after about 1 ms when electrons slide down from the level ($^4I_{11/2}$) to the metastable ($^4I_{13/2}$) energy level. As already mentioned, the electrons can live up to 11 ms at the metastable level before they radiate the energy and fall back to the ground level. The pumps operating at 980 nm have lower noise figure since signal and pump are separated in wavelengths by several hundred nanometers. In addition, they are rather reliable and require a simper and less expensive WDM coupler.

The 1,480-nm semiconductor pump lasers are also widely used since they are readily available and have better reliability than 980-nm pump lasers. In addition, they radiate at the wavelengths where optical fiber attenuation is lower than the attenuation at wavelengths around 980 nm, which makes them more suitable for applications where remote pumping might be needed. In this case, the population inversion is achieved in a single step since the electrons have been excited directly by the metastable energy levels. On the other hand, the noise figure related to the 1,480-pumps is higher than the noise figure associated with 980-nm pumps since the signal and pump wavelengths are relatively close to each other.

Some additional optical fiber amplifier schemes utilize specialty fibers that contain dopants different than erbium ions (such as praseodymium, neodymium, or thulium ions.) Such a design is suitable for amplification of the optical signal at wavelengths different than those around 1,550 nm. For example, both praseodymium and neodymium doped fiber amplifiers can be used in the wavelength region around 1,300 nm, while thulium doped fiber amplifiers (TDFAs) can be used at the wavelength region around 1,450 nm.

The second type of optical fiber amplifiers is based on stimulated Raman and stimulated Brillouin light scattering effects that occur in regular optical fibers. The physical nature of these effects is explained in Sections 3.3.9 and 3.3.10. While the Brillouin amplifiers can be used just for low bit rate applications, such as one presented in [45], the Raman amplifiers based on the SRS effect can find much wider practical applications [46].

The cost-effective design of Raman amplifiers was enabled by the availability of reliable high-power pump lasers. A simplified general scheme of a Raman amplifier is shown in Figure 2.20, while more detailed schemes of Raman amplifiers are presented in Section 6.1.1. The pump power is usually launched into the fiber in the opposite direction to the signal propagation. Two orthogonally polarized pump-signals are combined to provide a polarization-independent pumping scheme. As a result, the forward-propagating optical signals are enhanced since they get some energy through distributed SRS effect (refer to Section 3.3.9 to learn more about the SRS effect). The optical gain achieved by the Raman amplifier helps to keep the signal further above the noise level. A fairly flat gain profile over a wide

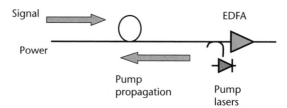

Figure 2.20 The Raman optical amplifier scheme.

range of optical wavelengths can be achieved by combining several pumps operating at different wavelengths.

The main advantage of Raman amplifiers is related to the fact that they do not require a specialty fiber. In addition, the Raman amplification is better suited to handle the impact of different nonlinearities that occur in the optical fiber since it lowers the level of the launched signal power. The Raman amplifiers can be used either as sole amplifiers, or to operate in combination with both EDFAs and TDFAs.

There is yet another type of optical amplifier that could find application in the near future. This type, known as a fiber parametric amplifier (PA), performs wavelength conversion in parallel with the signal level restoration [47]. Parametric amplifiers utilize so-called phase conjunction to perform functions mentioned above [please refer to (3.119) and (6.42)].

As a summary note, we can say that EDFAs can effectively cover both C and L wavelength bands. However, they cannot be used for other wavelength regions, such as the second wavelength window around 1,300 nm, or the S band from Figure 1.2. The praseodymium and neodymium doped fiber amplifiers are good candidates for application in the 1,300-nm wavelength region, although they did not yet find a wider application. As for the S band, it seems that TDFAs can be effectively used. A major obstacle for wider application of optical amplifiers that are based on specialty fibers, and operate at wavelength regions below 1,500 nm, is their relatively high cost and the lack of good pump lasers.

Advanced optical amplifiers for high-capacity long-haul transmission systems should provide enough gain for more than 100 optical channels, which means that the aggregate optical power should be in excess of 20 dBm. In addition, the noise figure should be as low as possible, while the gain profile should be equalized along the entire amplifier bandwidth. Typical parameters of optical amplifiers available on the market are summarized in Table 2.3.

Table 2.3 Typical Values of the Optical Amplifier Parameters

Parameter	SOA	EDFA	Raman Amplifier
Operating wavelength (nm)	1,280–1,350 1,530–1,610	1,528–1,610	1,200–1,700
Peak gain (dB)	10–25	17–45	10–25
Maximum output power (dBm)	Up to 15	Up to 37	Up to 40
Noise figure (dB)	Around 8	5–7	N/A (see Section 6.1.1)

2.5 Photodiodes

The photodiode is an integral part of an optical receiver (refer to Figures 2.1 and 5.1). The main role of the photodiode is to absorb photons of the incoming optical signal and convert them back to the electrical level through the process just opposite to one that takes place in semiconductor lasers. All incoming photons with energy greater than the bandgap of the semiconductor p-n structure can generate the electron-hole pairs. The electron-hole pairs, which are separated by the strong electrical field across the p-n junction created by the bias voltage, drift very rapidly toward electrodes (refer to Section 7.6 to learn more about physics of photodetection process). The photocurrent that is generated is consequently amplified and processed by electrical circuits.

All photodiodes belong either to the group of the PIN photodiodes, or the APDs [48–51]. The detection process within PIN photodiodes is characterized by a probability that each photon will eventually generate the electron-hole pair. On the other hand, each primary generated electron is accelerated in the avalanche photodiode by a strong electric field, which than causes a generation of several secondary electron-hole pairs through the effect of impact ionization. This process in APD is random in nature and avalanche-like. There are three major semiconductor materials that are used for photodiode manufacturing:

- Silicon (Si), which is used for photodiodes that have a total bandwidth of up to 200 nm, centered around 800 nm (wavelength peak responsivity);
- Germanium (Ge), which is used for photodiodes that have the total bandwidth of up to 400 nm, centered around 1,400 nm;
- Indium-Gallium-Arsenide (InGaAs), which is used for photodiodes that have the total bandwidth of up to 600 nm, centered around the wavelength peak of 1,500 nm.

The PIN photodiode is structured as a sandwich with lightly doped I-region (I stands for intrinsic) placed between p-type and n-type semiconductor layers, as shown in Figure 2.21(a). Therefore, we can say that p-type sits on I- type, which is placed on the top of n-type. (The name PIN associates with the nature of this positioning.) The PIN photodiode is reverse biased, with very high internal impedance, which means that it acts as a current source and generates the photocurrent that is proportional to the incoming optical signal. The following photodiode parameters are important from the systems engineering perspective:

- The quantum efficiency η, which is the ratio between the number of the electrons detected in the process and the number of the photons that arrived. This parameter is always lower than 100%.
- The responsivity R, which is the ratio of the output current generated in PIN photodiode, and the input optical power coming to the PIN. There is the following relationship between the responsivity and quantum efficiency:

$$R = \frac{q\eta}{h\nu} \qquad (2.10)$$

where q is the electron charge equal to 1.6×10^{-19} Coulombs, h is the Plank's constant equal 6.63×10^{-34} Js, and v is the optical frequency expressed in hertz. (Please refer to Chapter 7 for the meaning and the values of physical variables and units used in this book.)

- Frequency response of PIN photodiode and frequency bandwidth (characterized by the cutoff frequency f_c), which are the parameters that characterize the response and switching speed of the PIN photodiode. The limiting factor of the frequency response is the capacitance of the reverse-biased p-I-n structure. The width of the I-layer can be decreased in order to decrease the total capacitance, but the responsivity of the photodiode would also decrease. Therefore, the responsivity of PIN photodiodes used in high-speed optical receivers is lower than the responsivity of photodiodes used in low-speed optical receivers.
- The total noise generated during detection process. The total noise contains the dark current, quantum noise, and the thermal noise components. These components will be discussed in mode detail in Section 3.2.

The structure of the APD is shown in Figure 2.21(b). The structure is optimized to support the amplification of generated electron-hole pairs before they reach the photodiode electrodes through an internal gain. The strong electrical field increases the kinetic energy of electrons by accelerating them, so they became capable of generating new electron-hole pairs. The newly generated electrons are further accelerated by the electric field to generate additional electron-hole pairs through the process of impact ionization. The avalanche breakdown, which might occur if the number of generated electron-hole pairs grows quickly and without a real correlation with an incident optical signal level, is prevented by adjusting the bias voltage to be below a critical value that would produce a breakdown condition. The breakdown bias voltage can be found in a manufacturer's product data sheet.

The structure of APD is slightly different than the structure of the PIN photodiode, since an additional layer is added to enhance the impact ionization process. This is the active layer where the avalanche multiplication and current signal gain occur. The reverse bias voltage applied to APD can vary from tens to hundreds of

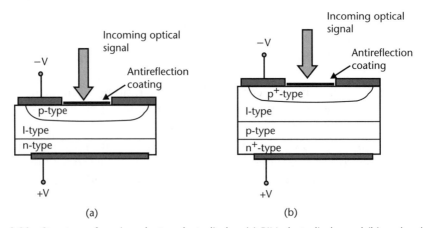

Figure 2.21 Structure of semiconductor photodiodes: (a) PIN photodiode, and (b) avalanche photodiode.

volts, as opposed to the PIN bias voltage, which is only several volts. In addition to higher bias voltage, n-type and p-type layers in APD are suitably doped to increase the carriers' density at the junction. The following parameters are used to characterize the APD:

- Responsivity R, which is the ratio between the output current generated in the APD photodiode and the input optical power coming to APD. It is about the same as the responsivity of the PIN photodiode. However, the generated photocurrent is much higher in APD since an internal amplification is involved.
- Frequency response of the APD photodiode and frequency bandwidth (characterized by cut-off frequency f_c), which are the parameters that characterize the response and switching speed of APD photodiode. The capacitance of the reverse-biased APD is a factor that limits the speed of the APD response. Accordingly, the widths of two internal layers (I-layer, and p-layer) can be decreased in order to decrease the capacitance. However, this will produce a side effect since the responsivity of the photodiode will be decreased. The cut-off frequency of APD is generally lower than the frequency associated with PIN photodiodes, which limits applications of APD to bit rates up to 10 to 15 Gbps.
- The instantaneous avalanche signal gain $M(t)$, which is a stochastic parameter that fluctuates around the average value. This internal gain is very beneficial for applications related to bit rates up to 2.5 Gbps.
- The total noise generated during the detection of an incoming optical signal. There is the noise component associated with the avalanche character of the amplification process in APD that comes in addition to the noise components found in PIN photodiodes. This is known as the shot noise, which is proportional to the deviation of the gain $M(t)$ from its average value. The APD noise parameters will be discussed in more detail in Sections 3.2 and 5.2.

Typical parameters associated with the photodetection process in PIN and ADP are summarized in Table 2.4. Photodiode parameters should be considered in connection with the parameters of the front-end amplifier, which amplifies input photocurrent signal and converts it to a voltage signal (refer to Figure 5.1). It is worth mentioning that the photodiode and front-end amplifier can be integrated on the same substrate, which is very often done in practice [52, 53].

2.6 Key Optical Components

There are a number of optical components that can be deployed in an optical transmission system (refer to Figure 2.1). This section provides a description of

Table 2.4 Typical Values Photodiode Parameters

Parameter	PIN	APD
Responsivity (A/W)	0.7–0.95	0.7–0.9
Cutoff frequency (GHz)	Up to 75	Up to 15
Internal gain	1	Up to 100

design and operational principles of key optical components. It also introduces the physical parameters that are relevant from the systems engineering perspective. More details with respect to the key optical elements can be found in literature [54, 55].

All components used in an optical transmission system can be divided in two groups: active and passive. The active components need some voltage, either as a power supply or for setting up an operational regime, while passive components can operate without any external electrical power. The most important active components are lasers, optical modulators, photodiodes, optical amplifiers, wavelength converters, and optical switches. The main passive optical components are optical couplers, optical multiplexers, optical isolators, and optical filters. In some cases an optical component can be either active or passive, such as optical filters, for example.

The optical component parameters that are important from the systems engineering perspective are insertion losses, return coupling losses, polarization-dependent losses, channel signal crosstalk, inserted chromatic dispersion, and inserted polarization mode dispersion.

There are several technologies that can be used in the components manufacturing process, such as fiber fusing, the combination of graded-index (GRIN) rods and optical filters, employment of planar optical waveguides, and the application of optical fiber Bragg gratings. The next section contains a description of the operational principles of the most important optical components. This will not include lasers, optical modulators, optical amplifiers, and photodiodes, since they have been treated separately.

2.6.1 Optical Couplers, Isolators, Variable Optical Attenuators, and Optical Circulators

The *optical coupler* is used to combine, or split, optical signals at different points along the lightwave path. There are three major manufacturing schemes of optical couplers, which are based on fused optical fibers, planar optical waveguides, and the combination of the GRIN roads and filters [54]. The first two schemes are illustrated in Figure 2.22.

The fused tapered optical couplers from Figure 2.22(a) are produced when two optical fibers are stripped from their claddings, and two fiber cores are brought together. The fibers are then heated and stretched, which results in a waveguide structure that can exchange the energy between the branches. As a result, the optical power from input 1 is split between two outputs [outputs 1 and 2 from Figure 2.22(a)]. The same happens with the optical power coming from input 2. Such an optical coupler type is known as 2×2 optical coupler. The coupler effectively becomes 1×2 type if just one input port is active. The coupler structure with planar optical waveguides is similar to the one just described, but with the planar optical waveguides used instead of optical fibers.

The directional coupler is characterized by the coupling region with an effective length L, as shown in Figure 2.22(a). The optical power coupling capability is characterized by so-called coupling coefficient α, which is a function of the optical signal wavelength, coupling length, cross-section parameters of the optical fibers or

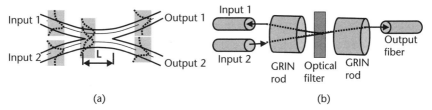

Figure 2.22 Optical couplers: (a) fused tapered coupler, and (b) GRIN-rod plus optical filter.

planar waveguides that are used, the difference in refractive indexes between the waveguide region and surrounding layers, and the proximity of two axes related to either fibers or planar waveguides.

It is possible to change both the ratio of the optical powers directed to outputs 1 and 2, and the wavelength content of the signals appearing at these outputs, by changing aforementioned parameters. The parameters are usually adjusted in such a way that the peak of the coupling coefficient coincides with a specified wavelength. The coupling region should be protected from any external effects that might cause the change in the coupling coefficient, which is done by a proper packaging. In addition, the temperature control might be applied if necessary.

The combination of GRIN rod and optical filter can be effectively used to produce the optical coupler in a way illustrated in Figure 2.22(b). The pair of GRIN rod lenses is employed to collimate and transfer the light from the input to the output fiber port, while the optical filter selects the wavelengths that should be directed to a specified port. There are three distinct optical signal flows that can be recognized in this design, and they are related to the input signal, the part that goes through the filter, and the reflected portion of the input signal. Therefore, the characteristics of the filter determine the amount and the content of the transmitted and reflected optical signals. A variety of optical coupler designs can be achieved by inserting the optical filters with specified characteristics.

There are several types of optical couplers used in optical communication systems. They are usually recognized as the optical taps and optical directional couplers. The optical taps are 1×2 optical couplers that are used for signal monitoring, while directional optical couplers present 2×2 structures that are used for power sharing. The typical coupler splitting ratio, which is the ratio between the powers at two outputs, can vary, but the following values are commonly found in practice: 1%/99%, 5%/95%, 10%/90%, and 50%/50%. Both the fused fiber couplers and GRIN rod devices are often used as optical taps and directional couplers. The major application issue with respect to optical taps and directional couplers is related to the inserted polarization-dependant loss (PDL), which is around 0.1 dB.

More complex optical coupler structures are known as $N \times M$ optical couplers, where N and M are numbers that can be larger than 1 or 2. Such couplers are used either for power distribution among number of users or for wavelength-specific functions in optical systems. The wavelength-specific couplers, which are also known as the WDM optical couplers, are a special group of directional 1×2 couplers designed for different applications, such as the pump power introduction, the wavelength band separation, the course wavelength multiplexing, and the introduction and separation of the supervisory channel.

Both the optical couplers based on fused optical fiber technology and GRIN-rod devices can be used for wavelength-specific applications. The fused optical fiber couplers are more suitable for applications where wavelengths, which are about to be separated, are not close to each other (such as the separation of signals placed around 1,300 nm from signals that are around 1,550 nm, or separation between contents that occupy 980- and 1,480-nm wavelengths). It is important to notice that the PDL in fused fiber couplers can be up to 0.2 dB, and it varies with the operational wavelength. On the other hand, the GRIN rod–based couplers provide a flat wavelength response across the wavelength band, while the PDL is relatively small (usually less than 0.1 dB).

The *optical isolator* is another optical component that is very often used in optical transmission systems, mainly to prevent the impact of the back-reflected optical signal. The two main causes of back-reflection are optical connectors and Rayleigh scattering in fiber spans. The back-reflection effect arises in optical connectors due to reflection from the front surface, while the Rayleigh scattering occurs in a distributed manner, as explained in Section 2.3.

Semiconductor lasers are particularly sensitive to any back-reflections since it will cause additional noise and severe degradations of the system performance (refer to Section 3.2 to find more details related to the impact of back-reflection signal to the transmission system performance). Next, an optical amplifier needs an optical isolator to prevent the lasing effect and to improve the amplifier performance by isolating the amplifier stages. The operation of commonly used optical isolators is based on utilizing a nonreciprocal change in the state of polarization of the incoming optical signal, which occurs in some materials under the presence of the magnetic field. This effect is also known as Faraday rotation [20]. Accordingly, the polarization is changed for the forward propagating signal that passes through the Faraday material. If there is a backward propagating light that passes through the Faraday material again, the material will turn the polarization state once more, rather than undoing the forward-induced polarization shift.

The Faraday material is placed between polarizer and analyzer, as illustrated in Figure 2.23. The polarization state of the forward propagating signal will be rotated by 45° after passing through the Faraday material. The same will happen with the back-reflected backward propagating signal. Therefore, the total rotation of the backward propagating signal will be 90°. In such a case, the polarizer will not recognize the polarization state of the back-reflected signal, and it will be denied the further throughput.

The key system parameters of optical isolators are the return isolation loss (which should be as high as possible), polarization mode dispersion (which should be as low as possible since an optical path can contain as many as 20 to 40 isolators), and the insertion loss (which should be as low as possible). The typical and high-performance values of optical isolator parameters are shown in Table 2.5.

Variable optical attenuators (VOAs) have become very important components, especially in multichannel optical systems and networks, where optical signals might travel through different optical transmission paths before being combined and processed together. Optical powers of different channels should be equalized at

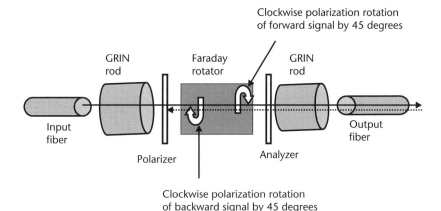

Figure 2.23 Operational principle of an optical isolator.

some reference point in order to assure the equal system performance for individual optical channels.

The equalization of optical powers is related to a very precise adjustment of the optical signal powers. It is highly desirable that the process of power adjustment can be remotely controlled. Variable optical attenuators serve that purpose very well, and they can be manufactured as separate optical components or combined together with other devices, such as optical couplers or optical filters. The operational principle of the commonly used variable optical attenuator is based on changing either the polarization state of the incoming polarized light or the signal loss in the material. Variable optical attenuators are not quite passive optical devices since some control voltage is needed to change the state of the material that is used. Optical attenuators are characterized by dynamic attenuation range and the increments in attenuation during the adjustment process. The typical dynamic range of optical attenuators is between 0 and 40 dB, while it can be adjusted in increments that range from 0.1 to 20 dB [29].

Optical circulators are optical devices that are very often used within optical subsystems, such as optical add/drop multiplexers and dispersion compensators. The operation of an optical circulator is similar in nature to the operation of a revolving door in terms that the signal is suppose to exit at the next port, while propagation is unidirectional. In essence, an optical circulator is a cascade of several optical isolators that form a closed circle. The signal can only go through a single Faraday rotator, since the entrance to the following isolator stage is effectively denied.

A hybrid integration of multiple functions into a single device can greatly enhance the performance of the individual optical components. Such integration

Table 2.5 Optical Isolator Parameters

Parameter	Typical Value	High-Performance Value
Insertion loss (dB)	0.6	0.5
Isolation loss (dB)	35	45
PDL (dB)	0.15	0.05
PMD (ps)	0.07	0.05

into a single package reduces the insertion loss, decreases both PDL and PMD, and increases the package reliability. The GRIN-rod technology that is commonly used for both optical couplers and optical isolators is quite suitable for different types of hybrid optical devices. Some examples of multifunctional optical components are as follows: WDM coupler + isolator, optical filters + isolators, WDM coupler + isolator + tap coupler, and WDM coupler + isolator + bandwidth filter [49].

2.6.2 Optical Switches

Optical switches are distinguished optical components used for changing the direction of optical signal by switching it from one lightwave path to the other. They can be used for different applications scaling from provisioning of lightpaths and protection switching (where there is no need for a fast switching time), to the optical packet switching and optical signal modulation (where a high switching speed is required). An optical switching matrix can have either a basic 1×2 or 2×2 forms, or may be more complex in terms of the number of input and output ports. Although a large switching matrix can be designed as a single entity, it is very often just a combination of basic 2×2 forms [29, 49].

It is important for all optical switches that the extinction ratio, defined as the ratio of the optical power corresponding to signal flow to the power level related to idle state, is as high as possible. In addition, a good isolation is required between lightwave paths traveling between different input and output ports. Finally, optical switches should have smaller insertion loss and be insensitive to the polarization state of the incoming optical signal.

There are several major types of optical switches used today, such as mechanical switches, electro-optic switches, thermo-optical switches, and semiconductor amplifier-based switches. Mechanical switches perform lightpath redirection through some mechanical action, such as by moving a mirror in and out of the lightwave path, or by using a flexible directional coupler that changes the coupling ratio if it is bent or stretched. The microelectromechanical switch (MEMS) design is a well-known example of optical switches using movable mirrors. Mechanical optical switches have low insertion loss and high isolation of the lightwave paths. In addition, the sensitivity to the polarization state is also relatively small. On the other side, the long-term reliability is of some concern since there is mechanical movement involved. Mechanical switches can switch a wavelength path within a time of several milliseconds.

Electro-optical switches are based on optical couplers in which the coupling ratio is changed by appropriate voltage applied to the electrodes. The voltage takes two discrete values producing "on" and "off" states of the switch. Electro-optical switches can operate relatively fast and switch the light path in less than 1 ns, which makes them good candidates for high-speed applications. Although the basic 2×2 electro-optical switches can be integrated in more complex structures on a single substrate, they tend to have higher insertion loss. In addition, they also insert PDL.

Thermo-optic switches are based on the application of the MZ interferometer, where the temperature is applied to change the refractive index in the interferometer arms. The phase difference between signals in two arms is changed by varying the refractive index in the arms, which results in constructive or destructive interference at the output ports. The switching process usually takes several milliseconds to select the output port where the constructive interference takes place. The isolation between two distinguished lightpaths is lower than the isolation associated with the mechanical switches. Thermo-optical switches are relatively simple and cheap, which makes them suitable for applications related to protection and restoration of the lightwave path.

Finally, there is a group of optical switches based on SOAs, where the switching process is performed by varying the bias voltage. If the bias voltage is low enough, the device absorbs the incoming optical signal, which corresponds to the off state of the switch. However, the input optical signal is amplified by increasing the bias voltage since the population inversion is achieved, and this situation corresponds to the on state of the switch. The change from the absorption to the amplification state can be done very quickly, usually taking less than 1 ns. Although the SOA-based optical switches have amplification capability and can be integrated in larger switching matrix, the noise accumulation and lightpath isolation are of some concern in practical applications.

2.6.3 Optical Filters

The role of optical filters has become very important in multichannel WDM transmission systems and the optical networking environment. The operation of an optical filter is based either on optical spectral interference of two or more monochromatic coherent optical waves, or on absorption of the specified wavelength band by some special material [56–58]. From the design perspective, all optical filters can be either fixed or tunable [59].

The spectral interference occurs when two or more electromagnetic waves, which originate from the same source, travel along the paths with slightly different lengths before they are reunited again. The difference in lengths will produce the difference in the phases, which will have an impact on the summation result after the waves are rejoined again. The interference effect implies that there are at least two versions of the same optical signal that are combined together after passing different paths. There are two main interferometer schemes that are used for design of optical filters: FP and MZ interferometers. These names were already mentioned several times with respect to some other elements (laser diodes, optical modulators), which means that the same physical phenomenon is a base for different optical components.

The MZ interferometer consists of two directional optical couplers connected by optical waveguides that have different lengths, as shown in Figure 2.24(a). The waveguides can be either optical fibers or planar optical waveguides that are produced on the structure such as silica on silicon. In general, there is a difference Δl between two-waveguide arms, which causes the signal delay $\Delta \tau$. There is the following relationship between the power of the optical signal at output 1 and the input optical signal entering input 1 in Figure 2.24(a) [49]

$$P_{out,1}(v) = P_{in,1}(v)\sin^2(\Delta\tau\pi v) \qquad (2.11)$$

where v is the frequency of the optical wave. (Please recall that the frequency v and wavelength λ of an optical signal are connected by the relationship $\lambda = c/v$.) At the same time, the optical power from output 2 is connected with the power at input 1 by the relation

$$P_{out,2}(v) = P_{in,1}(v)\cos^2(\Delta\tau\pi v) \qquad (2.12)$$

Therefore, the transfer functions that correlate the powers at input and output ports are raised sine functions, which are out of phase with one another. The periodical structure of the function given by (2.12) is the reason why the MZ filter belongs to the group of so-called periodic optical filters [54].

Several MZ elementary structures can be cascaded to form a multistage MZ interferometer. The total path difference between the two portions of the signal that travel through the chain is calculated as a sum of differences associated to individual stages. If there are M interferometers in the chain, the signal from output 2 can be expressed by the signal at the input 1 as

$$P_{out,2}(v) = P_{in,1}(v)\prod_{i=1}^{M}\cos^2(\Delta\tau_i\pi v) \qquad (2.13)$$

An optical filter that is based on the chain of MZ interferometers can be designed by adjusting optical path parameters from (2.13). Such an optical filter will be able to isolate one of $N = 2^M - 1$ optical channels, which are mutually separated by frequency spacing Δv, if the delay of the ith interferometer in the chain is

$$\Delta\tau_i = \frac{1}{2^i\Delta v} \qquad (2.14)$$

As an example, the chain that consists of six stages is capable of distinguishing 1 out of 64 channels, while the chain that consists of seven stages will provide selection among 128 channels. The optical filter function presented by (2.13) has a shape that contains the main lobe and periodical arcades, which are more or less suppressed, as illustrated in Figure 2.24(b).

The FP interferometer, often known as the "etalon," is another basic interferometer structure. In consists of resonant cavity established by two parallel mirrors, as shown in Figure 2.25(a). Light enters the cavity through the outer side of mirror 1, which is transparent to the incoming signal. The light signal that entered the cavity

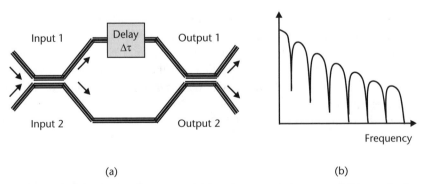

Figure 2.24 Mach-Zehnder interferometer: (a) interferometer scheme, and (b) transfer function shape of a cascaded design.

2.6 Key Optical Components

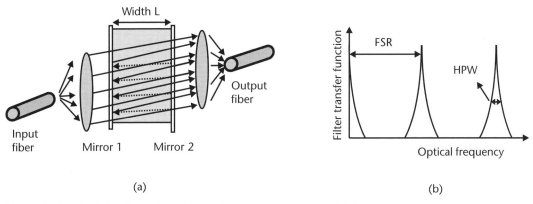

Figure 2.25 The Fabry-Perot filter: (a) interferometer structure, and (b) filter response.

will then bounce back and forth between mirrors. There will be the forward and backward propagating waves that will either contribute to each other or cancel each other, which depends on the characteristics of the resonator.

The resonator is tuned to its resonant position if the cavity length is adjusted to take the value

$$L = \frac{i\lambda}{2n} \qquad (2.15)$$

where n is the refractive index within the cavity, and i is an integer known as the filter order. A number of reflections contribute constructively to the filter response, and they occur before the light signal decays and eventually vanishes following a zig-zag path. The number of reflections that occur before the light intensity decreases to approximately $1/e$ of its original value is proportional to the parameter called filter "finesse" (please recall that e is the natural logarithm number). The full constructive interference will no longer take a place if the resonator length is detuned from the value L given by (2.15). Accordingly, the light output through mirror 2 will be suppressed.

The powers of optical signals that come to mirror 1 and exit through mirror 2 will be related to each other by function [54]

$$P_{out}(v) = P_{in}(v) \frac{(1-\alpha-r)^2}{(1-r)^2 - 4r\sin^2(2\pi\tau v)} \qquad (2.16)$$

where v is optical frequency, τ is the one-way propagation time across the resonant cavity, r is the power reflectivity of each of two mirrors, and α is the optical power absorption coefficient. (Recall that the reflectivity coefficient equals $r = 1 - t$, where t is so-called transmission coefficient [54].) The right-hand side of (2.16) defines a periodic function known as the Airy function [56]. The frequency at which the function transmission peaks periodically repeat themselves is known as the free spectral range, defined as

$$FSR = \frac{1}{2\tau} = \frac{c}{2nL} \qquad (2.17)$$

The FP interference structure is also characterized by the filter finesse F, which is defined as

$$F = \frac{FSR}{\Delta v_{FP}} \approx \frac{\pi\sqrt{r}}{1-r} \tag{2.18}$$

where Δv_{FP} is related to the width of the transmission peaks, as shown in Figure 2.25(b). This width, which determines the filter bandwidth, is defined by the points at which the transfer function decreases to the half of the transmission peak value. The relation between the filter finesse by the mirror reflectivity in (2.18) is obtained by assuming that the optical power absorption coefficient α can be neglected. If the FP filter is used for channel selection in an optical multichannel system, the FSR of the filter should be larger than the combined bandwidth of the multichannel signal. The number of optical channels that can be effectively resolved in such a multichannel environment is measured by the filter finesse.

The simplest way to produce a high-finesse optical filter based on the FP interferometer is by cascading several stages. There are two approaches to achieve the cascade of FP interferometers. The first one is to employ several interferometers as a simple chain, where the output signal from preceding resonator becomes the input signal to the following stage. In the second scheme, the optical signal from the output mirror [mirror 2 from Figure 2.25(a)] is reflected back in order to enter the cavity again. In such a scheme, light passes through the same cavity twice. The total number of signal passes through the cavity can be additionally increased if the output signal (this time from mirror 1) is directed back to enter the cavity one more time.

The effective FSR of a cascaded FP structure increases proportionally to the number of times the light passes through the resonator, while the filter bandwidth decreases in proportion to the numbers of cavities that are cascaded. The cascaded structure of FP filters that are used in practical applications consists of several dielectric layers. Such a design does not need classical mirrors since the difference in refractive indexes between neighboring layers is used to reflect the light signals back and forth.

Optical fiber gratings are special optical filters that are imprinted in optical fibers [57]. Due to such design, they are relatively cheap and easy for packaging and coupling with other optical fibers. These gratings are usually written in conventional germanium doped silica fibers that are more photosensitive than other fiber types. The gratings are formed by changing the refractive index in the fiber core, which occurs under the impact of ultraviolet radiation. A permanent grating can be written if the fiber core that is exposed to two interfering ultraviolet signals based on the fact that the light intensity of the resulting wave varies periodically along the fiber length. The grating is formed by an increase in the refractive index at the places where the resultant wave has maximums, since the index stays unchanged at the other places. The increase in the refractive index is in the range of 0.005% to 0.01%.

There are two groups of optical fiber gratings. The first group, known as fiber Bragg gratings (FBG), is based on Bragg reflection since the distance between the grating lines is comparable with the wavelength of the incoming optical signal. Such design is also known as short-period gratings (refer to Section 6.1.3). Such fiber gratings, or the optical filters based on FBR, have low insertion loss (usually lower than 0.1 dB), and a relatively sharp transition between the passband and the rest of the spectrum. In addition, these filters have a flat top of the passband and low sensitivity to the polarization states of the incoming optical signal. The FBG-based optical filters are commonly used in chromatic dispersion compensation schemes, and for optical signal filtering in optical add-drop multiplexers.

The other group of optical fiber gratings has the grating period that is much longer than the signal wavelength. The optical signal energy is not reflected back and forth as in fiber Bragg gratings, rather it is coupled to the vanishing fiber cladding modes. These long-period fiber gratings are widely used to flatten the spectral gain profile in different kinds of multichannel optical amplifiers, or as band-rejection filters, since their spectral profile can be precisely shaped [54].

Tunable optical filters are very useful optical components since they can dynamically select a specified range of optical wavelengths [58]. In a general case, it is desirable that tunable optical filters have wide tuning range, a high filter finesse, a flat-top of the transfer function peaks, and can be tuned as quickly as possible.

Both MZ and FP optical filters structures are potentially tunable. A tunable MZ structure is achieved through an active control of the optical signal delay $\Delta\tau$, which is done by changing the refractive index in the waveguides (refer to Figure 2.24). The simplest way to control the refractive index is by using the temperature change imposed by the thin film heater, which is often referred as the thermo-optic effect. The refractive index can also be altered by an external electrical field applied to the waveguide, which is referred as the electro-optical effect, already mentioned in Section 2.2.2. Therefore, the resonant wavelengths, which are associated with the transmission passband peaks, can be dynamically shifted by changing the refractive index.

The passband tuning of the FP filter can be achieved by either changing the cavity length or by varying the refractive index within the cavity. Both methods lead to a change in propagation delay τ, and to a shift of the resonant wavelength away from its initial position. The simplest way to change the cavity length is by mechanical movement of one of the mirrors. It can be done, for example, by an all-fiber design that changes the air gap between two optical fibers. The two fibers face each other in an enclosed piezoelectric chamber [54]. The cavity length, which is the air space between two polished fiber ends that act as mirrors, is changed electronically through piezoelectric contraction. The finesse of the all-fiber FP filters is higher than 100 and can be additionally increased by putting two filters in cascade [29]. The tuning range of such filters is up to 20 nm, while the tuning speed is relatively slow and can exceed 1 sec.

The tunable FP filters that use the refractive index change, rather than mechanical movement of the mirrors, are based on special materials, such as liquid crystals or semiconductors. The resonant cavity is formed by placing some of these materials between two mirrors, while the tuning is done electronically by changing the refractive index of the material in the resonant cavity. The finesse of these filters can be more than 300, the tuning range can exceed 50 nm, and the tuning time is about 1 ms. The cascaded structure of FP filters that consists of several dielectric layers can be also tuned, either thermally or electronically, with the tuning range of up to 40 nm [29, 58].

Acousto-optic tunable filters (AOTF) are very promising candidates for different applications related to the optical wavelength selection. The operation of the AOTF is also based on the Bragg reflection. The refractive index grating is imprinted by generating acoustical frequencies through the transducer, which is driven by an external RF signal, as shown in Figure 2.26. The acoustic frequencies form a standing wave, which determines the character of the refractive index change. The maximums in the refractive index coincide with the peaks of the standing acoustic wave, while the minimums coincide with the nodes of such a wave.

The acoustic transducer is applied to a waveguide structure based on a highly birefringent material. There are two polarization modes of the incoming signal, commonly referred as TM and TE modes [20], which see different refractive indexes during the propagation through the material (refer to Section 3.3.5 to find more details about polarization modes). The coupling, or the energy exchange between TM and TE modes, occurs if the associated refractive indexes satisfy the Bragg condition

$$n_{TM} = n_{TE} \pm \frac{\lambda}{\Lambda} \qquad (2.19)$$

where λ is the wavelength of the optical signal, Λ is the period of the Bragg grating created by an acoustic wave, and n_{TM} and n_{TE} are the refractive indexes associated with TM and TE modes, respectively.

The energy exchange is unidirectional since the optical signal energy around the wavelength λ is being transferred from TE to TM mode. Therefore, this scheme also needs the TE-mode polarizer at the front of the filter and the TM-mode polarizer at the filter's end. The length of the interaction determines the bandwidth of the filter, which is narrower for longer interaction lengths. However, the tuning speed is decreased with an increase in the interaction length, which means that there are different designs tailored for specific applications [59, 60]. The tuning range of AOTFs is more than 100 nm, while the tuning speed can be less than 10 μs.

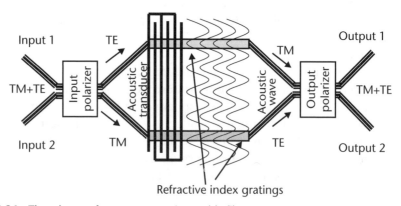

Figure 2.26 The scheme of an acousto-optic tunable filter.

2.6.4 Optical Multiplexers and Demultiplexers

Optical multiplexers and demultiplexers are optical devices that are used to either combine several distinct wavelength channels into a composite signal or split a composite signal into its channel constituents. Generally, the same device can be used in both roles, and just the direction of the signal will determine what function is performed. There are several types of optical multiplexers that are commonly used today, and they are based on either the diffraction or the interference effect.

Diffraction-based demultiplexers are related to the Bragg diffraction effect [20], which is usually enforced by employing by some angularly dispersive elements, such as diffraction gratings. The incident light signal is reflected from the grating and dispersed spatially into a number of wavelength components, which are then focused by lenses and introduced into individual optical fibers. The diffraction grating should be properly designed to create a wavelength-specific reflection angle. The operation principle of an optical multiplexer based on diffraction grating is the same as for an optical demultiplexer. In fact, an optical multiplexer can play the role of demultiplexer, and vice-versa. For that purpose, all that is needed is to switch the roles of the input and output ports.

The focusing lens is usually the GRIN rod since this is the most suitable from the design perspective. Furthermore, there is a possibility that the GRIN rod can be integrated with the diffraction grating. Another possibility to simplify the design is to use concave diffraction gratings, so that there is no need for a focusing lens. The concave diffraction grating can be made on the planar waveguide structure and eventually integrated with planar waveguides that serve as input or output ports [54]. However, there is a practical issue of how to couple multiplexers with optical fibers for a larger number of optical wavelengths, and this issue still needs to be effectively resolved.

The second group of optical multiplexers, which is based on the interference effect, uses the optical couplers and optical filters to combine two or more wavelength channels into a composite signal. The two commonly used types of optical multiplexers that are based on interferometric effect are dielectric thin-film filter multiplexers [54, 55], and the arrayed waveguide gratings (AWG) [61, 62].

AWGs are widely used for wavelength multiplexing and for wavelength routing purposes. The AWG multiplexer is a generalized version of the Mach-Zehnder modulator, which consists of two optical couplers interconnected by optical waveguides—as shown in Figure 2.27(a). Optical waveguides, which form multiple arms of the MZ interferometer, have distinctly different lengths in order to introduce a phase shift between corresponding optical signals. On the other hand, in the optical demultiplexer function, there is just one input and several output ports. The interferometric process that governs demultiplexing is the same process as the one associated with the signal splitting in a standard version of the MZ interferometer, which forces each output wavelength to take just one output port. The optical multiplexer functionality is obtained by switching the direction of the optical signals shown in Figure 2.27(a). In this capacity there are multiple inputs and just one output that accommodate a composite optical signal.

The "branch and three" structure of MZ interferometers, shown in Figure 2.27(b), can serve as an alternative to the AWG design that was just described. However, the structure from Figure 2.27(b) will have higher insertion

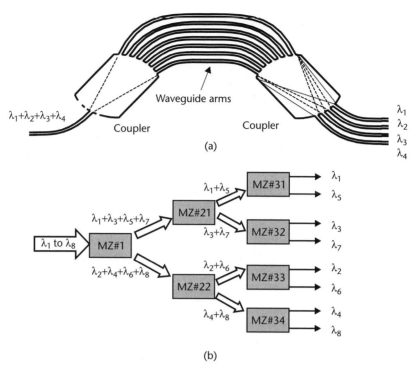

Figure 2.27 Optical multiplexers: (a) arrayed waveguide gratings, and (b) Mach-Zehnder filter chain.

loss and a nonflat response in the wavelength passband as compared to the AWG design. In addition, it is much easier to realize AWG as an integrated waveguide structure placed on the waveguide substrate [62]. The waveguide substrate material used for AWG manufacturing is usually silicon, while either pure silica, or silica mixed with some dopants, can be used as the waveguide structure.

The AWG structure with multiple input and output ports can also serve as a wavelength router with predetermined routing paths. The routing patterns can be established by proper adjustment of MZ parameters in the AWG interferometric structure.

The second type of interferometric optical multiplexer is based on a cascade of Fabry-Perot filters, where each of them is constructed with multiple-layer dielectric thin-films, as shown in Figure 2.28. Each filter contains several resonant cavities in order to flatten the passband and to provide the steeper slope of the passband edges. As for the filter cascade shown in Figure 2.28(b), each of individual filters selects a different wavelength from the composite signal. For example, the first filter passes just one wavelength, and directs the rest to the second filter in the cascade, where another wavelength is selected before the rest is directed to the third filter, and so on.

The optical multiplexers described above can provide relatively small crosstalk between neighboring optical channels, and reasonably flat top of the passband. In addition, they are relatively stable with respect to temperature changes and insensitive to the polarization state of the incoming optical signal. Typical values of parameters that characterize the commonly used optical multiplexers are summarized in Table 2.6 [29].

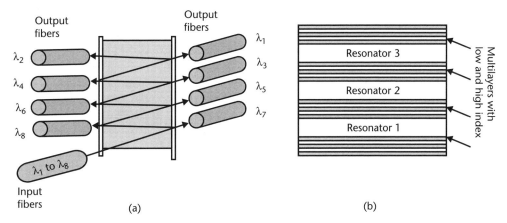

Figure 2.28 Optical multiplexers based on (a) the Fabry-Perot filter, and (b) the thin-film filter design.

Table 2.6 Typical Values of Optical Multiplexer Parameters

Parameter	AWG	FP Thin Film
Insertion loss (dB)	3.5–6	0.3–0.8
Crosstalk attenuation (dB)	25–45	15–25
Channel spacing (GHz)	12.5–200	50–200

2.7 Summary

In summary, we can say that the reader should now be familiar with the basic operational principles and key engineering parameters of optical components employed along the lightwave path. Typical values of these parameters were presented in Tables 2.1 to 2.6, and these can be used as reference points in the system engineering process.

References

[1] Chuang, S. L., *Physics of Optoelectronic Devices*, New York: Wiley, 1995.
[2] Yariv, A., *Quantum Electronics*, 3rd ed., New York: Wiley, 1989.
[3] Henry, C. H., "Theory of the Linewidth of the Semiconductor Lasers," *IEEE J. Quantum Electron*, Vol. QE-18, 1982, pp. 259–264.
[4] Lee, T. P., "Recent Advances in Long-Wavelength Lasers for Optical Fiber Communications," *IRE Proc.*, Vol. 19, 1991, pp. 253–276.
[5] Arakawa, Y., and A. Yariv, "Quantum-Well Lasers—Gain, Spectra, Dynamics," *IEEE Journal Quant. Electron.*, Vol. QE-22, 1986, pp. 1887–1899.
[6] Agrawal, G. P., and N. K. Duta, *Semiconductor Lasers*, 2nd ed., New York: Van Nostrand Reinhold, 1993.
[7] Morton, P. A., et al., "25 GHz Bandwidth 1.55mm GaInAsP p-doped Strained Multiquantum-well Lasers," *Electron Letters*, Vol. 28, 1992, pp. 2156–2157.
[8] Osinski, M., and J. Buus, "Linewidth Broadening Factor in Semiconductor Lasers," *IEEE Journal Quant. Electron.*, Vol. QE-23, 1987, pp. 57–61.

[9] Chang-Hasnain, C., "Monolithic Multiple Wavelength Surface Emitting Laser Arrays," *IEEE Journal Lightwave Techn.*, Vol. LT-9, 1991, pp. 1665–1673.

[10] Suematsu, Y., et al., "Advanced Semiconductor Lasers," *Proc. of IEEE*, Vol. 80, 1992, pp. 383–397.

[11] Margalit, N. M., et al., "Vertical Cavity Lasers for Telecom Applications," *IEEE Communication Magazine*, Vol. 35, 1997, pp. 164–170.

[12] Verdeyen, J. T., *Laser Electronics,* 2nd ed., Upper Saddle River, NJ: Prentice Hall, 1990.

[13] Saleh, B. E., and A. M. Teich, *Fundamentals of Photonics*, New York: Wiley, 1991.

[14] Ohstu, M., *Frequency Control of Semiconductor Lasers*, New York: Wiley, 1996.

[15] Kobayashi, K., and I. Mito, "Singular Frequency and Tunable Laser Diodes," *IEEE/OSA Journal of Lightwave Techn.*, Vol. LT-6, 1988, pp. 1623–1633.

[16] *Lightwave Optical Engineering Sourcebook, 2003 Worldwide Directory*, Lightwave 2003 Edition, Nashua, NH: PennWell, 2003.

[17] Hong, J., et al., "Matrix-Grating Strongly Gain Coupled (MG-SGC) DFB Lasers with 34 nm Continuous Wavelength Tuning Range," *IEEE Photon. Technol. Lett.*, Vol. 11, 1999, pp. 515–517.

[18] Kudo, K., et al., "1.55 mm Wavelength Selectable Microarray DFB-LD's with Integrated MMI Combiner, SOA, and EA Modulator," *Proc. of European Conf. on Optical Comm.*, ECOC 2000, TuL 5.1, Munich, 2000.

[19] Libatique, N. J. C, and K. J. Jain, "A Broadly Tunable Wavelength Selectable WDM Source Using a Fiber Sagnac Loop Filter," *IEEE Photon. Technol. Lett.*, Vol. 13, 2001, pp. 1283–1285.

[20] Born, M., and E. Wolf, *Principles of Optics*, 7th ed., New York: Cambridge University Press, 1999.

[21] Kim, H., and A. H. Gnauck, "Chirp Characteristics of Dual-Drive Mach-Zehnder Modulator with a Finite DC Extinction Ratio," *IEEE Photon. Technol. Lett.*, Vol. 14, 2002, pp. 298–300.

[22] Li, G. L., et al., "High Saturation High Speed Traveling-Wave InGaAsP-InP Electroabsorption Modulator," *IEEE Photon. Technol. Lett.*, Vol. 13, 2001, pp. 1076–1078–517.

[23] Mason, B., et al., "40 Gbps Tandem Electroabsorption Modulator," *IEEE Photon. Techn. Lett.*, Vol. 14, 2002, pp. 27–29.

[24] D., Marcuse, *Light Transmission Optics*, New York: Van Nostrand Reinhold, 1982.

[25] Gloge, D., "Propagation Effects in Optical Fibers," *IEEE Trans. Microwave Theor. Trans.*, Vol. MTT-23, 1975, pp. 106–120.

[26] Rudolph, H. D., and E. G. Neumann, "Approximation of the Eigenvalues of the Fundamental Mode of a Step Index Glass Fiber Waveguide," *Nachrichtentechnischen Zeitschrift*, Vol. 29, 1976, pp. 328–329.

[27] Adams, M. J., *An Introduction to Optical Waveguides*, New York: Wiley, 1981.

[28] Gloge, D., "Weakly Guided Fibers," *Applied Optics*, Vol. 10, 1971, pp. 2252–2258.

[29] *Lightwave Optical Engineering Sourcebook, 2003 Worldwide Directory*, Lightwave 2003 Edition, Nashua, NH: PennWell, 2003, pp. 5–6.

[30] Snyder, A. W., "Understanding Monomode Optical Fibers," *Proc. IEEE*, Vol. 69, 1981, pp. 6–13.

[31] Jeager, R. E., et al., "Fiber Drawing and Control," in *Optical Fiber Telecommunications*, S. E. Miller and A. G. Chynoweth, (eds.), New York: Academic Press, 1979.

[32] MacChesney, J. B., "Materials and Processes for Preform Fabrication—Modified Chemical Vapor Deposition and Plasma Chemical Vapor Deposition," *Proc. IEEE*, Vol. 68, 1980, pp. 1181–1184.

[33] Izawa, T., and N. Inagaki, "Materials and Processes for Fiber Preform Fabrication—Vapor Phase Axial Deposition," *Proc. IEEE*, Vol. 68, 1980, pp. 1184–1187.

[34] Murata, H., *Handbook of Optical Fibers and Cables*, New York: Marcel Dekker, 1996.

[35] Miller, C. M., et al., *Optical Fiber Splices and Connectors*, New York: Marcel Dekker, 1986.

[36] Ito, T., et al., "Extremely Low Power Consumption Semiconductor Optical Amplifier Gate for WDM Applications," *Electron. Letters*, Vol. 33, 1997, pp. 1791–1792.

[37] Mikkelsen, B., et al., "High Performance Semiconductor Optical Amplifiers as In-line and Pre-amplifiers," *Europ. Conf. on Optical Communications, ECOC'94*, Vol. 2, pp. 710–713.

[38] Mukai, T., et al., "5.2 dB Noise Figure in a 1.5 mm InGaAsP Traveling Wave Laser Amplifier," *Electron Letters*, Vol. 23, 1987, pp. 216–217.

[39] Mayers, R. J., et al., "Low Noise Erbium Doped Fiber Amplifier Operating at 1.54 mm," *Electron. Letters*, Vol. 23, 1987, pp. 1026–1028.

[40] O'Mahony, M. J., "Semiconductor Laser Optical Amplifiers for Use in Future Fiber Systems," *IEEE/OSA J. of Lightwave Techn.*, Vol. LT-6, 1988, pp. 531–544.

[41] Desurvire, E., *Erbium Doped Fiber Amplifiers*, New York: Wiley, 1994.

[42] Miniscalco, W. J., "Erbium Doped Glasses for Fiber Amplifiers at 1500 nm," *IEEE/OSA J. of Lightwave Techn.*, Vol. LT-9, 1991, pp. 234–250.

[43] Keiser, G. E., *Optical Fiber Communications*, Third Edition, New York: McGraw Hill, 2000.

[44] Clesca, B., et al., "Gain Flatness Comparison between Erbium Doped Fluoride and Silica Fiber Amplifiers with Wavelength Mixed Signals," *IEEE Photonics Techn. Letters*, Vol. 6, 1994, pp. 509–512.

[45] Atkins, C. G., et al., "Application of Brillouin Amplification in Coherent Optical Transmission," *Electron. Letters*, Vol. 22, 1986, pp. 556–558.

[46] Mochizuki, K., et al., "Amplified Spontaneous Raman Scattering," *IEEE/OSA J. of Lightwave Techn.*, Vol. LT-4, 1986, pp. 1328–1333.

[47] Radic, S. et al., "Continuous-Wave Parametric Gain Synthesis Using Nondegenerate Pump Four-Wave Mixing," *IEEE Photonics Techn. Letters*, Vol. 14, 2002, pp. 1406-1408.

[48] Kressel, H., (ed.), *Semiconductor Devices for Optical Communications*, New York: Springer-Verlag, 1980.

[49] Digonnet, M. J., (ed.), *Optical Devices for Fiber Communications*, Bellingham, WA: SPIE Press, 1998.

[50] Keyes, R. J., *Optical and Infrared Detectors*, New York: Springer, 1997.

[51] Yuan, P., et al., "Avalanche Photodiodes with an Impact Ionisation Engineered Multiplication Region," *IEEE Photon. Technol. Lett.*, Vol. 12, 2000, pp. 1370–1372.

[52] Kuebart, W., et al., "Monolithically Integrated 10 Gbps InP-based Receiver OEIC, Design and Realization, European Conf of Optical Commun.," *Proc. of European Conf. On Optical Comm., ECOC 1993*, TuP6.4, pp. 305–308.

[53] Bitter, M., et al., "Monolitic InGaAs-InP p-i-n/HBT 40 Gbps Optical Receiver Module," *IEEE Photon. Technol. Lett.*, Vol. 12, 2000, pp. 74–76.

[54] Kashima, N., *Passive Optical Components for Optical Fiber Transmission*, Norwood, MA: Artech House, 1995.

[55] Special Issue on Multiwavelength Technology and Networks, *IEEE/OSA J. of Lightwave Techn.*, Vol. LT-14, 1996.

[56] Abramovitz, M., and I. A. Stegun, *Handbook of Mathematical Functions*, New York: Dover, 1970.

[57] Bennion, I., et al., "UV-written in Fibre Bragg Gratings," *Optical Quantum Electronics*, Vol. 28, 1996, pp. 93–135.

[58] Kobrinski, H., and K. W. Cheung, "Wavelength Tunable Optical Filters: Applications and Technologies," *IEEE Commun. Magazine*, Vol. 27, 1989, pp. 53–63.

[59] Song, G. H., "Toward the Ideal Codirectional Bragg Filter with an Acousto-Optic Filter Design," *IEEE/OSA J. of Lightwave Techn.*, Vol. LT-13, 1995, pp. 470–481.

[60] Cheung, K. W., "Acoustooptic Tunable Filters in Narrowband WDM Networks; System Issues and Network Applications," *IEEE J. of Selected Areas in Commun.*, Vol. 8, 1990, pp. 1015–1025.

[61] Takada, K., et al., "Low-Loss 10-GHz Spaced Tandem Multi-Demultiplexer with More Than 1000 Channels Using 1x5 Interference Multi/Demultiplexer as a Primary Filter," *IEEE Photon. Technol. Lett.*, Vol. 14, 2002, pp. 59–61.

[62] Takahashi, H., et al., "Transmission Characteristics of Arrayed n×n Wavelength Multiplexer," *IEEE/OSA J. of Lightwave Techn.*, Vol. LT-13, 1995, pp. 447–455.

CHAPTER 3
Optical Signal, Noise, and Impairments Parameters

This chapter identifies all key parameters involved in optical transmission systems engineering. The most important parameters (such as chromatic dispersion and spontaneous emission noise) are given detailed mathematical treatment, and a more general approach based on basic mathematical formulas is given for other parameters, which are either well understood (such as the extinction ratio) or do not have a significant impact in most practical situations (such as modal noise).

Optical transmission systems engineering involves accounting for all effects that can alter the optical signal during modulation, propagation, and detection processes. Different impairments will degrade and compromise the integrity of the signal before it arrives at the decision point in the optical receiver to be recovered from corruptive additives. The transmission quality is measured by the received SNR, which is defined as the ratio of the signal level to the noise level at the decision point. This ratio is related to the receiver sensitivity, which is defined as the minimum optical power needed to keep SNR at the specified level. The receiver sensitivity is additionally degraded by other impairments, such as pulse dispersion or signal crosstalk, which will add to the impact of the noise and make it more difficult to recover the signal.

The received signal level at the decision point should be as high as possible to keep the distance between the signal and noise, and to provide a margin necessary to compensate for other corruptive effects. In this chapter we will deal with three types of parameters that are important from the systems engineering viewpoint:

1. Optical signal parameters, which define the signal level;
2. Optical noise parameters, which define the SNR;
3. Impairment parameters, which determine the power margin that should be allocated to compensate for the impact of impairments.

These parameters are related to individual optical components and modules within an optical transmission system. This is shown in Figure 3.1, which indicates the origin of the most relevant parameters.

3.1 Optical Signal Parameters

An optical signal, generated in an optical transmitter, is characterized by its output power and the extinction ratio. The signal is periodically amplified in order to recover the power level, which is needed to keep SNR above a specified value. The

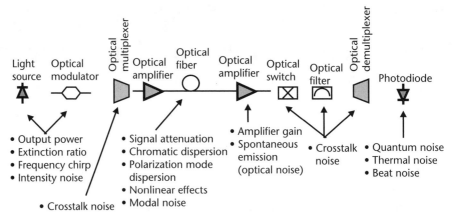

Figure 3.1 The origin of the most relevant parameters in an optical transmission system.

signal amplification process is characterized by the optical amplifier gain. The signal level, but this time in an electrical form, is also impacted by the value of the photodiode responsivity.

Accordingly, there is a set of parameters that characterize the received signal at the decision point:

- *Output signal power*, which is a power level from the laser/modulator, or from an optical amplifier coupled to the fiber pigtail;
- *The extinction ratio*, which is the ratio between the optical power related to 1 bits and 0 bits;
- *Optical amplifier gain*, which determines the level of the optical signal that is being amplified;
- *Photodiode responsivity*, which defines the conversion efficiency through the ratio of generated photocurrent and incoming optical signal. The responsivity is closely related to the ratio between the number of electrical carriers produced during the photodetection process and the number of incoming photons.

3.1.1 Output Signal Power

Output signal power from the laser/modulator is coupled to the fiber pigtail, which means that power from the fiber pigtail is a parameter that should be used in engineering considerations. The output power is defined for each individual optical channel and depends on the laser/modulator that was used. It is usually expressed in decibels per milliwatt (dBm) units, defined as dBm =10 log (P), where P is the output power expressed in milliwatts. The total output power P_{tot} of an aggregated optical signal that consists of several multiplexed optical channels is

$$P_{tot} = \sum_{i=1}^{M} P_i \qquad (3.1)$$

where $P_i (i = 1, 2, ..., M)$ represents the power of the ith optical channel, and M is the total number of optical channels.

The total power can be expressed as a product of an individual channel power and the number of optical channels, but only if the powers of all channels are equalized. Therefore, the total output power in decibels that correspond to aggregation of several channels with equal powers is

$$P_{tot}[dBm] = P + 10\log(M) \tag{3.2}$$

For example, the total output power is P_{tot} = 20 dBm if there is 100 optical channels and each of them has the optical power of 1 mW (or 0 dBm).

3.1.2 The Extinction Ratio

The extinction ratio R_{ex} is expressed as

$$R_{ex} = 1/r_{ex} = \frac{P_1}{P_0} \tag{3.3}$$

where P_1 is the power associated with the 1 bit, and P_0 is the power related to the 0 bit. In an ideal case, the extinction ratio would be indefinitely large. In reality, however, most sources and modulators generate a nonzero optical power output for 0 bits, and the extinction ratio takes a finite value. The optical power carried by the 0 bits will increase the probability that these bits will be mistaken for the 1 bits at the decision point. The SNR can be increased by increasing the extinction ratio, but at the cost of additional penalties in the modulation speed and the laser frequency chirp (refer to Section 3.3.3).

It is worth mentioning that parameter $r_{ex} = 1/R_{ex} = P_0/P_1$, which is also introduced by (3.3), is used more often than R_{ex} in different calculations, such as those related to the receiver sensitivity degradations (see Section 5.3.1).

3.1.3 Optical Amplifier Gain

Optical amplifiers enhance the level of the input optical signal through the process of stimulated light emission, or the stimulated scattering process. In this section we will deal with the gain related to the stimulated emission process, while the gain properties related to the stimulated light scattering will be explained in Section 6.1.1.

The optical gain can be realized after an inverse population is achieved through the pumping process, as explained in Section 7.5. The amplification coefficient that characterize the stimulated emission process can be expressed as [1, 2]

$$g(\nu) = \frac{g_0}{1 + P_{in}/P_{sat} + \left[2\pi T_2 (\nu - \nu_0)\right]^2} \tag{3.4}$$

where ν is frequency of the incident optical signal, ν_0 is the so-called atomic transition frequency related to the two-level energy diagram (as shown in Figure 7.1), g_0 is the value of the amplification peak, P_{sat} is the saturation power, and T_2 is known as dipole relaxation time that takes subpicosecond values. The saturation power P_{sat} is a medium-specific parameter related to the population relaxation time, which is the time that carriers spend on the upper energy level (see Section 7.5). The relaxation time can be in the range of 0.1 to 10 ms for commonly used optical amplifiers discussed in Chapter 2.

The optical signal amplification is proportional to the amplification coefficient and the total length L_a of the amplifier medium. The amplifier medium is often a doped fiber, such as an EDFA. The gain factor G, which determines the level of the optical signal that is being amplified, can be expressed as

$$G(\nu) = \exp[g(\nu)L_a] = \exp\left\{\frac{g_0 L_a}{1 + P_{in}/P_{sat} + \left[2\pi T_2 (\nu - \nu_0)^2\right]}\right\} \quad (3.5)$$

The gain factor, or simply the gain, has a maximum $G_0 = \exp(g_0 L_a)$ related to the gain peak g_0 and frequency ν_0. Therefore, the gain depends on the level of the incident optical power, and it degrases as incident optical power becomes comparable with the saturation power P_{sat}. This reduction in amplification capability is known as the gain saturation. The amplification process covers effectively some optical frequency bandwidth $\Delta\nu_a$, which is defined as the FWHM of the gain function $G(\nu)$; that is,

$$\Delta\nu_a = \frac{1}{\pi T_2}\sqrt{\left[\frac{\ln 2}{\ln(G_0/2)}\right]} \quad (3.6)$$

The output optical power from an optical amplifier is determined by the amplifier gain and can be expressed as

$$P_{out} = GP_{in} \quad (3.7)$$

where P_{in} and P_{out} are the input and output powers, respectively. The enhancement in the output power can also be expressed as $P_{out} = P_{in} + G$, if all parameters in (3.7) are given in decibels.

Equations (3.5) and (3.7) can be used to express the gain G in the following form:

$$G = G_0 \exp\left(-\frac{G-1}{G}\frac{P_{out}}{P_{sat}}\right) \quad (3.8)$$

Therefore, the gain gradually decreases from its maximum value G_0 if the amplifier output power approaches the level P_{sat} determined by the population relaxation time. The functional dependence given by (3.8) is plotted in Figure 3.2 for several values of the maximum gain value G_0, which is also known as a small signal gain.

Another parameter of practical interest is known as the output saturation power $P_{o,sat}$, which is defined as the output power where the gain G drops to just half of its maximum value. By replacing G with $G_0/2$, and solving (3.8), it is

3.1 Optical Signal Parameters

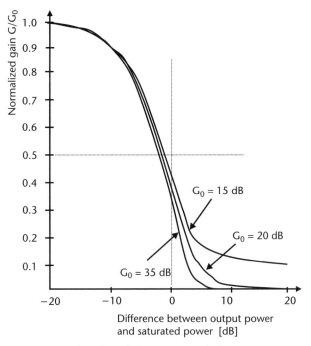

Figure 3.2 Gain parameter as a function of the output optical power.

$$P_{o,sat} = \frac{G_0 \ln 2}{G_0 - 2} P_{sat} \approx P_{sat} \ln 2 \tag{3.9}$$

The saturated value of the gain parameter, which is associated with the saturated optical power, satisfies the following implicit mathematical equation

$$G_{sat} = 1 + \frac{P_{sat}}{P_{in}} \ln \frac{G_0}{G_{sat}} \tag{3.10}$$

3.1.4 Photodiode Responsivity

Photodiode responsivity R, which is expressed in amperes of the output electric current per watt of the input optical power (i.e., as A/W), defines the optoelectronic conversion rate. The photocurrent I_p is related to the incident optical power by the equation

$$I_p = RP_{in} \tag{3.11}$$

The responsivity R can also be expressed through the quantum efficiency η, which defines the ratio between the number of generated electrical carriers (electrons) and the number of incoming photons; that is

$$R = \frac{\eta q}{h\nu} \approx \frac{\eta \lambda}{1.24} \tag{3.12}$$

where h is the Planck's constant, ν is the optical frequency, $\lambda = c/\nu$ is the optical wavelength, and q is the electron charge. Please note that the wavelength in (3.12) should be expressed in microns.

The photodiode responsivity can be increased by increasing the size of the area at which the incident light falls. On the other side, the increase in size will slow the response process and limit the photodiode bandwidth. In addition, the responsivity is wavelength dependent, as shown by (3.12). Such an unequal response over the wavelength band should be corrected through the subsequent signal processing in order to provide the same signal-to-noise ratio for all optical channels. The responsivity of commonly used PIN photodiodes is in the range of 0.4 to 0.6 A/W for silicon-based PIN photodiodes, 0.5 to 0.7 A/W for germanium-based PIN photodiodes, and 0.6 to 0.85 A/W for InGaAs-based PIN photodiodes, as shown in Figure 3.3.

3.2 Noise Parameters

It was mentioned earlier that different corruptive additives would degrade and compromise the integrity of the signal. The noise, as a major corruptive additive to the signal, can originate from different places within an optical transmission system, as shown in Figure 3.4. Semiconductor lasers are the source of the relative intensity noise (RIN), laser phase noise, and mode partition noise. Optical fibers are responsible for modal noise generation, while optical splices and optical connectors are the origin of the reflection-induced noise. Optical amplifiers generate spontaneous emission noise, which is subsequently amplified by the chain of optical amplifiers, and becomes ASE noise. In addition to the amplification of the spontaneous emission noise, each amplifier generates its own spontaneous emission noise. Finally, there are several noise components generated during optoelectronic conversion in photodiodes, all of them related either to the thermal behavior of electrons or the quantum nature of the detected light signal. Accordingly, there are quantum and thermal noise components if the PIN photodiode is used, and thermal, quantum, and shot noise components if the avalanche photodiode is used.

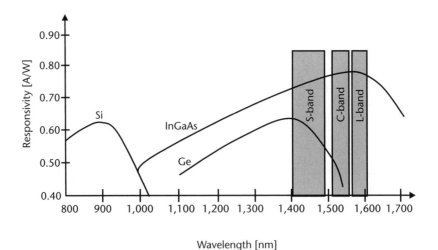

Figure 3.3 Typical responsivity curves of different PIN photodiodes.

3.2 Noise Parameters

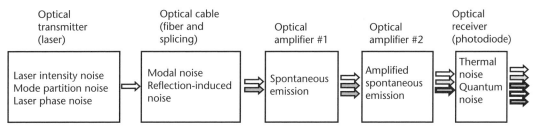

Figure 3.4 Originators of the noise in optical transmission system.

All noise components mentioned above could be divided into ones generated at the optical level, and the electrical components added to the signal while it is in an electrical form. Any optical noise component carries some optical power, which is proportional to the square of the electric field. Since the photodetector generates a photocurrent that is proportional to the incoming optical power rather than to the incoming electric field associated with the power, there will be several components the total photocurrent. These components arise since the noise field beats against the signal, against itself, and against the other optical noises. Although the number of beat components can be substantial, most of them are relatively small as compared with two dominant components, which are related to the noise field beating against the signal, and against itself, as explained in Section 3.2.8. The reader is advised to refer to Table 4.1, which shows the typical values of noise components generated during the photodetection process.

The most relevant noise components can be sorted at the receiving side in a manner shown in Figure 3.5. The scheme presented in Figure 3.5 refers to a general case that contains an optical preamplifier, photodiode, and the front-end amplifier at the receiving side (see Figure 4.1). The most relevant optical noise components coming to the optical preamplifier are the intensity noise and spontaneous emission noise. The spontaneous emission noise contains the spontaneous emission photons originating from the preceding in-line amplifier, and the amplified spontaneous emission noise from the other amplifiers placed along the lightwave path. (The amplified components from Figure 3.5 are denoted by lowercase letters if they are

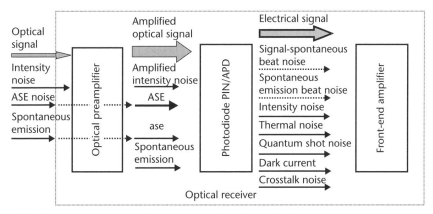

Figure 3.5 Noise components at the receiving side of an optical transmission system.

related to one-stage amplification, and by capital letters if they are related to multi-stage amplification.)

An optical amplifier will enhance all optical inputs in proportion to the amplifier gain. At the end, both signal and noise components are converted to the electrical level through the photodetection process. In addition to the optoelectronic conversion, the noise beat components have been created. Therefore, the total number of noise components that come to the input of the front-end amplifier has been increased. The most important noise beat components are the signal-spontaneous emission beat noise and the spontaneous emission beat noise.

The total noise is a stochastic process that has both multiplicative and additive components. The multiplicative noise components are produced only if the signal is present, while the additive components remain even if the signal is not present. The multiplicative noise components are:

- *Mode partition noise*, which occurs in multimode lasers due to nonuniform distribution of the total power among longitudinal modes and microvariations in the intensity of each longitudinal mode.
- *Laser intensity noise*, which occurs due to microvariations in the laser output power intensity. This noise is characterized through the RIN parameter.
- *Laser phase noise*, which is related to microvariations in phase of generated photons. This is the main reason why the output optical signal, which is a collection of individual photons, exhibits a finite nonzero spectral width.
- *Modal noise*, which arises in multimode fibers through the random process of excitation of transversal modes.
- *Quantum shot noise*, which is caused by the quantum nature of light and the random distribution in time of the electrons generated during the photodetection process.
- *Avalanche shot noise*, which is caused by random nature of the amplification of primary electron-hole pairs through the effect of impact ionization in avalanche photodiodes.

The additive noise components are:

- *Dark current noise*, which is generated in photodiodes due to the thermal process.
- *Thermal noise*, which is created in the resistive part of the input impedance that belongs to the combination of photodiode and front-end amplifier.
- *ASE noise*, which is generated by optical amplifiers placed along the lightwave path.
- *Crosstalk noise*, which occurs in multichannel WDM systems. Optical components that introduce the crosstalk in the WDM systems include optical filters, optical multiplexers and demultiplexers, optical switches, semiconductor optical amplifiers, and optical fibers themselves (through fiber nonlinearities.) The crosstalk can either be intrachannel or interchannel. The intrachannel (or inband) crosstalk, occurs when another signal of the same wavelength interferes with the signal in question; interchannel (out-of band) crosstalk occurs

when some portion of a neighboring channel has been spread out and has been detected by the specified signal's receiver. The crosstalk is often analyzed separately since it is different in nature than other noise components.

3.2.1 Mode Partition Noise

Noise generated in semiconductor lasers is related to fluctuations of the output power parameters (intensity, phase, and frequency), which occur even under fairly constant bias current. It is caused by the spontaneous emission of photons that accompanies the process of stimulated radiation in the laser cavity.

The spontaneous emission is a stochastic process. The spontaneously emitted photons supplement the coherent optical power by contributing randomly to the signal amplitude and phase, and by causing random perturbations in amplitude and phase of the output power. Therefore, these fluctuations in intensity and phase of the emitted light are the physical origin of the overall laser noise.

Mode partition noise is associated with multimode semiconductor lasers, due to uncorrelated emissions among longitudinal modes that are confined within the laser spectrum [refer to Figure 2.4(a)]. The difference in intensities of any pair of longitudinal modes from Figure 2.4(a) fluctuates randomly, even if we assume that the total output power is constant. These intensity fluctuations will be transferred all the way to the optical receiver since chromatic dispersion in the optical fiber will force all longitudinal modes to travel with different speeds. Consequently, the mode partition noise will be converted to the electrical noise and will corrupt the signal at the decision point.

Mode partition noise is relevant for multimode semiconductor lasers and transmission systems where the product BL is relatively low (B is the signal bit rate and L is the transmission distance). The intensive study of the mode partition noise was carried out in the literature some time ago in order to estimate the power penalty related to the noise impact [3, 4]. It was concluded that the impact of the mode partition noise could be almost entirely suppressed by satisfying the following condition:

$$BLD\sigma_\lambda \leq 0.075 \quad (3.13)$$

where D is the chromatic dispersion coefficient expressed in ps/km-nm, while σ_λ is the spectral linewidth of the multimode semiconductor laser, which refers to the root mean square (RMS) width of the spectral envelope from Figure 2.4(a). It is often assumed that spectral envelope has a Gaussian shape, which means that $\sigma_\lambda \sim 0.426\Delta\lambda$, where $\Delta\lambda$ is the FWHM of the spectral envelope (refer to Table 2.1 for numerical values).

The BL product can be maximized by selecting an operational wavelength within the zero dispersion region, or at the region with chromatic dispersion is lower than 1 ps/km-nm. In such a case, the signals with 1-Gbps bit rate can be transmitted over about 30 km, while 10-Gbps signals can be effectively transmitted over 3 km.

Semiconductor lasers designed to operate in the single-mode regime do not produce mode partition noise. However, the existence of remaining side-modes in the laser spectrum may be of some concern. The strength of side-modes is characterized by the mode suppression ratio (MSR), which is defined as a difference in powers between the governing longitudinal mode and the most dominant sidemode [refer to Figure 2.4(b)]. We can assume that lasers that have MSR > 100 will cause a negligible mode partition noise effect since the power penalty will be lower than 0.1 dB [3].

3.2.2 Laser Intensity and Phase Noise

The laser intensity noise, which has the same nature as the mode partition noise, is associated with single-mode lasers. The intensity fluctuations created at the transmitter side will eventually experience both the attenuation in the optical fiber and amplification through the chain of optical amplifiers. The laser intensity noise will be converted to electrical noise by a photodiode and will corrupt the signal at the decision circuit point.

Laser intensity noise can be estimated by RIN, for which the spectrum is defined as

$$RIN(\omega) = \int_{-\infty}^{\infty} \frac{\langle \delta P(t) \delta P(t+\tau) \rangle}{\langle P \rangle^2} \exp(-j\omega t) dt \qquad (3.14)$$

where <P> represents the average value of the laser output power, and $\delta P = P(t) - <P>$ represents small power fluctuations around the average value. The RIN value can be calculated by solving generalized laser rate equations that contain a noise term, known as Langevin forces [see (7.55) to (7.57)]. The function $RIN(\omega)$, which is usually expressed in decibels per hertz, has a peak at the relaxation-oscillation frequency, which was introduced by (2.4). That peak RIN value for a typical semiconductor laser operating at 1,550 nm is in the range of –130 to –120 dB/Hz. However, it decreases gradually at a rate of about 5 to 10 dB/Hz per gigahertz at frequencies further away from the relaxation frequency [5].

The parameter of practical interest is the $(SNR)_{RIN}$ due to the RIN effect, which is defined as

$$SNR_{RIN} = 1/r_{int} = \left[\frac{\langle P \rangle^2}{\langle \delta P(t) \delta P(t+0) \rangle} \right]^{1/2} \qquad (3.15)$$

The coefficient r_{int} is often used as a measure of the receiver sensitivity degradation (refer to Section 5.3.2). Although the $(SNR)_{RIN}$ value is proportional to the output optical power, it was shown that it eventually saturates with an increase of the optical power. We can assume that the $(SNR)_{RIN}$ value is in the range of 20 to 30 dB for output powers above a few milliwatts [5].

The reflection-induced noise is related to the appearance of the back-reflected optical signal due to refractive-index discontinuities at optical splices, connectors, and optical fiber ends. The amount of reflected light can be estimated by the refection coefficient r_{ref}, defined as

$$r_{ref} = \left[\frac{n_a - n_b}{n_a + n_b} \right]^2 \qquad (3.16)$$

where n_a and n_b are the refractive-index coefficients of materials facing each other. The amount of the reflected light is directly proportional to the coefficient r_{ref}. Therefore, r_{ref} is higher for a larger difference in refractive indexes, and vice-versa. The strongest reflection occurs at the glass-air interface. Assuming that $n_a = 1.46$ (for silica), and $n_b = 1$ (for air), the refection coefficient becomes $r_{ref} \sim 3.5\%$ (or -14.56 dB). This value can be even bigger if the optical fiber ends are polished. The amount of reflected light can be reduced below 0.1% if some index-matching oils or gels are used at the fiber-air interface, or if the fiber ends are cut at an angle to deviate the reflected light from the fiber axis. Both methods are used extensively in high-speed optical transmission systems.

A considerable amount of back-reflected light can come back and enter the semiconductor laser resonant cavity, which would negatively affect the laser operation and lead to excess intensity noise at the laser output. That is the main reason why the laser is often separated from the optical fiber link by an optical isolator, which will eventually suppress the impact of the reflected light. The relative intensity noise can be increased by as much as 20 dB if the back-reflected light exceeds an absolute level of -30 dBm.

The impact of the reflection-induced noise is not limited just to the laser source, since multiple back and forth reflections between optical splices and connectors can be the source of an additional intensity noise. Multiple reflections will eventually create multiple copies of the same signal traveling forward. These copies will be shifted in phase, which means that the total signal will experience an additional phase noise. Such a phase noise is eventually converted to the intensity noise by chromatic dispersion and enhanced by optical amplifiers along optical fiber link. In addition to chromatic dispersion, phase noise can be converted to intensity noise at any two reflecting surfaces along the optical fiber links, since they act as the mirrors of the FP interferometer. The end result of the conversion of phase noise to intensity noise will be the increase in the total RIN. Therefore, it is extremely important to suppress the back-reflections along the entire optical transmission line by careful selection of optical connectors that minimize reflections.

The second type of noise produced by lasers is the laser phase noise, which is directly related to the spectral linewidth. The spectral linewidth of a single-mode laser is enhanced by the amplitude-phase coupling of the output optical signal [see (3.49) and (3.50)]. Laser phase noise is one of the most important sources of signal degradation in coherent lightwave systems [6], while its direct impact is much smaller in systems that employ the intensity modulation and direct detection, and can be considered as a minor factor in transmission systems engineering presented by this book.

3.2.3 Modal Noise

Modal noise is related to multimode optical fibers. Please recall that the total optical power is nonuniformly distributed among a number of modes within a multimode optical fiber, as shown in Figure 2.10. Such modal distribution creates the so-called speckle pattern at the receiving side that contains brighter and darker spots in accordance to the mode distribution. The photodiode effectively eliminates the speckle pattern impact by registering the total power that is integrated over the photodiode area. However, if the speckle pattern is not stable but changes with time, it will induce fluctuations in the received optical power. Such fluctuations are referred to as the modal noise and will eventually be converted to photocurrent fluctuations.

The fluctuations in speckle pattern occur in optical fibers due to mechanical disturbances within optical cables, such as microbends and vibrations, and due to the impact of splices and connectors since they act as spatial optical filters and influence the power distribution over transversal modes.

The modal noise is inversely proportional to the spectral linewidth Δv of the light source. This comes from the fact that mode interference and speckle pattern changes are relevant only if coherence time ($t \sim 1/\Delta v$) is longer than intermodal dispersion in optical fibers (refer to Section 2.4). This condition is not satisfied if LEDs are used for signal transmission, since the LED spectral linewidth is quite large, as shown in Table 2.1. Therefore, it is a good idea to use LED sources in combination with multimode optical fibers whenever possible to avoid the impact of modal noise. On the other hand, it is important to mention that the modal noise also depends on the laser frequency chirp since the spectral linewidth of a single-mode laser is enhanced by the amplitude-phase coupling of the output optical signal [see (3.49) and (3.50)].

The situation is quite different if single-mode lasers are used in combination with multimode optical fibers, since the modal noise impact could be quite a serious problem. The impact of the modal noise is higher for a smaller number of modes propagating through an optical fiber, and the most serious situation occurs if the optical power at the receiving side is effectively shared by only several transversal modes. As for systems engineering, it is necessary to allocate some power margin to accommodate the modal noise effect in the case when single-mode lasers are used in combination with multimode optical fibers. That margin should be as high as 1 dB for a combination of single-mode lasers and multimode optical fibers.

Even sections of single-mode optical fibers that are up to few meters long can introduce the modal noise since a higher-order mode can be excited at one fiber discontinuity (connector or splice), and then converted back to the fundamental mode at the next discontinuity. It is good idea, therefore, to use optical fibers that are a little bit longer even if the distance is just 1m to 2m. The distance of 5m, for example, can effectively eliminate the impact of the modal noise since the higher-order mode cannot reach the second fiber discontinuity.

It is important to notice that VCSELs are often used in combination with multimode optical fibers for very short links (up to several kilometers). Although this combination is a cost-effective solution for gigabit-per-second signal rates over very short distances, it is also a place where modal noise may be a serious factor and cause a power penalty even higher than of 1 dB [7]. This penalty should be compensated for by proper power margin allocation (refer to Chapter 5).

3.2.4 Quantum Shot Noise

The optical signal arriving at the photodiode contains a number of photons that generate the electron-hole pairs through the photoelectric effect. The electron-hole pairs

3.2 Noise Parameters

are effectively separated by the inverse bias voltage, which produces a photocurrent, as shown in Figure 7.2. The probability of having n electron-hole pairs at the photo diode during the time interval Δt is expressed by the Poisson probability distribution [8, 9]

$$p(n) = \frac{N^n e^{-N}}{n!} \tag{3.17}$$

where N is the mean number of photoelectrons detected during the time interval Δt given as

$$N = \frac{\eta}{h\nu} \int_0^{\Delta t} P(t) dt \tag{3.18}$$

The parameters in (3.18) are: ν − optical frequency, h − the Plank's constant, and η − quantum efficiency (which is the ratio between the number of the generated electrons and the number of the arrived photons). The Poisson distribution approaches a Gaussian distribution for larger values of the mean N.

The mean intensity of the photocurrent, which has been generated by the stream of electrons, is given as

$$I = \langle i(t) \rangle = \frac{qN}{\Delta t} = \frac{qN}{T} \tag{3.19}$$

where q is the electron charge ($q = 1.6 \times 10^{-19}$ Coulombs). Note that it was assumed that the time interval Δt is equal to the duration T of either the 1 or 0 bits. The actual number of electrons generated during the bit duration will fluctuate around the mean value N due to the random nature of the photodetection process, while the generated photocurrent will fluctuate around the mean value I. It is useful to recall the property of Poisson distribution that the variance is equal to the mean value, in order to get the following equation connecting the actual number n of electrons with the mean value N:

$$\langle [n-N]^2 \rangle = N \tag{3.20}$$

where "<>" brackets indicate the average. The instantaneous current during the bit duration T is given as

$$i(t) = \frac{qn}{T} \tag{3.21}$$

Equations (3.20) and (3.21) can be used to estimate the fluctuations of the instantaneous current around the mean value. These fluctuations can be expressed through the mean-square value, which is

$$\langle i^2 \rangle_{sn} = \langle [i(t) - I]^2 \rangle = \frac{q^2 \langle [n(t) - N]^2 \rangle}{T^2} = \frac{q^2 N}{T^2} = \frac{qI}{T} \tag{3.22}$$

Equation (3.22) presents the power of the quantum shot noise, a multiplicative noise component that is caused by the quantum nature of light. We can correlate the bit duration T with the signal bandwidth Δf, as in [2], to get the following relation:

$$\langle i^2 \rangle_{sn} = 2qI\Delta f = S_{sn}(f)\Delta f \tag{3.23}$$

The same result as the previous one can be obtained by calculating the signal spectral density $S_{sn}(f)$ through the Fourier transform of the photocurrent correlation function [10, 11]. The spectral density of the quantum shot noise from (3.23) is constant and given as

$$S_{sn}(f) = \frac{d}{df}\langle i^2 \rangle_{sn} = 2qI \tag{3.24}$$

Equations (3.23) and (3.24) can be applied just to PIN photodiodes. That is because an internal amplification process in APDs increases the generated photocurrent and enhances the total quantum noise (recall the characteristics of APD from Section 2.5). Physical background behind this additional noise in APD is related to the fact that secondary electron-hole pairs are generated randomly through a stochastic process of impact ionization. It was shown that the avalanche shot noise can be characterized by the Gaussian probability density function, and it has the spectrum that is flat with the frequency [2]. The avalanche shot noise power is

$$\langle i^2 \rangle_{sn/APD} = S_{sn/APD}(f)\Delta f = 2q\langle M \rangle^2 F(M)I\Delta f \tag{3.25}$$

where <M> is the average value of the avalanche gain, and $F(M)$ is the excess noise factor, which measures the variations of the instantaneous avalanche gain M from its average value.

The excess noise factor can be expressed as [2, 10]

$$F(M) = k_N\langle M \rangle + (1 - k_N)\left[2 - \frac{1}{\langle M \rangle}\right] \tag{3.26}$$

where parameter k_N is known as the ionization coefficient. The ionization coefficient takes values in the range of 0 to 1 and measures the ability of a carrier to generate other carriers in the avalanche amplification process. This coefficient should be as small as possible in order to minimize the avalanche shot noise that is being generated. There is also an approximate form that is often used instead of (3.26), and it is given as

$$F(M) = \langle M \rangle^x \tag{3.27}$$

The noise coefficient x in (3.27) takes the values from the range (0 to 1), depending on the semiconductor compound used. Typical values of the noise coefficients k_N and x for commonly used APD compounds are shown in Table 3.1.

Table 3.1 Noise Parameters for Various Photodiodes

Semiconductor	x	k_N	Dark Current Noise, nA
InGaAs	0.5–0.8	0.3–0.6	Up to 20
Germanium	1.0	0.7–1.0	50–500
Silicon	0.4–0.5	0.02–0.04	Up to 10

The avalanche amplification process in APD will lead to an enhancement of both the signal and quantum noise. However, the SNR will be decreased since the noise factor $F(M)$ is higher than the unity value.

3.2.5 Dark Current Noise

The dark current is the current that flows through a biased photodiode even if no light is coming to the photodiode surface. The dark current consists of electron-hole pairs, which are thermally created in the photodiode p-n junction (see Figure 7.2). These carriers can also get accelerated in APD, and can contribute to the avalanche shot noise generation. The dark current noise power is

$$\langle i^2 \rangle_{dcn} = s_{dcn}(f)\Delta f = 2q\langle M \rangle^2 F(M) I_d \Delta f \tag{3.28}$$

where I_d is a primary dark current generated in the photodiode. The typical values of dark current for different semiconductor compounds are shown in Table 3.1. The power of dark noise generated in photodiodes is smaller than the power of the other noise components that arise during the photodetection process. That is the reason why the impact of dark current noise is often neglected.

The process of noise generation in the avalanche photodiode is illustrated in Figure 3.6. There are four time slots in Figure 3.6 to illustrate the Poisson statistics and the total noise generation. There is one incoming photon that generates one primary photoelectron within the first time slot. This primary electron will produce several secondary electron-hole pairs (there are seven pairs in Figure 3.6). There are no signal photons captured within the second time slot, but a dark current electron is generated, which was able to produce a pair of secondary electrons through the ionization process. Next, there are two incident photons within the third time slot that produce a number of secondary electrons. Finally, for time slot number four, there is a similar situation to that related to slot number two, but this time a smaller number of secondary electrons have been generated. As a result, both the total number of electrons and the generated current flow fluctuate in time around their average values. These fluctuations are associated with the noise at the photodiode output.

3.2.6 Thermal Noise

The photocurrent generated during the detection process is converted to voltage by a load resistor. The voltage at the load resistor is then amplified by the front-end stage (refer to Figure 5.1). The load resistor generates its own noise due to a random thermal motion of electrons. Such noise occurs as a fluctuating current that adds to the already-generated photocurrent. This additional noise component, also known

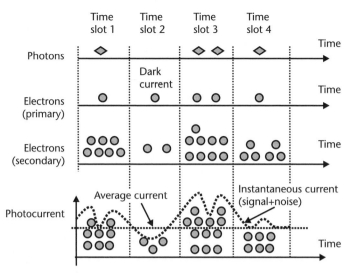

Figure 3.6 Noise generation in a photodiode.

as the Johnson noise [12], has a flat frequency spectrum that is characterized with the zero-mean Gaussian probability density function. The thermal noise spectral density, which is expressed in square amperes per hertz, is given as

$$S_{the}(f) = \frac{4k\Theta}{R_L} \quad (3.29)$$

where R_L is the load resistance, Θ is the absolute temperature in Kelvin, and k is the Boltzmann's constant ($k = 1.38 \times 10^{-23}$ J/K). The thermal noise power, which is represented by the mean-square of the current, contained in the receiver bandwidth Δf is

$$\langle i^2 \rangle_{the} = S_{the}(f)\Delta f = \frac{4k\Theta\Delta f}{R_L} \quad (3.30)$$

The thermal noise can be reduced by using a large value of the load resistance. Such a design, often referred to as the high-impendence front-end amplifier, also increases the receiver sensitivity. On the other hand, it limits the receiver bandwidth since the RC constant (C is the capacitance of the circuit) is also increased. Therefore, the high-impedance input requires an equalizer that will boost the high-frequency components and increase the receiver bandwidth. In the general case, the receiver bandwidth is increased by selecting a smaller value for the load resistance. Such a design, also known as low-impedance front-end, has a smaller receiver sensitivity than the version with the high-impendence front-end amplifier.

A distinct design approach, often used to achieve both the high receiver sensitivity and high-speed operation, is known as trans-impedance front-end. The trans-impedance front-end design also improves the dynamic range of the optical receiver, which is important in cases when significant variations of optical power can occur at the receiving side. Two commonly used optical receiver front-end schemes are shown in Figure 3.7.

Some components (mainly transistors) within the front-end amplifier will enhance the thermal noise generated in the load resistor. Such a noise contribution can be accounted for by the amplifier noise figure F_{ne}, which is the factor that measures the thermal noise enhancement at the front-end output. The total power of the thermal noise that also accounts the contribution of the front-end amplifier is given as

$$\langle i^2 \rangle_{the} = S_{the}(f)\Delta f = \frac{4k\Theta\Delta f F_{ne}}{R_L} \tag{3.31}$$

The noise figure F_{ne} (the index "e" stands for electrical) can vary from amplifier to amplifier, but for a low-noise front-end amplifier it is around 3 dB.

3.2.7 Spontaneous Emission Noise

The spontaneous emission of light occurs during optical signal amplification in an optical amplifier. That process is additive, which means that there is no correlation between the signal and the noise generated through the spontaneous emission. The noise induced by spontaneous emission also has a flat frequency spectrum characterized with the zero-mean Gaussian probability density function. The noise spectral density can be written as [1, 2]

$$S_{sp}(\nu) = (G-1)F_{no}\, h\nu/2 \tag{3.32}$$

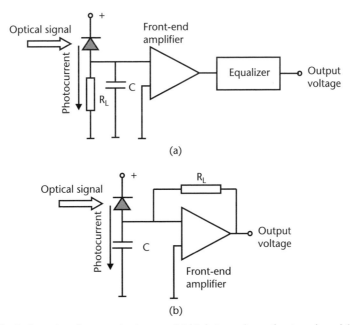

Figure 3.7 Optical receiver front-end schemes: (a) high-impedance front-end, and (b) transimpedance front-end.

where G is the optical amplifier gain, F_{no} is optical amplifier noise figure (the index "o" stands for optical), h is the Planck's constant ($h = 6.63 \times 10^{-34}$ J/Hz), and v is the optical frequency. Please note that we will temporarily use two notations for frequency: f for the frequency of an electrical signal and v for the frequency of an optical signal. However, the variables f and v refer to the same physical parameter, which is expressed in hertz. This distinction will be used for few more times (e.g., to distinguish between an optical and an electrical filter bandwidth, or to clearly point to optical and electrical SNRs).

The optical amplifier noise figure is related to the spontaneous emission factor n_{sp}, defined as

$$n_{sp} = \frac{N_2}{N_2 - N_1} \tag{3.33}$$

where N_1 and N_2 are the populations of the atoms at the ground level and at the upper energy levels, respectively (see Figure 7.1). The spontaneous emission factor will become unity if all atoms are excited and moved in energy to the upper level. This is just a theoretical case since such an inversion population cannot be achieved in practice. That is the reason why the spontaneous emission factor will always take values higher than one. There is the following relation between the noise figure and the spontaneous emission factor [1, 2]:

$$F_{no} = \frac{2n_{sp}(G-1)}{G} \approx 2n_{sp} \geq 2 \tag{3.34}$$

In most practical cases it will be $2 < F_{no} < 5$, or in decibels, 3 dB $< F_{no} < 7$ dB.

The effective noise figure of the chain of cascaded optical amplifiers can be calculated as

$$F_{no,eff} = F_{no,1} + \frac{F_{no,2}}{G_1} + \frac{F_{no,3}}{G_1 G_2} + ... \frac{F_{no,k}}{G_1 G_2 ... G_{k-1}} \tag{3.35}$$

where $F_{no,eff}$ is the effective noise figure of the amplifier chain that contains the total number of k optical amplifiers. The first amplifier in the chain is the most important one in terms of the noise impact. That is the reason why multistage optical amplifiers should be designed to have the first stage with lower noise figure. Accordingly, any decrease in the effective value of the amplifier's noise figure will bring a significant benefit in the overall system performance (refer to Section 5.2).

The total power of the spontaneous emission noise is

$$P_{sp}(v) = 2|E_{sp}|^2 = 2S_{sp}(v)B_{op} = (G-1)F_{no}hvB_{op} \tag{3.36}$$

where E_{sp} is the electric field of the spontaneous emission, and B_{op} is the effective bandwidth of spontaneous emission determined either by the optical amplifier bandwidth or by an optical filter. Please note that factor 2 in (3.36) accounts for the

3.2.8 Noise Beat Components

Optical amplifiers are employed along the transmission line to boost the power of the optical signal, but the process is accompanied by the generation of spontaneous emission. The spontaneous emission noise generated in a preceding amplifier stage will eventually be amplified in the following stage, thus becoming the ASE noise. In addition, the spontaneous emission noise is also generated at the any specific amplifier in question, which means that the amplifier output is noisier than the input. The power of the ASE noise is converted from an optical to electrical level in parallel with the optical signal conversion. The total photocurrent generated at the photodiode output, in cases when optical amplifiers are employed, can be written as

$$I_p = I + i_{noise} = R\left|E\sqrt{G} + E_{sp}\right|^2 + i_{sn} + i_{the} \tag{3.37}$$

where $E = (P)^{1/2}$ and $E_{sp} = (P_{sp})^{1/2}$ are the electrical fields associated with optical signal power P and amplified spontaneous emission power P_{sp}, respectively, I is the mean value of the signal current ($I = RP$), R is photodiode responsivity, G is the amplifier gain, i_{sn} is the quantum shot noise component, and i_{the} is the thermal noise component. The three noise components from (3.37) are the dominant ones and the most relevant from the systems engineering perspective. The reader can refer to Section 4.2 to find more details with respect to comparison of different noise components.

The ASE is not simply converted to the corresponding electrical noise since there is a beating process between the ASE and the signal, which results in the appearance of several components that can be classified as beat noise components. It is necessary to express the incoming signal and ASE by the corresponding electric fields, as in (3.37), in order to evaluate these beat noise components. Please note that (3.37) contains just half of the noise power from (3.36), which belongs to a component of the ASE that has the same polarization as the signal. This is because the orthogonally polarized components cannot beat effectively, and only the component that has the same polarization with the signal is a relevant factor.

The total noise current associated with the optical ASE noise occurs due to beating of the ASE field E_{sp} with the signal field E, and due to beating of the field E_{sp} with itself. The total variance of such a fluctuating current can be found by expressing all electrical fields in (3.37) by a basic definition from

$$E = \sqrt{P}\exp(-j\omega t + \phi) \tag{3.38}$$

(where ω and ϕ are the frequency and the phase of the electric filed), and by averaging the field products over random phases as in [10, 11]. This process leads to the equation

$$\langle i^2 \rangle = \langle i^2 \rangle_{the} + \langle i^2 \rangle_{sn} + \langle i^2 \rangle_{sig-sp} \langle i^2 \rangle_{sp-sp} \tag{3.39}$$

where the first term on the right side represents the power of the thermal noise, while the three remaining terms are associated with the optical noise. The noise components from (3.39) are given as

$$\langle i^2 \rangle_{the} = \frac{4k\Theta F_{ne} \Delta f}{R_L} \tag{3.40}$$

$$\langle i^2 \rangle_{sn} = 2qR\left[GP + S_{sp}B_{op}\right]\Delta f \tag{3.41}$$

$$\langle i^2 \rangle_{sig-sp} = 4R^2 GPS_{sp} \Delta f \tag{3.42}$$

$$\langle i^2 \rangle_{sp-sp} = 2R^2 S^2_{sp}\left[2B_{op} - \Delta f\right]\Delta f \tag{3.43}$$

From a practical viewpoint, it is useful to compare the noise powers of components presented in (3.40) to (3.43), which is done in Section 4.2. It is easy to verify that the thermal noise and the shot noise from (3.40) and (3.41) are generally smaller than the beat noise components from (3.42) and (3.43). In addition, it is possible to reduce the spontaneous-spontaneous beat noise by optical filtering and make it smaller than the signal-spontaneous beat noise.

3.2.9 Crosstalk Noise Components

Crosstalk occurs in multichannel optical transmission systems. There are two types of crosstalk noise, and they are known as intrachannel (out-of-band) and interchannel (inband) components.

Out-of-band crosstalk occurs when the power from a neighboring channel crosses the border between channels and mixes with the power of the specified optical channel, as shown in Figure 3.8(a). Optical receiver bandwidth captures the interfering power and converts it to the electrical current. This often happens when optical filters and optical multiplexers are deployed in an optical transmission system.

Out-of band crosstalk is different from the random noise in an advantageous way. Namely, various noise types have amplitude probability distribution with a gradually decaying tail, which determines the decision threshold position (refer to Figure 5.2). On the other hand, in the out-of-band crosstalk case the undesired crosstalk power that may appear at the decision point is bounded since there are a finite number of contributing sources. The worst case scenario in terms of crosstalk impact would be if all channels were bit-synchronized, and when the channel in question carried 0 bit, while all other channels carried 1 bits. However, in reality, the incoming crosstalk optical power is not correlated with the signal, which means that the out-of band noise is incoherent in nature. The out-of-band crosstalk is dominated by two immediately adjacent channels, as shown in Figure 3.8(a).

The portion of the photocurrent generated due to conversion of the out-of-band crosstalk to electrical signal can be considered as a noise and treated similarly as the dark current noise was treated. The noise current generated by conversion of the out-of-band crosstalk to the electrical signal is

3.2 Noise Parameters

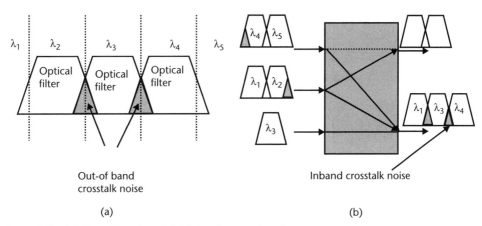

Figure 3.8 (a) Out-of band, and (b) inband crosstalk noise components.

$$i_{cross,out} = \sum_{n \neq m}^{M} RX_n \qquad (3.44)$$

where R is photodiode responsivity, and X_n is a portion of the nth channel power that has been captured by the optical receiver of the mth optical channel (i.e., the channel in question). We assumed that there are M optical channels altogether, and that the largest contribution to the crosstalk noise level comes from the neighboring channels. The impact of the out-of-band crosstalk can be minimized by optimizing the optical channel spacing and by selecting optical filters with steeper bandwidth characteristics.

The so-called inband or interchannel crosstalk occurs when a signal at the same optical wavelength as the wavelength of the optical channel in question comes to the optical receiver to be converted to electrical current. This can happen if there are multiple optical multiplexing/demultiplexing stages between optical transmitter and optical receiver, or if the optical routing process takes place along the lightwave path.

Namely, any separation of optical channels by optical filtering can cause interchannel crosstalk. Such crosstalk can be enhanced by an increased mismatch between multiplexing and demultiplexing filters, which is usually caused by the temperature change, or by aging. This will create an out-of-band noise, as mentioned above, if demultiplexing is followed by the photodetection process. However, if the signal continues its propagation along a specified lightwave path, the total power will now consist of the signal power and out-of-band crosstalk power. Any additional multiplexing/filtering process, which can eventually occur before photodetection, will also produce out-of-band crosstalk. It might happen that crosstalk power from the neighboring channel brings back a part of power that originally belonged to the channel in question. Although that portion contains the same data as the channel in question, they are not in phase with each other any more since they experienced different delays before reunion occurred. The same process (separation and reunion) might happen at an optical switch due to nonideal isolation between switch ports, as shown in Figure 3.8(b).

Inband crosstalk will eventually produce the beat noise components by a process similar to one related to the ASE noise. The only real difference is that the crosstalk, rather than the ASE noise, will beat with the signal and with itself. The total beat photocurrent in the mth channel is

$$i_{cross,in}(t) = R|E_m(t)|^2$$
$$= R\left|E_m(t)\exp[-j\omega_m t + \phi_m(t)] + \sum_{n=1,n\neq m}^{M} X_n E_n(t)\exp[-j\omega_n t + \phi_n(t)]\right|^2 \quad (3.45)$$

where R is photodiode responsivity, E_m is the electric field of the optical signal in question, and X_n and ϕ_n are the amplitudes and phases the electric fields associated with inband crosstalk components. Parameter M refers to the potential number of inband contributions. Please notice that the exponential term denotes the coherent nature of the total electric field. By using substitution $E = (P)^{1/2}$, and by performing the multiplication in (3.45), the total current becomes

$$i_{cross,in}(t) = RP_m + 2R\sum_{n=1,n\neq m}^{M}\sqrt{P_n}\cos[\phi_m(t) - \phi_n(t)] = R(P_m + \Delta P_{cross,in}) \quad (3.46)$$

It becomes evident from (3.46) that the inband crosstalk has the same character as the intensity noise and can be treated as a component of the total intensity noise.

3.3 Signal Impairments

There are a number of impairments that will either corrupt or distort an optical signal during its propagation along the wavelength path. The impact of these impairments can be calculated either through the SNR, or through optical power penalty related to the receiver sensitivity degradation. The impact of some impairments, such as the attenuation, is very straightforward and can be accounted directly through the SNR. The impact of most other impairments is more complex and needs to be evaluated differently. The major signal impairments will be introduced throughout the rest of this chapter.

3.3.1 Fiber Attenuation

Optical fiber attenuation, also known as the fiber loss, is defined as the ratio between the output and input optical powers associated with the specified optical fiber section. The change in the optical signal power P, which propagates through the optical fiber, is given as

$$\frac{dP}{dz} = -\alpha P \quad (3.47)$$

where α is the attenuation coefficient expressed in km^{-1} (or in m^{-1}). Optical powers P_2 and P_1, which are related to the output and the input of the link, respectively, are connected by relation

$$P_2(t) = P_1(t)\exp(-\alpha L) \tag{3.48}$$

where L is the length of the optical fiber link.

Optical attenuation coefficient α is commonly expressed in decibels per kilometer. The decibel is defined as dB = 10 log (P_2/P_1). The decibel value of the coefficient α can be simply obtained from its numeric value as: α [dB/km] = 4.343α. Recall from Section 2.3 that there are several factors contributing to the total value of the attenuation coefficient (such as absorption and scattering). The reader can also recall that the minimal attenuation coefficient corresponding to the 1,300-nm wavelength window is about 0.35 dB/km, while it is around 0.2 dB/km in the 1,550-nm wavelength window.

3.3.2 Insertion Losses

Fiber losses are not the only cause of optical signal attenuation along the transmission line. There are fiber joints (fiber splices and fiber connectors), which also cause the signal attenuation. Fiber splices can either be fused or joined together by some mechanical means, as mentioned in Section 2.3. Typical mean values of the attenuation inserted by fused optical splices is somewhere between 0.05 and 0.1 dB, while mechanical splices insert signal loss comparable (or slightly above) to 0.1 dB.

Optical connectors are designed to be removable, thus allowing many repeated connections and disconnections. The insertion loss for high-quality single-mode optical fibers should not be higher than 0.25 dB. A special connector design is very often applied to minimize reflection of the incoming optical signal from the surface of the receiving part of the connector. Such a design can include angled fiber-end surfaces, or perhaps some index matching fluid applied at the fiber surfaces. The matching fluid will minimize the reflection coefficient given by (3.16) by decreasing the refractive index differences when the optical signal crosses from one fiber to the other.

Optical connectors with angled fiber-end surfaces are the most convenient ones to minimize *connector return loss*, which is a fraction of the optical power reflected back into the fiber at the connection point. In some cases, such as with bidirectional transmission through the same fiber, the reflection power should be suppressed to more than 60 dB below the input level. This requirement is satisfied by using connectors that provide a physical contact of angled fiber-ends.

The number of optical splices and connectors depends on the length of the lightwave path, and this should be taken into account during the optical transmission systems engineering. The engineering is greatly simplified if the total attenuation due to fiber joints is distributed over the overall fiber transmission line, and added to the optical fiber attenuation. In such case, some per-kilometer attenuation value can be assigned to account for signal attenuation.

3.3.3 Frequency Chirp

A finite spectral linewidth can be attached to each individual longitudinal mode of a multimode Fabry-Perot laser [1]. The same is applicable for the remaining mode of single-mode lasers (DFB, DBR). The linewidth of individual longitudinal modes is between 10 and 100 MHz, which is much larger than the linewidth of an individual longitudinal mode in gas lasers [13] (the enlargement factor is somewhere between 5 and 40).

The linewidth enhancement was originally explained by C. H. Henry in [14]. This effect occurs due to a change of the lasing frequency, which is affected by a change of the optical gain in the laser resonant cavity. The optical gain changes due to fluctuations in carrier density, which is caused by the spontaneous emission process [refer to laser rate equations (7.49) to (7.51)]. On the other hand, fluctuations in carrier density will produce fluctuations in both the refractive index and fluctuations in the optical signal phase, since the carrier density also determines the refractive index. All these random fluctuations are accounted through terms related to the Langevin forces in (7.55) to (7.57).

Therefore, the random fluctuations in carrier density will be transferred to the random fluctuations of the frequency, which leads to the frequency noise and spectral linewidth enhancement. The linewidth enhancement is proportional to the factor $(1 + \alpha^2_{chirp})$, where α_{chirp} is an amplitude-phase coupling parameter that determines the ratio between the refractive index change and the gain change. It is defined as

$$\alpha_{chirp} = \frac{dn/dN}{dG/dN} \quad (3.49)$$

where n, N, and G are refractive index in the laser cavity, number of carriers (electrons), and the laser gain, respectively. Parameter α_{chirp} is also associated with the frequency chirp of the output optical signal. That is the reason why this parameter is often called simply "the chirp factor." It takes the values ranging from 2 to 8 for different types of semiconductor lasers [5, 14]. The chirp factor can also be defined for the external optical modulators (i.e., for the EA and MZ modulators), in which case it can take not just positive, but negative values as well. The absolute values of the chirp factor, which characterizes an external modulator parameter, are typically in the range of 0 to 1.

The frequency chirp effect, which occurs during an amplitude modulation of the optical signal, is proportional to the chirp factor. This process occurs through several steps. First, the change in carrier density during the amplitude modulation causes the change in both the optical gain and the refractive index. Secondly, changes in the optical gain lead to the intensity modulation of the output optical signal, while changes in the refractive index lead to the detrimental optical phase modulation. Finally, the phase modulation causes a shift in instantaneous frequency from the steady-state value v_0. Such an instantaneous frequency shift can be calculated by using the formula derived in [15]

$$\delta v(t) = \frac{\alpha_{chirp}}{4\pi} \left[\left(\frac{d}{dt} \ln P(t) \right) + \chi P(t) \right] \quad (3.50)$$

where $P(t)$ is the time variation of the output optical power, and χ is a constant related to the material and design parameters. Parameter χ can vary from values close to zero to up to several tens.

The first term in the brackets on the right-hand side of (3.50) is referred to as the transient (or instantaneous) chirp, and the second term defines adiabatic (or steady-state) frequency chirp. There is an offset between the adiabatic and transient chirp, as shown in Figure 3.9, which is sometimes used in direct modulation schemes for their partial mutual cancellation.

Generally, the temporary shape of the modulation electrical signal plays an important role in the total value of the frequency chirp, which will be discussed in the next section. Please note from Figure 3.9 that the leading edge of the pulse undergoes the shift towards higher frequencies (so-called blue shift), while the trailing edge shifts towards lower frequencies (so-called red shift).

As a summary note, we can say that the frequency chirp causes signal bandwidth increase. It also interacts with the chromatic dispersion in the optical fiber, thus contributing to the deviation of the pulse shape. The character of the pulse change during the propagation through an optical fiber will be discussed in the next section.

3.3.4 Chromatic Dispersion

An optical signal experiences an intensive distortion as it travels through the optical fiber. Both the intermodal delay effect and chromatic dispersion cause the distortion in multimode optical fibers, while chromatic dispersion is the only cause of the signal distortion in single-mode optical fibers.

It was mentioned in Chapter 2 that multimode optical fibers transfer light through multiple spatial or transversal modes, and that each mode takes a slightly distinguished path along the optical fiber. The difference in mode path lengths causes a difference in arrival times at the receiving point, which is a phenomenon known as multimode dispersion.

Single-mode optical fibers effectively eliminate multimode dispersion by limiting the number of modes to just one through a much smaller core diameter. However, there is still signal distortion in single-mode optical fibers due to chromatic dispersion, which is caused by the difference in velocities among different wavelength components within the same pulse. Chromatic dispersion is a result of two contributing factors called material and waveguide dispersion. The material dispersion is caused by the wavelength dependence of the refractive index on the

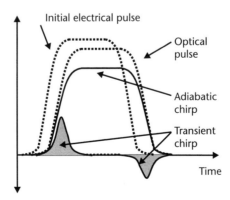

Figure 3.9 Transient and adiabatic frequency chirp.

fiber core material, while the waveguide dispersion occurs due to dependence of the mode propagation constant on the fiber parameters (core radius, and difference between refractive indexes in fiber core and fiber cladding) and signal wavelength.

The multimode optical fibers are usually used for point-to-point signal transmission over shorter distances, which are measured by kilometers or tens of kilometers. In such cases, the transmission engineering is a straightforward process related to optical power balance calculation and verification of the available bandwidth, as explained in Sections 5.4.1 and 5.4.2.

The intermodal dispersion effect can be expressed through the optical fiber bandwidth B_{fib} that is related to the 1-km long optical fiber length. This parameter is specified in bandwidth-length units, and expressed in GHz-km, or in MHz-km. Optical fiber bandwidth is specified by the fiber manufacturer, and presented in the product data sheet. It is usually measured at wavelengths around 1,310 nm to make sure that there is not any impact of the chromatic dispersion to the measured results. The bandwidth B_{fib} varies from several tens of MHz-km for so-called step-index multimode optical fibers, to more than 2 GHz-km for so-called graded-index optical fibers.

The bandwidth B_{fib} can be used to calculate the bandwidth of any specific fiber length. The bandwidth of the specified fiber length can be calculated by using the formula [16]

$$B_{fib,L} = \frac{B_{fib}}{L^{\mu}} \tag{3.51}$$

where L is the length of the fiber in question, and μ is the coefficient that can take values in the range of 0.5 to 1 (this is around $\mu = 0.7$ for most field applications).

The bandwidth parameter does not account for the chromatic dispersion impact. That impact should be included separately, as shown in Section 5.4.2. Therefore, the total dispersion that occurs in multimode optical fibers has two independent components that are related to intermodal and chromatic dispersion. The impact of the modal dispersion component is predominant if transmission is done around 1,310 nm, while the impact of chromatic dispersion can be neglected. However, the impact of chromatic dispersion cannot be neglected if transmission is done in some other wavelength region.

Chromatic dispersion, also known as the intramodal dispersion, is related to each individual optical frequency. Chromatic dispersion is often associated just with single-mode optical fibers, since there is no intermodal dispersion effect. At the same time, it adds to the intermodal dispersion in multimode optical fibers and puts more stringent requirements in terms of the available system bandwidth.

Chromatic dispersion that occurs in single-mode optical fibers is the result of the fact that the group velocity of optical signal is a function of wavelength. Therefore, chromatic dispersion would not exist if a monochromatic wave (i.e., a single wavelength) propagated through the optical fiber. However, since the optical source is not an ideal monochromatic source, each pulse in its time domain contains different spectral components that travel at different velocities through the fiber. The amount of the dispersion is proportional to the spectral width of the optical source.

Chromatic dispersion induces pulse broadening when the neighboring pulses cross their allocated time slot borders, as illustrated in Figure 3.10. Such a spreading can severely limit the signal bit rates that can be effectively transmitted. The effect presented in Figure 3.10 is known as the intersymbol interference (ISI). The amount of the signal spread outside the time slot will degrade the optical receiver performance. That degradation is expressed through the power penalty (refer to Section 4.3). It is important to mention that chromatic dispersion is also a cumulative effect that increases with optical fiber length.

The signal pulses forming the digital stream in Figure 3.10 provide envelopes in the time domain for the spectral content of the light source. Each spectral component propagates independently through the fiber across the axial coordinate z. A specific spectral component, characterized by the angular optical frequency $\omega = 2\pi\nu$, will arrive at the output end of the fiber after some delay τ_g, which is given as

$$\tau_g = \frac{L}{v_g} = L\frac{d\beta}{d\omega} = \frac{L}{c}\frac{d\beta}{dk} = -\frac{L\lambda^2}{2\pi c}\frac{d\beta}{d\lambda} \tag{3.52}$$

where L is the fiber length, $\lambda = 2\pi c/\omega = c/\nu$ is the wavelength, ν is the linear frequency, c is the light speed in vacuum, β is the propagation constant that was introduced by (2.7), and

$$v_g = \left(\frac{d\beta}{d\omega}\right)^{-1} = -\frac{2\pi c}{\lambda^2}\left(\frac{d\beta}{d\lambda}\right)^{-1} \tag{3.53}$$

is the group velocity defined as a speed at which the energy of an optical pulse travels through the medium [13].

As a result of the difference in time delays, the optical pulse disperses after traveling a certain distance before arriving at the output end of the optical fiber. The following relation characterizes the quantity of the pulse broadening due to chromatic dispersion:

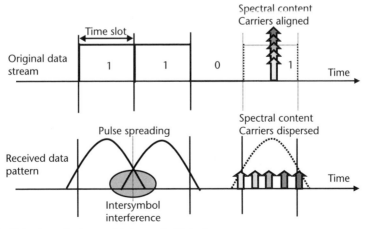

Figure 3.10 Pulse broadening and intersymbol interference.

$$\Delta\tau_g = \frac{d\tau_g}{d\omega}\Delta\omega = \frac{d\tau_g}{d\lambda}\Delta\lambda \qquad (3.54)$$

where $\Delta\omega$ and $\Delta\lambda$ represent the range of frequencies and wavelengths emitted by the optical source, respectively. By inserting (3.53) into (3.54), it becomes

$$\Delta\tau_g = L\frac{d^2\beta}{d\omega^2}\Delta\omega = -\frac{L}{2\pi c}\left(2\lambda\frac{d\beta}{d\lambda} + \lambda^2\frac{d^2\beta}{d\lambda^2}\right)\Delta\lambda \qquad (3.55)$$

The factor

$$D = -\frac{1}{2\pi c}\left(2\lambda\frac{d\beta}{d\lambda} + \lambda^2\frac{d^2\beta}{d\lambda^2}\right) \qquad (3.56)$$

is known as the chromatic dispersion coefficient, or just the chromatic dispersion. It is expressed in ps/(nm-km).

The chromatic dispersion coefficient from (3.56) is a result of two contributing factors called material and waveguide dispersion. The material dispersion D_m arises due to wavelength dependence of the refractive-index on the fiber core material, which causes a wavelength dependence of the group delay. The wavelength dependence of the refractive-index is well approximated by the Sellmeier equation [see (2.7)]. The waveguide dispersion D_w occurs since propagation constant β is a function of the fiber parameters (core radius, and difference between refractive indexes in fiber core and fiber cladding), and it is also function of the wavelength λ.

The factors that affect both the material and wavelength dispersion are interrelated since the dispersive properties of the refractive-index have an impact on the waveguide dispersion, which makes it more difficult to evaluate them separately. However, a simplified approach is often used in many practical situations, in which the material and waveguide dispersion components are calculated separately, while the total chromatic dispersion is calculated as a sum of these components [17]. By using such an approach, (5.56) becomes

$$\Delta t_g \approx \frac{d(t_m + t_w)}{d\lambda}\Delta\lambda = (D_m + D_w)L\Delta\lambda \approx \left(\frac{\lambda}{c}\frac{d^2n}{d\lambda^2} + \frac{n_{cl}\Delta n}{c\lambda}V\frac{d^2(Vb)}{dV^2}\right)L\Delta\lambda \qquad (3.57)$$

where n_{cl} is the refractive-index in the optical fiber cladding, Δn is the difference between the refractive indexes at the fiber axis and in the fiber cladding, and V and b are the V parameter and normalized propagation constant, respectively, which were introduced by (2.6) and (2.7).

Material and waveguide dispersion components from (3.57) are shown in Figure 3.11 as functions of wavelength. The dispersion curves are plotted by using approximations given by (2.8) and (2.9), and for typical values of fiber parameters ($V = 2.1$, $\Delta = 0.25\%$). Please note that material dispersion passes through the zero value in the wavelength region around 1,300 nm (1,270 nm for pure silica core and 1,310 nm for doped silica). Since material dispersion is not a linear function of

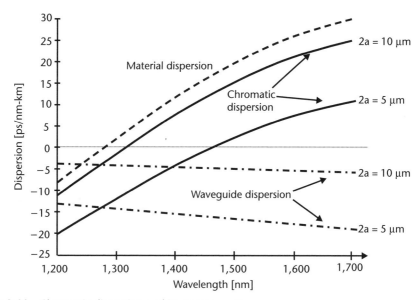

Figure 3.11 Chromatic dispersion and its components.

wavelength, there is a slope associated with the dispersion curve. That slope is often referred to as the chromatic dispersion slope.

Waveguide dispersion remains negative in the wavelength region above 1,000 nm, while its absolute value increases with the wavelength. The waveguide component of the chromatic dispersion is much smaller than the material dispersion component if considering the wavelength region from 800 to 900 nm. However, they become comparable in the wavelength region around 1,300 nm. The waveguide dispersion can be enhanced to become much larger than material dispersion if a special fiber design is applied. This approach is used to manufacture dispersion compensating fibers (DCF) (refer to Section 6.1.2).

It is possible to reduce the total chromatic dispersion by mutual cancellation of material and waveguide dispersions. A particular wavelength where chromatic dispersion is reduced to zero value can vary, but it is always higher than the wavelength value where material dispersion curve crosses the x axis. Mutual cancellation of chromatic dispersion components at a specified wavelength can be done through the proper selection of dopants (by changing the material dispersion component) or by controlling the waveguide effects through the fiber core diameter and the refractive index profile.

This approach for mutual cancellation was used to produce several types of single-mode optical fibers that are different in design than the standard single-mode fibers. The characteristics of different optical fiber types are standardized by the International Telecommunication Union (ITU-T) in documents [18–20]. In addition to standard SMF, there are two major fiber types:

1. Dispersion shifted fibers (DSF) defined by ITU-T Recommendation G.653, with dispersion minimum shifted from the 1,310-nm wavelength region to the 1,550-nm wavelength region;

2. Nonzero-dispersion shifted fibers (NZDSF) defined by ITU-T Recommendation G. 655, with dispersion minimum shifted from the 1,310-nm wavelength region to anywhere within C or L bands. There are several commercial types of optical fibers from this group that are optimized for the DWDM transmission, such as TrueWave fiber or LEAF fiber.

Chromatic dispersion characteristics of three types of single-mode optical fibers are illustrated in Figure 3.12.

As mentioned earlier in this section, chromatic dispersion is fully determined by frequency dependence of the propagation constant $\beta = \beta(\omega)$. However, the effect of the signal pulse broadening is not determined only by waveguide properties and material structure of optical fiber, but also by the frequency chirp imposed by the optical source. The pulse broadening in a single-mode optical fiber can be evaluated by analyzing the behavior of each individual frequency component within the signal spectrum. The analysis should be done with an assumption that a specific frequency chirp is induced at the input of the optical fiber.

It is also convenient to assume that so-called chirped Gaussian-like optical pulses have been generated at the fiber input [10, 11, 16]. Such an approach will be followed in this section. We will also utilize general equations presented in Chapter 7 (refer to Section 7.3).

Each axial component of the monochromatic electromagnetic wave can be presented by its complex electric field function as

$$E(z,t) = E_a(z,t)\exp[j\beta(\omega)z]\exp(-j\omega_0 t) \qquad (3.58)$$

where $E_a(z,t)$ is the pulse envelope that changes with time t and distance z. Parameter $\omega_0 = 2\pi\nu_0$ is the radial optical frequency of the monochromatic wave, ν_0 is linear

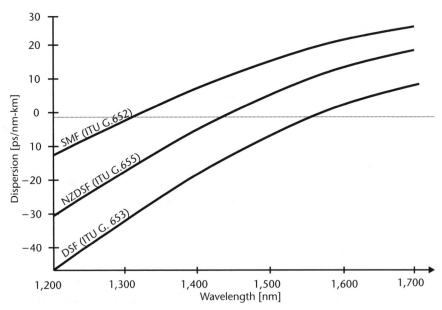

Figure 3.12 Chromatic dispersion of different types of single-mode optical fibers.

optical frequency, and $\beta(\omega)$ is the propagation constant of the monochromatic optical wave.

The pulse broadening, which occurs along the optical fiber, can be evaluated by exploring the frequency dependence of the propagation constant $\beta = \beta(\omega)$. Each spectral component within the launched optical pulse will experience a phase shift proportional to $\beta(\omega)z$. The pulse spectrum, which is observed at the distance z, is given in frequency domain as

$$\tilde{E}_a(z,\omega) = \tilde{E}_a(0,\omega)\exp[j\beta(\omega)z] \qquad (3.59)$$

Recall that superscript (~) denotes the frequency domain of a specified function.

The pulse shape in the time domain can be obtained by the inverse Fourier transform of (3.59); that is,

$$E_a(z,t) = \frac{1}{2\pi}\int_{-\infty}^{\infty}\tilde{E}_a(0,\omega)\exp[j\beta(\omega)z]\exp(-j\omega t)d\omega \qquad (3.60)$$

The exact calculation of the inverse Fourier transform cannot be generally carried out since the function $\beta = \beta(\omega)$ is not known in most cases. It is useful, therefore, to expand the propagation constant $\beta = \beta(\omega)$ in a Taylor series around the carrier frequency $\omega_0 = 2\pi\nu_0$. The expansion can be done only if condition $(\omega - \omega_0) = \Delta\omega << \omega_0$ is satisfied, which is true even if the bit rate of the optical signal goes up to several terabits. By applying Taylor's expansion, function $\beta = \beta(\omega)$ becomes

$$\beta(\omega) \approx \beta(\omega_0) + (\omega - \omega_0)\frac{d\beta}{d\omega}\Big|_{\omega=\omega_0} \\ + \frac{(\omega-\omega_0)^2}{2}\frac{d^2\beta}{d\omega^2}\Big|_{\omega=\omega_0} + \frac{(\omega-\omega_0)^3}{6}\frac{d^3\beta}{d\omega^3}\Big|_{\omega=\omega_0} \qquad (3.61)$$

The first derivative at the right side of previous equation is the group delay with respect to a unity length, which was introduced by (3.52). This parameter is sometimes denoted as $\beta_1 = d\beta/d\omega$, which is just the inverse value of group velocity v_g [see (3.53)].

Parameter $\beta_2 = d^2\beta/d\omega^2$ is commonly known as the group velocity dispersion (GVD) coefficient, and this determines the extent of the pulse broadening during the propagation. It is easy to relate this parameter with the dispersion parameter D introduced through (3.56). It can be done by using relationships $\omega = 2\pi c/\lambda$ and $\Delta\omega = -2\pi c\Delta\lambda/\lambda^2$ between optical frequency ω and wavelength λ, which are given by (7.35) and (7.36), so it becomes

$$D = -\frac{2\pi c}{\lambda^2}\beta_2 \qquad (3.62)$$

where c is light speed in vacuum. Parameters D and β_2, which are opposite in sign, are used to recognize two distinctive wavelength regions:

- The normal dispersion wavelength region characterized by $D < 0$ and $\beta_2 > 0$.
- The anomalous dispersion wavelength region characterized by $D > 0$ and $\beta_2 < 0$.

Finally, parameter $\beta_3 = d^3\beta/d\omega^3$ from (3.61) is known as the differential dispersion parameter, which determines the chromatic dispersion slope over a specified wavelength range. This parameter plays an important role if operation is done at wavelengths where the chromatic dispersion changes the sign (the zero-dispersion region).

Equation (3.61) can be used to justify a concept of slowly varying amplitude $A(z,t)$ of the pulse envelope that can be introduced to express the pulse field function from (3.58) as

$$E(z,t) = E_a(z,t)\exp[j\beta(\omega)z]\exp(-j\omega_0 t) = A(z,t)\exp(j\beta_0 z - j\omega_0 t) \qquad (3.63)$$

The slowly varying amplitude is the most important function from the pulse propagation perspective, and it can be found by inserting (3.61) and (3.63) into (3.60):

$$A(z,t) = \frac{1}{2\pi}\int_{-\infty}^{\infty}\tilde{A}(0,\omega)e^{\left[j\beta_1 z(\omega-\omega_0)+\frac{j}{2}\beta_2 z(\omega-\omega_0)^2+\frac{j}{6}\beta_3 z(\omega-\omega_0)^3\right]}e^{-jt(\omega-\omega_0)}d\omega \qquad (3.64)$$

Equation (3.64) can be rewritten in the form of a partial differential equation, just by calculating the partial derivative per axial coordinate z, and by recalling that $(\omega - \omega_0)$ can be considered as a partial derivative of amplitude per time coordinate t. The partial differential equation obtained from (3.64) has a form [11]

$$\frac{\partial A(z,t)}{\partial z} = -\beta_1\frac{\partial A(z,t)}{\partial t} - \frac{j\beta_2}{2}\frac{\partial^2 A(z,t)}{\partial t^2} + \frac{\beta_3}{6}\frac{\partial^3 A(z,t)}{\partial t^3} \qquad (3.65)$$

This equation is the basic one that governs the pulse propagation through a dispersive medium, such as single-mode optical fiber.

It is common to analyze the pulse propagation assuming that slowly varying amplitude takes the Gaussian function shape, given as

$$A(0,t) = A(0)\exp\left(-\frac{t^2}{2\tau_0^2}\right) \qquad (3.66)$$

where $A(0) = A_0$ is the peak amplitude, while $2\tau_0 = T_{FWEM}$ represents the full-width at $1/e$ intensity point. The full width at the $1/e$ intensity point is related to the full-width at half-maximum ($T_{FWHM} = 2T_0$) by relation $T_{FWEM} = 2(ln2)^{1/2}T_{FWHM}$, which means that there is the following relation between half-widths τ_0 and T_0:

$$T_0 = \tau_0\sqrt{\ln 2} \approx 1.67\tau_0 \qquad (3.67)$$

The frequency spectrum of the modulated optical wave at the fiber input is determined by the Fourier transform of (3.66), which is given as

$$\tilde{A}(0,\omega) = A_0 \int_{-\infty}^{\infty} \exp\left(-\frac{t^2}{2\tau_0^2}\right) \exp[j(\omega-\omega_0)] dt = A_0 \tau_0 \sqrt{2\pi} \exp\left[-\frac{\tau_0^2(\omega-\omega_0)^2}{2}\right] \quad (3.68)$$

The spectrum has a Gaussian shape centered around frequency $\omega_0 = 2\pi\nu_0$. The spectral half-width at the 1/e intensity point is

$$\Delta\omega_0 = 1/\tau_0 \quad (3.69)$$

The pulses that satisfy (3.68) are referred to as the transform-limited pulses.

Equation (3.69) represents the case when there is no frequency chirp imposed on the optical pulse during its generation. However, in most practical cases there is a frequency chirp induced during the optical pulse generation or modulation. It can be expressed through an initial chirp parameter C_0 in (3.66), which then becomes

$$A(0,t) = A_0 \exp\left(-\frac{(1+jC_0)t^2}{2\tau_0^2}\right) \quad (3.70)$$

The above function corresponds to chirped Gaussian pulse. The chirp parameter C_0 defines the instantaneous frequency shift, and can be associated with the negative value of the linewidth enhancement factor α_{chip} introduced by (3.49). In addition, the chirp parameter C_0 can be changed by applying some special modulation methods (refer to Section 6.1.2). The instantaneous frequency increases linearly from the leading to the trailing pulse edge for positive values of the parameter C_0. The opposite situation occurs for negative values of the chirp parameter since instantaneous frequency decreases linearly from the leading to trailing edge.

The amount of frequency chirp imposed through the parameter C_0 is measured by the frequency deviation $\delta\omega$ from the carrier frequency ω_0, and it can be found by making a derivative of the phase in (3.70). It is given as

$$\delta\omega = \frac{C_0}{\tau_0^2} t \quad (3.71)$$

The spectrum of the chirped Gaussian pulse can be found by taking the Fourier transform of (3.70) so that it becomes

$$\tilde{A}(0,\omega) = A_0 \int_{-\infty}^{\infty} \exp\left[-\frac{(1+jC_0)t^2}{2\tau_0^2}\right] \exp(j\omega t) dt$$
$$= A_0 \left[\frac{2\pi\tau_0^2}{1+jC_0}\right]^{1/2} \exp\left[-\frac{(\omega\tau_0)^2}{2(1+jC_0)}\right] \quad (3.72)$$

The spectrum has a Gaussian shape, with the spectral half-width at the 1/e intensity point given as

$$\Delta\omega = (1+C_0^2)^{1/2} / \tau_0 = \Delta\omega_0 (1+C_0^2)^{1/2} \qquad (3.73)$$

where $\Delta\omega_0$ is the spectral half-width of the chirp-free pulse, given by (3.69). Therefore, the spectral width of the chirped Gaussian pulse is enhanced by factor $(1+C_0^2)^{1/2}$.

A closed-form expression for slowly varying pulse amplitude can be found by inserting (3.72) into (3.64) and by performing the analytical integration afterwards. The contribution of the term associated with coefficient β_3 can be neglected by assuming that the carrier wavelength is far away from the zero-dispersion wavelength region. In addition, we can omit the contribution of the term associated with β_1, since it does not impact the pulse shape. (Recall that term β_1 contributes just to the pulse delay.) After all these transactions, the expression for the output pulse envelope becomes

$$A(z,t) = \frac{A_0 \tau_0}{\sqrt{\tau_0^2 - j\beta_2 z + C_0\beta_2 z}} \exp\left[-\frac{(1+jC_0)t^2}{2\tau_0^2 + 2C_0\beta_2 z - j2\beta_2 z}\right] = |A(z,t)|e^{j\Phi(z,t)} \qquad (3.74)$$

where $|A(z,t)|$ and $\Phi(z,t)$ are the magnitude and the phase of the complex pulse envelope, respectively. They can be expressed as

$$|A(z,t)| = \frac{A_0}{\left[\left(1+\frac{C_0\beta_2 z}{\tau_0^2}\right)^2 + \frac{\beta_2^2 z^2}{\tau_0^4}\right]^{1/4}} \exp\left[-\frac{t^2}{2\tau_0^2 + 2C_0\beta_2 z + \frac{2\beta_2^2 z^2}{\tau_0^2}}\right] \qquad (3.75)$$

$$|\Phi(z,t)| = -\frac{1}{2}\frac{\beta_2 z t^2}{(\tau_0^2 + C_0\beta_2 z)^2 + \beta_2^2 z^2} + \frac{1}{2}\tan^{-1}\left[-\frac{\beta_2 z}{\tau_0^2 + C_0\beta_2 z}\right] \qquad (3.76)$$

Equation (3.75) shows that the pulse shape remains Gaussian, but with modified amplitude due to an impact of the chirp parameter. The chirp parameter and the pulse width (expressed though its half-width at the 1/e intensity point) change from their initial values C_0 and τ_0, respectively, and after distance z become equal to

$$C(z) = C_0 + \frac{(1+C_0^2)\beta_2 z}{\tau_0^2} \qquad (3.77)$$

and

$$\tau(z) = \tau_0 \left[\left(1+\frac{C_0\beta_2 z}{\tau_0^2}\right)^2 + \left(\frac{\beta_2 z}{\tau_0^2}\right)^2\right]^{1/2} = \tau_0\left[\left(1+\frac{C_0 z}{L_D}\right)^2 + \left(\frac{z}{L_D}\right)^2\right]^{1/2} \qquad (3.78)$$

where $L_D = \tau_0^2/|\beta_2|$ is the dispersion length, as defined in [11]. The time-dependent pulse phase from (3.76) means that there is an instantaneous frequency variation around the carrier frequency ω_0. That frequency variation is given as

$$\delta\omega(z,t) = -\frac{\partial \Phi(z,t)}{\partial t} = \frac{\beta_2 z}{\left(\tau_0^2 + C_0\beta_2 z\right)^2 + \beta_2^2 z^2} t \qquad (3.79)$$

This instantaneous frequency shift is again referred to as a linear frequency chirp, since it scales in proportion with time. There are two factors contributing to the sign and the slope of a linear function presented by (3.79). They are the GVD parameter β_2 and the initial chirp parameter C_0.

The frequency change along the pulse width, caused only by the impact of chromatic dispersion, is illustrated in Figure 3.13. This figure shows the case when the initial chirp parameter is zero. As we can see, the instantaneous frequencies at the leading edge are lower than the carrier frequency $\omega_0 = 2\pi\nu_0$, if referred to the normal dispersion region (for $D < 0$, and $\beta_2 > 0$). At the same time, the frequencies at the trailing edge are higher than the carrier frequency. The opposite situation occurs in the anomalous dispersion region (for $D > 0$, and $\beta_2 < 0$).

The impact of the initial chirp parameter can be evaluated through the ratio $\tau(z)/\tau_0$, which reflects the broadening of the pulse. It is shown in Figure 3.14 as a function of normalized distance z/L_D, where $L_D = \tau_0^2/|\beta_2|$ is the dispersion length parameter introduced in (3.78). The pulse broadening depends on the sign of the product $C_0\beta_2$. When $C_0\beta_2 > 0$, a monotonical broadening occurs. Since the broadening rate is proportional to the initial value of the chirp parameter, the smallest width increase occurs for an unchirped pulse.

It is worth noting that an initial narrowing occurs if $C_0\beta_2 < 0$, but it is followed by subsequent almost linear broadening. This situation can happen in either of the following cases:

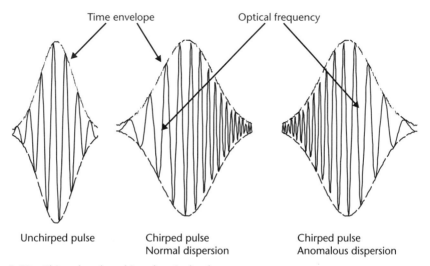

Figure 3.13 Chirped and unchirped optical pulses.

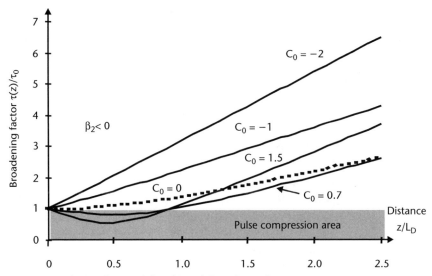

Figure 3.14 Broadening factor of the chirped Gaussian pulse.

- The pulse with a positive value of C_0 propagates in the anomalous dispersion region characterized by $\beta_2 < 0$ and $D > 0$.
- The pulse with a negative value of C_0 propagates in the normal dispersion region (for $\beta_2 > 0$, and $D < 0$).

In either case the frequency chirp introduced by chromatic dispersion counteracts the initial frequency chirp. The pulse undergoes narrowing until these two chirps cancel each other, which can be identified as a minimum in the curve from Figure 3.14. For longer propagation distances, these two frequency chirps are out of balance and the pulse broadens monotonically. The pulse broadening at the point where $z = L_D = \tau_0^2/|\beta_2|$ is identical to the broadening of a chirpless Gaussian pulse.

The pulse broadening given in (3.78) does not include a contribution of the higher-order dispersion terms (i.e., term β_3, β_4 and up). Although this equation can be used in most practical situations, it does not produce good results in some cases where a more precise estimate is needed, such as transmission in a wavelength region with low chromatic dispersion.

The analytic solution of (3.64) can be obtained even if the term β_3 is included, but the pulse shape will not be a pure Gaussian function any more. Since neither the FWHM, nor the full-width at the 1/e intensity point (FWEM) can be used to characterize pulse broadening, the width of the pulse can be expressed by its RMS, defined as

$$\sigma(z) = \left\{ \frac{\int_{-\infty}^{\infty} t^2 |A(z,t)|^2 dt}{\int_{-\infty}^{\infty} |A(z,t)|^2 dt} - \left[\frac{\int_{-\infty}^{\infty} t|A(z,t)|dt}{\int_{-\infty}^{\infty} |A(z,t)|dt} \right]^2 \right\}^{1/2} \quad (3.80)$$

The amount of pulse broadening at distance z can be estimated by the variance ratio $\sigma^2(z)/\sigma_0^2$, where $\sigma_0^2 = \tau_0^2/2$ is determined by the half-width of the input Gaussian pulse [11, 16],

$$\frac{\sigma^2(z)}{\sigma_0^2} = \left(1 + \frac{C_0 \beta_2 z}{2\sigma_0^2}\right)^2 + \left(\frac{\beta_2 z}{2\sigma_0^2}\right)^2 + \left(1 + C_0^2\right)\left(\frac{\beta_3 z}{4\sigma_0^3 \sqrt{2}}\right)^2 \quad (3.81)$$

This equation is very useful from the systems engineering perspective, since it helps to estimate the pulse broadening due to chromatic dispersion in a single-mode optical fiber and for a specified initial chirp parameter. Equation (3.81) is applicable for all cases where spectral width of the light source is much smaller than the signal spectral width $\Delta\omega_0$, which was introduced by (3.73). Since this condition is not satisfied in most cases where a direct modulation is applied, (3.81) should be modified to account for the source spectral width. It could be done through a broadening factor $\sigma_c = 2\sigma_0\sigma_s$, where σ_0 is the spectral width of the Gaussian input pulse, and σ_s is the source spectral width measured in gigahertz. Equation (3.81) now becomes [11]

$$\frac{\sigma^2(z)}{\sigma_0^2} = \left(1 + \frac{C_0 \beta_2 z}{2\sigma_0^2}\right)^2 + \left(1 + \sigma_c^2\right)\left(\frac{\beta_2 z}{2\sigma_0^2}\right)^2 + \left(1 + C_0^2 + \sigma_c^2\right)\left(\frac{\beta_3 z}{4\sigma_0^3 \sqrt{2}}\right)^2 \quad (3.82)$$

Equations (3.81) and (3.82) can be used to analyze the chromatic dispersion impact for two cases that are the most interesting from the systems engineering perspective:

- The case when $\sigma_c = 2\sigma_0\sigma_s > 1$, which means that the source spectrum is much larger that the signal spectrum. The effect of the frequency chirp can be neglected here (i.e., $C_0 = 0$). In addition, if the carrier wavelength is far away from the zero-dispersion region, the term containing β_3 can be neglected as well. In this case (3.82) takes a simplified form:

$$\sigma^2(z) = \sigma_0^2 + \sigma_s^2 \beta_2^2 z^2 = \sigma_0^2 + \sigma_\lambda^2 D^2 z^2 = \sigma_0^2 + \sigma_{D,1}^2 \quad (3.83)$$

If the carrier wavelength is within the zero-dispersion region, the term containing β_2 can be neglected, so that (3.82) leads to

$$\sigma^2(z) = \sigma_0^2 + \frac{\sigma_s^4 \beta_3^2 z^2}{2} = \sigma_0^2 + \frac{\sigma_\lambda^4 S^2 z^2}{2} = \sigma_0^2 + \sigma_{D,2}^2 \quad (3.84)$$

Note that wavelength-related parameters σ_λ (the source spectral width expressed in nanometers), D (chromatic dispersion parameter expressed in ps/nm·km), and $S = dD/d\lambda$ (chromatic dispersion slope expressed in ps/nm²km) have also been introduced in (3.83) and (3.84).

- The case when $\sigma_c = 2\sigma_0\sigma_c \ll 1$, which means that the source spectrum is much smaller that the signal spectrum. This implies that the impact of frequency chirp is dominant and that (3.81) is more suitable than (3.82). If the carrier wavelength is far away from the zero-dispersion region, the term containing β_3 can be neglected, so that (3.81) leads to

$$\sigma^2(z) = \sigma_0^2 + C_0 \beta_2 z + \left(1 + C_0^2\right)\left(\frac{\beta_2 z}{2\sigma_0}\right)^2 = \sigma_0^2 + \sigma_{D,3a}^2 \quad (3.85)$$

or, for the chirpless pulse case,

$$\sigma^2(z) = \sigma_0^2 + \left(\frac{\beta_2 z}{2\sigma_0}\right)^2 = \sigma_0^2 + \sigma_{D,3b}^2 \quad (3.86)$$

The term containing β_2 in (3.81) can be neglected if the carrier wavelength is within the zero-dispersion region, so it transforms to

$$\sigma^2(z) = \sigma_0^2 + \left(1 + C_0^2\right)\left(\frac{\beta_3 z}{4\sigma_0^2 \sqrt{2}}\right)^2 = \sigma_0^2 + \sigma_{D,4a}^2 \quad (3.87)$$

or, for the chirpless pulse case,

$$\sigma^2(z) = \sigma_0^2 + \left(\frac{\beta_3 z}{4\sigma_0^2 \sqrt{2}}\right)^2 = \sigma_0^2 + \sigma_{D,4b}^2 \quad (3.88)$$

Please note that parameter $\sigma_{D,i}$ ($i = 1, 2, 3, 4$) was introduced to be a measure of the pulse broadening due to chromatic dispersion. This index i was used to distinguish four cases that are the most interesting from the systems engineering perspective. The pulse broadening for different engineering cases is summarized in Table 3.2.

In summary, we can say that pulse broadening is determined by optical source and optical fiber parameters if the source spectral width is much larger than the signal spectrum. This is basically applicable to direct modulation schemes, when either Fabry-Perot lasers or LEDs are used. For cases where source spectrum is much smaller than the signal spectrum, the pulse broadening is also dependent on the initial pulse width. This situation can be associated with external modulation schemes. Therefore, the pulse broadening can be minimized by choosing an optimum value of the initial pulse width.

Table 3.2 Pulse Broadening Parameter σ_p for the Most Interesting Practical Cases

Transmission Case	Out of Zero Dispersion Point	Close to Zero Dispersion Point
Source with large spectral width	$\sigma_{D,1} = \sigma_s \beta_2 z = \sigma_\lambda D z$	$\sigma_{D,2} = \dfrac{\sigma_s^2 \beta_3 z}{\sqrt{2}} = \dfrac{\sigma_\lambda^2 S z}{\sqrt{2}}$
Source with small spectral width	for initially chirped pulse $\sigma_{D,3b} = \sqrt{C_0 b_2 z + \left(1 + C_0^2\right)\left(\dfrac{\beta_2 z}{2\sigma_0}\right)^2}$	for initially chirped pulse $\sigma_{D,4a} = \sqrt{\sigma_0^2 + \left(1 + C_0^2\right)\left(\dfrac{\beta_3 z}{4\sigma_0^2 \sqrt{2}}\right)^2}$
	for chirpless pulse $\sigma_{D,3b} = \dfrac{\beta_2 z}{2\sigma_0}$	for chirpless pulse $\sigma_{D,4b} = \dfrac{\beta_3 z}{4\sigma_0^2 \sqrt{2}}$

It is worth noting that the zero-dispersion region appears to be very forgiving with respect to chromatic dispersion effects. That is particularly true for a single-channel optical transmission. However, in multichannel systems, such as DWDM transmission, this is just one part of the total equation, since a potential benefit of the transmission in the zero-dispersion wavelength region vanishes due to the impact of nonlinear effects.

3.3.5 Polarization Mode Dispersion

There is no polarization mode dispersion in an ideal single-mode optical fiber with a perfectly cylindrical core of uniform diameter, which is composed of perfectly isotropic material. It appears in real optical fibers due to variations in the shape of their core along the fiber length. Any deviations from an ideal structure will cause birefringence effect and the creation of two distinct polarization modes [13]. In such a case, single-mode fibers have effectively become bimodal.

Two modes that are developed in real optical fibers due to birefringence effect are related to two orthogonal polarizations of the optical signal. These polarizations "see" either different core sizes due to elliptical core, or different material densities due to asymmetric internal stress, as shown in Figure 3.15. These differences are transferred to differences in refractive indexes, which are associated with two polarization states. The difference in refractive indexes (or birefringence) cause the difference in speeds of two orthogonally polarized modes, and that implies that the propagation constants of two modes are also different.

The degree of birefringence is measured by parameter $B = |n_x - n_y|$, where n_x and n_y are the refractive indexes experienced by signal portions polarized along orthogonal axes x and y axes, respectively. The total signal is a vector sum of two polarization modes, as shown in Figure 3.16(a). Optical fiber will keep the residual birefringence along its length if there is not any intrinsic rotation of the axes x and y, and if there is not any external perturbation applied. This case is known as the deterministic birefringence, and it is characterized by well-defined polarization states. These states, which are also known as the *eigenstates*, depend neither on fiber length nor on optical wavelength [21–23]. The difference in refractive indexes and propagation constants between two polarization modes will lead to a phase shift among them, all during the propagation process. The phase shift is observed through a differential group delay (DGD) between the slow and fast axis, which increases linearly with the fiber length. Consequently, the total width of the optical pulse will be dispersed, as shown in Figure 3.16(b).

In reality, two polarization modes do not propagate independently along the optical fiber, but exchange the energy through the mode coupling process. The

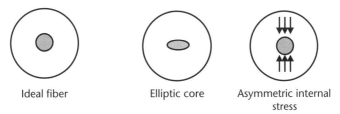

Figure 3.15 An ideal optical fiber and deformations that cause PMD.

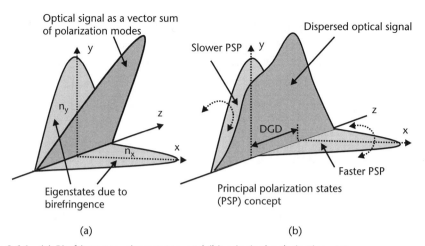

Figure 3.16 (a) Birefringence, eigenstates, and (b) principal polarization states.

polarization mode coupling is caused by the external perturbation due to bending, twisting, lateral stress, and temperature variations. In this case the degree of birefringence changes randomly along the fiber, and polarization eigenstates are no longer preserved. In other words, we have a stochastic rotation of the polarization states and a stochastic distribution of energy among them. Due to the random character of the energy exchange, neither the phase difference nor differential group delay between two polarization modes will scale linearly with the distance. It was shown that they would increase in proportion with the square root of the fiber length [22].

The random process that was just described causes a stochastic pulse dispersion known as polarization mode dispersion (PMD), which is extensively studied in the literature and through practical experiments [10, 11, 21–25]. In addition to stochastic rotation of the polarization states and stochastic distribution of energy between the principal states, different wavelength components will behave differently, which will add to the overall complexity of the polarization mode dispersion. An exact analytical treatment of the PMD effect is rather difficult due to the random nature and complexity of the overall process.

Real optical fibers do not generally exhibit well-defined orthogonally polarized eigenstates that are stable with respect to the optical frequency and fiber length. Instead, at any specific optical frequency ω, there are two orthogonally polarized input states that produce the output polarization states with minimal frequency dependence. These states are known as the principal states of polarization (PSP). It is important to notice that all other output polarization states will experience frequency dependence.

The DGD between two principal polarization states is the main contributor to the pulse broadening, as shown in Figure 3.16. The DGD, also known as first-order PMD, does not include any frequency dependence. The frequency dependence of the PMD effect is included through second-order PMD. In fact, the second-order PMD measures not only the wavelength dependence of the first-order PMD coefficient, but also quantifies the rotation of the principal polarization states associated with each individual wavelength. The second-order PMD is often called *chromatic polarization mode dispersion* since it accounts for frequency dependence of both DGD and PSP.

The first-order PMD is characterized by coefficient D_{P1}, expressed in ps/(km)$^{1/2}$, which is a statistical parameter that varies with time and operating conditions. The total delay between principal polarization states accumulates randomly and in proportion to the square root of the optical fiber length L, so it is [22]

$$\Delta \tau_{P1} = D_{P1} \sqrt{L} \qquad (3.89)$$

The pulse broadening due to first-order PMD depends on the differential group delay and on the power splitting between two principal polarization states. The following relation between RMS widths of input and output pulses can be established [22–24]:

$$\sigma_{out}^2 = \sigma_{in}^2 + 2\Delta\tau_{P1}^2 \zeta(1-\zeta) \qquad (3.90)$$

where σ_{out} and σ_{in} are RMSs of the output and input pulses, respectively, and ζ represents the power splitting of the signal between two principal polarization states, which can vary from 0 to 1.

There is a general agreement that a statistical nature of the first-order PMD coefficient can be characterized by the Maxwellian probability density distribution $p(D_{P1})$ given as

$$p(D_{P1}) = \sqrt{\frac{2}{\pi}} \frac{D_{P1}^2}{\alpha^3} \exp\left(-\frac{D_{P1}^2}{2\alpha^2}\right) \qquad (3.91)$$

The mean value $<D_{P1}>$ determined by the Maxwellian distribution equals $<D_{P1}> = (8/\pi)^{1/2} \alpha$. Coefficient α can be experimentally determined and has a typical value of about 30 ps [22, 23]. The probability function expressed by (3.91) is shown in Figure 3.17(a). This functional curve has a highly asymmetric character and decreases rapidly for an argument larger than $3<D_{P1}>$.

The overall probability $P(D_{P1})$ that coefficient D_{P1} will be larger than a specified value can be found by performing an integration in (3.91); that is,

$$P(D_{P1}) = \int_0^{D_{P1}} p(D_{P1}) d(D_{P1}) \qquad (3.92)$$

Function $P(D_{P1})$ is shown in Figure 3.17(b). The probability that the actual coefficient is three times larger than the average value $<D_{P1}>$ is 4×10^{-5}. This statistical evaluation means that the first-order PMD coefficient will remain larger than three times its average value for approximately 21 minutes per year. For example, if DGD varies significantly once a day, the first-order PMD coefficient will exceed three times the average value once every 70 years, but if it varies once a minute it will exceed three times the average value every 17 days.

The character of the distribution expressed by (3.91) has been the reason why three times the average value (i.e., $3<D_{P1}>$), rather that temporary value of the

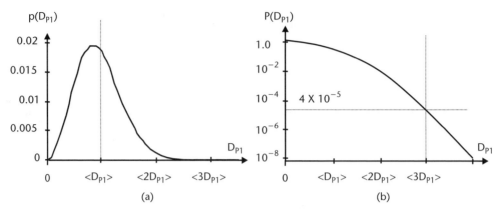

Figure 3.17 The first-order PMD: (a) probability density function, and (b) the overall probability function.

coefficient D_{P1}, has been used to characterize the first-order PMD, so that (3.89) takes more practical form

$$\Delta \tau_{P1} = 3 \langle D_{P1} \rangle \sqrt{L} \tag{3.93}$$

The value $\langle D_{P1} \rangle$ is usually measured after the manufacturing process takes place, and it listed in the fiber data sheet. The measured values can vary from 0.01 ps/(km)$^{1/2}$ to several ps/(km)$^{1/2}$. The older fibers that were installed 20 to 25 years ago have a relatively large $\langle D_{P1} \rangle$ coefficient due to imperfections in the manufacturing process, while $\langle D_{P1} \rangle$ is much smaller for newer optical fibers. It is usually lower than 0.1 ps/(km)$^{1/2}$, while manufacturers claim that typical values are around 0.05 ps/(km)$^{1/2}$. The specified value can increase after the installation is done since the cabling process and environmental conditions can add to the original number.

The second-order PMD is characterized by coefficient D_{P2}, and occurs due to frequency-dependent variations of both the DGD and PSP. It can be expressed as [23, 24]

$$D_{P2} = \sqrt{\left(\frac{1}{2}\frac{\partial D_{P1}}{\partial \omega}\right)^2 + \left(\frac{D_{P1}}{2}\frac{\partial \vec{S}}{\partial \omega}\right)^2} \tag{3.94}$$

where ω is the optical frequency, and vector **S** determines the position of principal states of polarization. This vector is also known as the Stokes vector. Therefore, the first term at the right side of (3.94) describes the frequency dependence of the differential group delay, while the second term describes the variations of the PSP and their dependence on optical frequency ω.

The statistical nature of the second-order PMD coefficient in real optical fibers can be characterized by probability density distribution $p(D_{P2})$ given as [26]

$$p(D_{P2}) = \frac{2\sigma^2 D_{P2}}{\pi} \frac{\tanh(\sigma D_{P2})}{\cosh(\sigma D_{P2})} \tag{3.95}$$

where σ is a parameter that can be experimentally determined, as in [26], and ranges around $\sigma = (240 \text{ ps}^2)^{-1}$.

The mean value $<D_{P2}>$ of the second-order PMD coefficient can be correlated to the mean value $<D_{P1}>$ of the first-order PMD coefficient. It was found that the following approximate relation can be established [26]:

$$\langle D_{P2} \rangle \approx \frac{\langle D_{P1} \rangle^2}{\sqrt{12}} \quad (3.96)$$

Equation (3.96) is related just to numerical values. (Please recall that the first order PMD is expressed in ps/km$^{1/2}$, while the second-order PMD is expressed in ps/km-nm). The probability density function given by (3.95) is shown in Figure 3.18. It is more asymmetrical and with a larger tail than the function related to the first-order PMD coefficient. The probability that the second-order PMD coefficient will take a value that is three times lager than to the mean value $<D_{P2}>$ is still countable. It becomes negligible for arguments that are larger than five times the mean value $<D_{P2}>$. Accordingly, we can assume that the actual value of the second-order PMD coefficient does not exceed five times of the mean value $<D_{P2}>$. Since the second-order PMD scales linearly with the fiber length L, the total pulse spreading due to the second-order PMD effect can be expressed as

$$\Delta \tau_{P2} = 5 \langle D_{P2} \rangle L \quad (3.97)$$

The second-order PMD, or better to say, the first term at the right side of (3.94), interacts with the total chromatic dispersion along the fiber length. It can induce either pulse broadening or pulse compression and could be treated as a component of chromatic dispersion. However, the portion of the second-order PMD that is represented by the second term at the right side of (3.94) is related to rotation of the principal polarization states and cannot be always treated as pure chromatic dispersion. This part usually dominates and determines the total character of the second-order PMD.

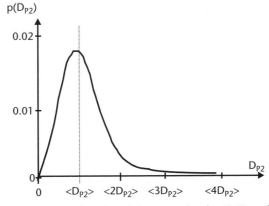

Figure 3.18 The probability density function of the second-order PMD coefficient.

As a summary, we can conclude that the first-order PMD is a dominant part of polarization mode dispersion, which causes the signal broadening and distortion. That is the reason why this term is considered in most analyses of the PMD effect, while the second-order term is sometimes neglected. Such an approach is quite beneficial in most cases if the bit rates do not exceed 10 Gbps. However, the second-order PMD effects are not negligible in high-speed optical transmission systems and should be included in overall engineering considerations. The penalties imposed by the PMD effect will be evaluated in Section 4.4. It is also important to mention that some other components employed along the lightwave path can contribute to the total value of the PMD effect. It is reasonable to assume that an additional PMD in excess of 0.5 ps can be expected if the lightwave path exceeds several hundred kilometers (see Section 4.4).

Different schemes have been considered so far in order to compensate for PMD effects [27]. The original idea of eliminating PMD through compensation has been abandoned for now due to PMD complexity and its statistical character. Instead, attention is concentrated to minimize the distortion effect caused by PMD. Such an approach is known as PMD mitigation. Some of PMD compensation schemes will be described in Section 4.4.

3.3.6 Self-Phase Modulation

The basic operational principles of optical transmission can be explained by assuming that optical fiber acts as a linear transmission medium with the following properties:

- Refractive index does not depend on optical power.
- The superposition principle can always be applied to several independent optical signals.
- Optical wavelength, or carrier frequency, of any optical signal stays unchanged.
- The light signal in question does not interact with any other light signal.

All of this is quite valid if, for example, optical power does not exceed several milliwatts in a single channel optical transmission system. However, the availability of high-power semiconductor lasers and optical amplifiers, together with deployment of DWDM technology, marked a new era in optical transmission. From a practical perspective, optical fiber has exposed its nonlinear properties, which are opposite to ones mentioned above. Nonlinear effects inoptical fibers are neither design nor manufacturing defects, but can occur regardless and can cause severe transmission impairments. On the other hand, in some special cases, they may be used to enhance the fiber transmission capabilities. Nonlinear effects are mainly related to single-mode optical fibers since they have a much smaller cross-sectional area of the fiber core. Therefore, all conclusions that refer to nonlinear effects are mainly related to single-mode optical fibers.

It is well known that any dielectric material makes its nonlinear properties more obvious if exposed to a strong electromagnetic field [13]. The same conclusion can be applied to optical fiber, since all nonlinear effects that can appear in the fiber are proportional to the intensity of the electromagnetic field of propagating optical signal. The total impact of nonlinear effects to characteristics of an optical transmission system is mostly negative since they cause signal crosstalk, limit the total optical signal power, and cause signal distortion. On the other hand, and under specific circumstances, they can be effectively utilized to perform some useful functions, such as chromatic dispersion suppression, or wavelength conversion. Nonlinear effects that can appear in optical fibers are shown in Figure 3.19.

There are two major groups of nonlinear effects related either to nonlinear refractive index, or to nonlinear optical signal scattering. The effects related to nonlinear refractive index are based on the Kerr effect [13], which occurs due to the dependence of the refractive index on light intensity. The following effects belong to this category:

- Self-phase modulation, which is an optical power dependent effect that leads to the pulse distortion. The SPM effect is related to a single optical channel.
- Cross-phase modulation, which is an optical power dependent effect that leads to pulse distortion and interchannel crosstalk. It is related to a composite optical signal that consists of several optical channels.
- Four wave mixing, which is optical power dependent effect that leads to the generation of new optical carriers and to signal crosstalk. In addition to optical power, FWM depends on the value of chromatic dispersion in an optical fiber.

On the other hand, stimulated scattering effects are caused by parametric interaction between light (i.e., photons) and material (i.e., crystal lattice or phonons). There are two types of scattering effects:

- Stimulated Raman scattering, which leads to energy transfer between different wavelengths, thus causing interchannel crosstalk;
- Stimulated Brillouin scattering, which leads to optical power coupling to backward traveling waves, thus limiting the available power per channel.

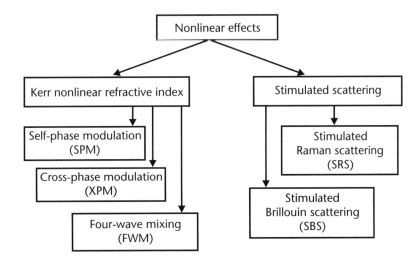

Figure 3.19 Classification of nonlinear effects.

It is important to mention that both SRS and SBS will have a spontaneous character if the optical power is lower than a certain threshold value (see Sections 3.3.9 and 3.3.10). However, it also important to say that such spontaneous character is not relevant from the signal transmission perspective.

The nonlinear effects mentioned above depend on the transmission length and cross-sectional area of the fiber. The nonlinear interaction will be stronger for a longer optical fiber length and for a smaller cross-sectional area of the fiber core. The nonlinear interaction will decrease along the transmission line due to decrease in the optical power. Therefore, the nonlinear effects have a local character, which makes an overall assessment of their impact more difficult. It is more practical from the engineering perspective to introduce an effective length L_{eff} as a parameter that characterizes the strength of nonlinear effects.

The effective length can be found by assuming that a constant optical power that acts over the effective length L_{eff} will produce the same effect as a decaying optical power $P = P_0 \exp(-\alpha z)$ that acts over the physical fiber length L. Therefore, the following relation can be established:

$$P_0 L_{eff} = \int_{z=0}^{L} P(z)dz = \int_{z=0}^{L} P_0 \exp(-\alpha z)dz \tag{3.98}$$

where P_0 presents an input optical power, and α is the fiber attenuation coefficient. Equation (3.98) leads to expression

$$L_{eff} = \frac{1 - \exp(-\alpha L)}{\alpha} \tag{3.99}$$

The value L_{eff} can be approximated by $1/\alpha$ if optical transmission is performed over distances at least several tens of kilometers long. As an example, $L_{eff} \sim 20$ km if a long-haul transmission is performed at wavelengths around 1,550 nm. The concept of the effective length is illustrated in Figure 3.20.

The optical power level will be periodically reset to higher values if there are optical amplifiers along the transmission line, which means that the total effective length for such a transmission line will include contributions of multiple fiber spans. If each fiber span has the same physical length, the total effective length will be

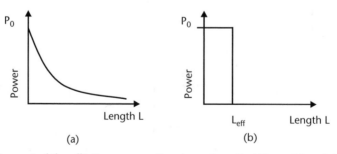

Figure 3.20 Concept of the effective cross-sectional area in optical fibers: (a) real situation, and (b) the effective cross-sectional area.

3.3 Signal Impairments

$$L_{eff,t} = \frac{1-\exp(-\alpha L)}{\alpha} M = \frac{1-\exp(-\alpha L)}{\alpha} \frac{L}{l} \quad (3.100)$$

where L is the total length of a transmission line, l is fiber span length that is equal to the optical amplifier spacing, and M is the number of fiber spans on the line. As we see from (3.100), the effective length can be reduced by increasing the span length, which means that the number of in-line amplifiers would be reduced. However, if we increase the amplifier spacing, we should also increase the power in proportion to $\exp(\alpha l)$ to compensate for additional fiber losses. On the other hand, any power increase will enhance nonlinear effects. Therefore, what matters in this case is the product between the launched power P_0 and the effective length $L_{eff,t}$. Since the product $P_0 L_{eff,t}$ increases with the span length l, the overall effect of nonlinearities can be reduced by reducing the amplifier spacing.

Nonlinear effects are inversely proportional to the area of the fiber core. This is because the concentration of the optical power per unit cross-sectional area, or the power density, is higher for a smaller cross-sectional area, and vice-versa. Recall from Section 2.3.1 that optical power in single-mode optical fibers is not uniformly distributed across the core section. It has a maximum at the optical fiber axes, and decays along the fiber diameter. The power still has some nonnegligible level at the core-cladding border and continues its decay through the optical fiber cladding area. Optical power distribution across the cross-sectional area is closely related to the overall refractive index profile. It is, therefore, convenient to introduce an effective cross-sectional core area A_{eff} by applying the same logic as one used above for the effective length. The effective cross-sectional area can be found by assuming that the effect of a constant optical power acting over the effective cross-sectional area is equal to the effect of decaying optical power acting over the entire fiber radius, as illustrated in Figure 3.21. The effective cross-sectional area is found as

$$A_{eff} = \frac{\left[\int_r \int_\theta r\, dr\, d\theta |E(r,\theta)|^2 \right]^2}{\int_r \int_\theta r\, dr\, d\theta |E(r,\theta)|^4} \quad (3.101)$$

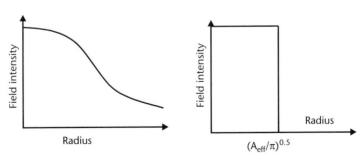

Figure 3.21 Concept of the effective cross-sectional area in optical fibers: (a) real situation, and (b) the effective cross-sectional area.

where r and θ are the polar coordinates, and $E(r, \theta)$ is the electric field of an optical signal. The distribution of the electric field is often approximated by Gaussian function:

$$E(r,\theta) = E_0 \exp\left(\frac{-r^2}{\varpi_0^0}\right) \quad (3.102)$$

where E_0 is the electric field at the fiber axes, and ϖ_0 is so-called mode radius. If we insert (3.102) in (3.101), the approximate value of the effective cross-sectional area becomes $A_{eff} = \pi\varpi_0^2$.

The Gaussian approximation cannot be applied to optical fibers having more complex refractive index profile. Such fibers include DSFs, NZDSFs), and DCFs. In such cases, the effective cross-sectional area can be approximated by $A_{eff} = k\pi\varpi_0^2$, where coefficient k is determined experimentally. The coefficient k can be either $k <1$ (for DCF, and some NZDSF) or $k > 1$ (for some newer NZDSFs that have large effective cross-sectional area). The size of the effective cross-sectional area for several types of commercially available optical fibers is shown in Table 3.3 in parallel with some other fiber parameters.

The refractive index acts as a function of the optical power density if the density is relatively high. (Please recall that the power density equals the optical power divided by effective cross-sectional area.) That is because the electric field effectively compresses molecules and increases the refractive index value. The effect where refractive index is dependent on the strength of the external electric field is known as the Kerr effect [13]. The effective value of the refractive index in a silica-based optical fiber can be now expressed as

$$n(P) = n_0 + n_2\Pi = n_0 + n_2\frac{P}{A_{eff}} \quad (3.103)$$

where Π is the power density per cross-sectional area, while n_2 is the so-called second-order refractive index coefficient (or the Kerr coefficient). The Kerr effect is very fast, with a response time of about 10^{-15} s, and can be considered as a constant with respect to both the signal wavelength and polarization state.

Table 3.3 Characteristics of Some Single-Mode Optical Fibers

Fiber Type	Effective Area A_{eff} [μm^2]	Zero-Dispersion Wavelength λ_0 [nm]	Chromatic Dispersion D [ps/nmkm] at 1,550 nm	Dispersion Slope S [ps/kmnm^2]	Nonlinear Coefficient γ [W^{-1} km^{-1}] at 1,550 nm
Corning SMF-28™	80	1,302–1,320	16–19	0.09	1.12–1.72
Lucent (AFS) AllWave™	80	1,300–1,322	17–20	0.088	1.12–1.72
Corning LEAF™	72	1,490–1,500	2–6	0.06	1.23–1.92
Corning Vascade™	101	1,300–1,310	18–20	0.06	0.9–1.36
Lucent (OFS) TrueWave-RS™	50	1,470–1,490	2.5–6	0.05	1.78–2.75
Alcatel Teralight™	65	1,440–1,450	5–10	0.058	1.37–2.11

The typical values of the second-order refractive index coefficient in silica-based optical fibers are in the range 2.2 to 3.4×10^{-20} m²/W (or 2.2 to 3.4×10^{-8} μm²/W). As an example, optical power $P = 100$ mW that propagates through a standard single-mode optical fiber with $A_{eff} = 80 \times 10^{-12}$ m² will induce the density $\Pi = P/A_{eff} = 1.25 \times 10^{9}$ W/m², and the index change $\Delta n = n_2 \Pi = 3 \times 10^{-11}$. Therefore, the nonlinear portion of the refractive index is much smaller than its constant part n_0 (please recall that n_0 is around 1.5 for silica-based optical fibers). However, even this small value will have a big impact in some cases of optical signal transmission.

The variations in refractive index due to the Kerr effect will change the propagation constant β (please recall that $\beta = 2\pi n/\lambda$, where n is the refractive index and λ is the signal wavelength). By using (3.103), the nonlinear propagation constant can be written as

$$\beta(P) = \beta_0 + \gamma P \qquad (3.104)$$

where β_0 is a linear portion related to linear refractive index n_0, and γ is a nonlinear coefficient given as

$$\gamma = \frac{2\pi n_2}{\lambda A_{eff}} \qquad (3.105)$$

Therefore, parameter γ is not just a function of nonlinear index n_2, but also depends on the effective cross-sectional area and signal wavelength. Typical values of parameter γ are in the range from 0.9 to 2.75 (W·km)⁻¹ for single-mode optical fibers operating at wavelengths around 1,550 nm.

The propagation constant β will vary along the duration of the optical pulse, since different points along the pulse will "see" different optical powers. Accordingly, the propagation constant associated with the leading edge of the pulse will be lower than the propagation constant related to the central part of the pulse. Such a difference in propagation constants will cause the difference in phases associated with different portions of the pulse. The central part of the pulse will acquire phase more rapidly than leading and trailing edges. The total nonlinear phase shift after some length L can be calculated by integrating the propagation constant over the distance z, so it becomes

$$\Delta\Phi(P) = \int_0^L [\beta(P_0) - \beta]dz = \int_0^L \gamma P(z)dz \qquad (3.106)$$

The phase shift can be expressed in an explicit form by using (3.98), so it becomes

$$\Delta\phi[P(t)] = \frac{\gamma P_0(t)[1-\exp(-\alpha L)]}{\alpha} = \gamma L_{eff} P_0(t) = \frac{2\pi}{\lambda} \frac{L_{eff}}{A_{eff}} n_2 P_0(t) \qquad (3.107)$$

If we introduce a *nonlinear length*, defined as [28]

$$L_{nel} = \frac{\lambda A_{eff}}{2\pi n_2 P_0} \qquad (3.108)$$

(3.107) becomes

$$\Delta\Phi[P(t)] = \frac{L_{eff}}{L_{nel}} \quad (3.109)$$

The time-dependent pulse phase from (3.107) will cause an instantaneous frequency variation $\delta\omega_{SPM}(z,t) = 2\pi\delta\nu_{SPM}(z,t)$ around the carrier frequency $\omega_0 = 2\pi\nu_0$. The frequency variation (i.e., the frequency chirp) can be found as

$$\delta\omega_{SPM}(t) = \frac{d[\Delta\Phi(t)]}{dt} = \frac{2\pi}{\lambda}\frac{L_{eff}n_2}{A_{eff}}\frac{dP_0(t)}{dt} \quad (3.110)$$

Equations (3.107) and (3.110) give us a clear view about the nature of the phase and frequency shift due to the Kerr effect. Please notice that we outlined the fact that the optical pulse power is time dependent, just as a reminder that the position along the pulse determines the instantaneous variation of the optical phase from its stationary value. It is also important to notice that frequency chirp induced by self-phase modulation is not linear in time, which is different from the case related to chromatic dispersion [see (3.71)].

The frequency chirping imposed by the SPM effect can be illustrated in the same fashion as the frequency chirping due to chromatic dispersion. The pulse time envelope and the frequency deviations along the pulse due to frequency chirping imposed by SPM are shown in Figure 3.22. The frequency chirp induced by SPM acts in conjunction with the group velocity dispersion that was described in Section 3.3.4. That conjunction can lead either to pulse broadening or to pulse compression.

More precise evaluation of the SPM impact on transmission system characteristics can be done by using the nonlinear Schrodinger equation [see (7.29) and (7.30)]. Such an approach was used in [28], with an assumption that the unchirped Gaussian pulse with envelope $A(0,t) = \exp(-t^2/2\tau_0^2)$ propagates through an optical fiber. The broadening of the pulse can be evaluated by the ratio of the input and output RMS as

$$\frac{\sigma^2(z)}{\sigma_0^2} = 1 + \frac{\sqrt{2}L_{eff}L\beta_2}{2L_{nel}\sigma_0^2} + \left(1 + \frac{4}{3\sqrt{3}}\frac{L_{eff}^2}{L_{nell}^2}\right)\frac{L^2\beta_2^2}{8\sigma_0^4} \quad (3.111)$$

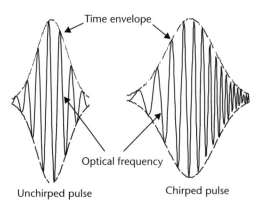

Figure 3.22 Frequency chirping due to self-phase modulation.

where $\sigma_0^2 = \tau_0^2/2$ is determined by the half-width of the input Gaussian pulse.

In general case, SPM causes the pulse broadening since it increases the frequency bandwidth of the optical signal. In some situations, however, the frequency chirp induced by SPM could be opposite in sign to the frequency chirp induced by chromatic dispersion. This can happen, for example, in the anomalous dispersion region, where GVD coefficient β_2 becomes negative. In such a case, SPM helps to reduce the impact of chromatic dispersion. The evaluation of SPM impact to transmission characteristics will be analyzed in Section 4.5.

The possibility of engineering the total frequency chirp has a broader scope since there are three parameters that determine the total frequency chirping: the initial frequency chirp, the GDV parameter, and the optical fiber nonlinear parameter. They can be connected together in a way that would be the most beneficial from a practical application perspective, as discussed in Section 6.1.

3.3.7 Cross-Phase Modulation

XPM is another effect caused by the intensity dependence of the refractive index, but it is related to multichannel transmission and occurs during the propagation of a composite optical signal through an optical fiber. The nonlinear phase shift of a specific optical channel is affected not just by the power of that channel, but also by the optical power of the other channels. Such impact of other optical channels on the channel in question can be evaluated using (3.107), which is modified to include the contributions of other channels, and becomes

$$\Delta\Phi_m(t) = \gamma L_{eff} P_{0m}(t) + 2\gamma L_{eff} \sum_{i \neq m}^{M} P_{0i}(t) = \frac{2\pi}{\lambda} \frac{L_{eff} n_2}{A_{eff}} \left[P_{0m}(t) + 2\sum_{i \neq m}^{M} P_{0i}(t) \right] \quad (3.112)$$

where m denotes the channel in question, and n_2, A_{eff}, L_{eff}, and M are the nonlinear index, the effective cross-sectional area, the effective fiber length, and the number of optical channels, respectively. The factor 2 in (3.112) shows that the cross-phase modulation effect is two times more effective than the self-phase modulation effect [25]. In addition, the phase shift is bit-pattern dependent, since just 1 bits will have an effective impact on the total phase shift. In the worst case scenario, when all channels contribute simultaneously by having 1 bits loaded with power P_m, (3.112) becomes

$$\Delta\Phi_m(t) = \frac{2\pi}{\lambda} \frac{L_{eff} n_2 P_{0m}(t)}{A_{eff}} (2M - 1) \quad (3.113)$$

Equations (3.112) and (3.113) can be used to estimate the XPM effect in a dispersionless medium, in which optical pulses from different optical channels propagate with the same group velocity. In real optical fibers, however, optical pulses in different optical channels will have different group velocities. The phase shift given by (3.112) and (3.113) can occur only during the overlapping time. The overlapping among neighboring channels is longer than the overlapping of channels spaced apart, and it will produce the most significant impact to the phase shift.

The process of the pulse overlapping and frequency shift due to cross-phase modulation is illustrated in Figure 3.23. When overlapping starts, the leading edge of pulse A experiences a decrease in optical frequency, or an increase in optical wavelength (often called red-shift), while the trailing edge of the pulse B experiences an increase in optical frequency, or a decrease in optical wavelength (often called blue-shift). After overlapping is over, the trailing edge of the pulse A has become blue-shifted, while the leading edge of the pulse B has experienced the red-shift. If pulses walk through one another quickly, the described effects on both pulses are diminished since distortion caused by the trailing edge undoes distortion caused by the leading edge. Such a case can be associated with the situation in which there is a significant chromatic dispersion or when interacting channels are widely separated. However, if the pulses walk through one another slowly, the effect that both pulses experience is similar to one illustrated in Figure 3.23.

Increasing the spacing between individual channels can reduce the effect of XPM. By doing so, the difference in propagation constants between these channels becomes large enough so that interacting optical pulses walk away from each other and cannot interact any further. That difference will be enhanced if there is a stronger impact of chromatic dispersion. On the other hand, the most unfavorable case will occur in the zero dispersion wavelength region, since optical channels will stay together and overlap for a significant amount of time.

In general, it is very difficult to estimate the real impact of XPM on the transmission system performance just by using (3.112) and (3.113). The impacts of both SPM and XPM can be studied more precisely by solving the nonlinear Schrodinger equations [refer to (7.32) to (7.34)]. This is often a part of a simulation software package that is used for design and engineering purposes.

3.3.8 Four-Wave Mixing

FWM is another nonlinear effect that occurs in optical fibers during the propagation of a composite optical signal, such as the WDM signal. The power dependence of the refractive index in optical fibers will not only shift the phases of the signals in individual channels as mentioned above, but will also give a rise to new optical signals

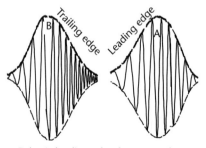
Pulse A: leading edge frequency decrease
Pulse B: trailing edge frequency increase

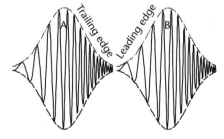
Pulse A: trailing edge frequency increase in addition to the leading edge frequency decrease

Pulse B: leading edge frequency decrease in addition to the leading edge frequency decrease

Figure 3.23 Pulse overlapping and frequency shift due to the cross-phase modulation effect.

through the process known as four-wave mixing. Four-wave mixing is an effect that originates from the third-order nonlinear susceptibility [13, 29] [see (7.5)].

Four-wave mixing is related to the fact that, if there are three optical signals with different carrier frequencies (v_i, v_j, and v_k; i,j,k = 1, ..., M) that propagate through the fiber, a new optical frequency ($v_{ijk} = v_i + v_j - v_k$) will be generated. The number of possible combinations with three optical frequencies grows rapidly with an increase of the total number M of optical channels propagating together. However, effective interaction between three frequencies also requires a phase matching of the corresponding propagation constants, which can be expressed as

$$\beta_{ijk} = \beta_i + \beta_j - \beta_k \tag{3.114}$$

(Please recall that propagation constant is defined as $\beta = 2\pi n \lambda / c$, where n is the refractive index, λ is the wavelength, and c is the speed of light in vacuum.) The value

$$\Delta \beta = \beta_i + \beta_j - \beta_k - \beta_{ijk} \tag{3.115}$$

presents a measure of the phase-matching condition with respect to optical waves involved in the four-wave mixing process. An effective interaction accompanied with new wave generation takes place only if $\Delta \beta$ approaches zero. The phase-matching condition can be easily understood if we look at the physical picture of the FWM process. The FWM process can be considered as a mutual interaction and annihilation of two photons with energies hv_i and hv_j (h is Planck's constant), resulting in the generation of two new photons with energies hv_k and hv_{ijk}. In such a case, the phase-matching condition can be considered just as requirement for momentum conservation.

The FWM process and generation of new optical frequencies is shown in Figure 3.24, which illustrates the cases where there are two and three interacting optical channels. If there are just two optical channels with frequencies v_1 and v_2, the two more optical frequencies will be created, which presents a degenerate case shown in Figure 3.24(a). If there are three optical channels with frequencies v_1, v_2, and v_3, then eight more optical frequencies can be created, as in Figure 3.24(b). The condition for phase matching can be easily understood for the degenerate case from

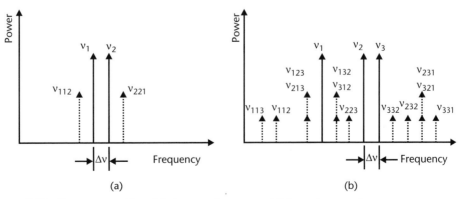

Figure 3.24 Four-wave mixing: (a) degenerate case, and (b) nondegenerate case.

Figure 3.24(a). In this situation we have that $v_2 = v_1 + \Delta v$, $v_{221} = v_2 + \Delta v$, and $v_{112} = v_1 - \Delta v$. If we use the expansion in Taylor series for each of the propagation constants in the degenerate case, the phase-matching condition becomes

$$\Delta \beta = \beta_2 \left(\frac{\Delta v}{2\pi} \right)^2 \tag{3.116}$$

where β_2 is the GVD coefficient [see (3.61) and (3.62)]. It is clear from (3.116) that a complete phase matching happens only if $\beta_2 = 0$, or at the zero chromatic dispersion point. However, a practical phase matching occurs either for very low values of the GVD coefficient, or for very narrow channel spacing.

The power of the newly generated optical frequency will be dependent on the powers of optical signals involved in the process, the intensity of the nonlinear Kerr effect, and the satisfaction of the phase-matching condition. This power is proportional to [30]

$$P_{ijk} \sim \left(\frac{2\pi v_{ijk} n_2 d_{ijk}}{3cA_{eff}} \right)^2 P_i P_j P_k L_{eff}^2 \tag{3.117}$$

where n_2 is nonlinear refractive index, A_{eff} is the effective cross-sectional area, L_{eff} is the effective length, and d_{ijk} is the degeneracy factor (d_{ijk} equals 3 and 6 for the degenerate and nondegenerate cases, respectively).

Four-wave mixing can produce significant signal degradation in WDM systems since several newly generated frequencies can coincide with any specific optical channel. The total number N of new frequencies that can be generated through the FWM process is

$$N = \frac{M^2(M-1)}{2} \tag{3.118}$$

where M is the total number of the optical channels propagating through the fiber. As an illustration, it can be mentioned that 80 WDM channels will produce 252,800 new frequencies. It is also important to notice that some of them will be very small and will have a negligible impact on the system performance. However, some of them can have a significant impact, especially if there are several newly generated frequencies that will coincide with a specified WDM channel. Namely, the sum of newly generated optical signals could be fairly significant if compared with the power level of the channel in question. The FWM effect can be minimized by either decreasing the power levels of interacting optical channels or by preventing a perfect phase matching. The prevention can be done by increasing the chromatic dispersion and by increasing the channel spacing. The impact of the FWM process to transmission system performance will be evaluated in Section 4.5.

It is worth mentioning here that the FWM process can be utilized to perform some functions that may enhance the overall system performance, such as the wavelength conversion (refer to Section 6.1.6). The wavelength conversion is done by generating a spectrally inverted signal through the process known as phase conjugation. The spectral inversion can be expressed through the following relation connecting the Fourier transform $F_{in}(\omega)$ of the input spectrum with the Fourier transform $F_{out}(\omega)$ of the output spectrum:

$$F_{in}(\omega_0 - \Delta\omega) = F_{out}(\omega_0 + \Delta\omega) \tag{3.119}$$

3.3.9 Stimulated Raman Scattering

There are two types of light scattering that can occur in optical fibers. The first one, known as linear or elastic scattering, is characterized by a scattered light signal that has the same frequency as the incident one. The second type, known as nonlinear or inelastic scattering, is characterized by a downshift in frequency of the scattered signal. Just for reference purposes, we can recall that the Rayleigh scattering is a classical example of the elastic process. On the other side, the Raman and Brillouin scatterings are two well-known examples of inelastic scattering [2].

Raman scattering is a nonlinear effect that occurs when a propagating optical power interacts with glass molecules in the fiber undergoing a wavelength shift. The result is a transfer of energy from some photons of the input optical signal to vibrating silica molecules, and a creation of new photons with lower energy than the energy of the incident photons. The incident optical signal is often referred as the pump signal. Newly generated photons form the Stokes signal, which is illustrated in Figure 3.25. Since energy of the Stokes photons is lower than the energy of incident pump photons, the frequency of the Stokes signal will also be lower than the frequency of the pump signal. A difference in frequencies, known as the Raman frequency shift ω_R, is expressed as $\omega_R = \omega_P - \omega_S$, where ω_P is optical frequency of the incident pump signal, while ω_S is the optical frequency of the scattered Stokes signal.

Scattered photons are not in phase with each other and do not follow the same scattering pattern, which means that the energy transfer from the pump to Stokes photons is not a uniform process. As a result, there will be some frequency band $\delta\omega_R$ that includes frequencies of all scattered Stokes photons. In addition, scattered Stokes photons can take any direction, which means that Raman scattering is an isotropic process. That direction can be either forward or backward with respect to the direction of the pump signal in the optical fiber.

If the pump power is lower than a certain threshold value, the Raman scattering process will have a spontaneous character, which is characterized by a relatively

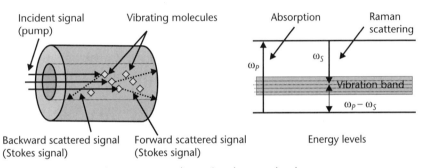

Figure 3.25 Raman scattering process and associated energy levels.

small number of pump photons that will be scattered and converted to Stokes photons. However, if the pump power exceeds the threshold value, Raman scattering becomes a stimulated process, and we are talking about stimulated Raman scattering. The stimulated scattering could be explained as a positive feedback process, where the pump signal interacts with Stokes signal and creates a beat frequency $\omega_{beat} = \omega_R = \omega_P - \omega_S$. The beat frequency then acts as a stimulator of molecular oscillations, and the process is enhanced, or amplified. If we assume that the Stokes signal propagates in the same direction as pump (i.e., in the forward direction), the intensity of the pump signal $\Pi_P = P_P/A_{eff}$ and the intensity of the Stokes signal $\Pi_S = P_S/A_{eff}$ (where A_{eff} is the effective cross-sectional area of the fiber core) can be characterized by the following pair of equations [31]:

$$\frac{d\Pi_P}{dz} = -g_R \left(\frac{\omega_P}{\omega_S}\right) \Pi_P \Pi_S - \alpha_P \Pi_P \qquad (3.120)$$

$$\frac{d\Pi_S}{dz} = g_R \Pi_P \Pi_S - \alpha_S \Pi_S \qquad (3.121)$$

where z is the axial coordinate, g_R is the Raman gain coefficient, and α_R and α_S are the fiber attenuation coefficients for the pump and Stokes signal, respectively.

As already mentioned, the scattered Stokes photons will not have equal frequencies, and they will occupy a certain frequency band. The number of photons corresponding to any specified frequency within the frequency band will determine the value of the Raman gain related to that specific frequency. Therefore, the Raman gain is not constant, but it is a function of the optical frequency. The spectrum of the Raman gain is related to the width of the energy band of silica molecules, and to the time decay associated to each energy state within the energy band. Although it is difficult to find an analytical representation for Raman gain spectrum, it can be roughly approximated by the Lorentzian spectral profile, which is given as

$$g_R(\omega_R) = \frac{g_R(\Omega_R)}{1 + (\omega_R - \Omega_R)^2 T_R^2} \qquad (3.122)$$

where T_R is the decay time associated with excited vibration states, while Ω_R is the Raman frequency shift corresponding to peak of the Raman gain. The decay time is around 0.1 ps for silica-based materials, which makes the gain bandwidth wider than 10 THz. The Raman gain peak $g_R(\Omega_R) = g_{Rmax}$ is between 10^{-12} and 10^{-13} m/W for wavelengths above 1,300 nm.

Figure 3.26 shows the approximation of the Raman gain with Lorentzian curve for silica fibers, and a typical shape of the actual gain profile. The actual gain profile extends over the frequency range of about 40 THz (which is approximately 320 nm), with a peak around 13.2 THz, as shown in Figure 3.26. There are also several smaller peaks that cannot be approximated by the Lorentzian curve. They are located around frequencies of 15, 18, 24, 32, and 37 THz [29, 32]. The gain profile can also be approximated by a triangle function [28], as

$$g_R(\omega_R) = \frac{g_R(\Omega_R)\omega_R}{\Omega_R} \tag{3.123}$$

This approximation is also shown in Figure 3.26.

It is important to estimate a threshold value of the pump power above which the Raman scattering takes a stimulated character. The threshold power, or the Raman threshold, is usually defined as the incident power at which the half of the pump power is eventually converted to the Stokes signal. The Raman threshold can be estimated by solving (3.120) and (3.121). For that purpose, value g_R from (3.120) and (3.121) should be approximated by the peak value $g_R(\Omega_R) = g_{Rmax}$. Accordingly, the amplification of the Stokes power along the distance L can be expressed as [31]

$$P_S(L) = P_{S0}\exp\left(\frac{g_{R\max}P_{S0}L}{2A_{eff}}\right) \tag{3.124}$$

The value P_{S0}, which corresponds to the Raman threshold P_{Rth}, as defined above, is

$$P_{Rth} = P_{S0} \approx \frac{16A_{eff}}{g_{R\max}L_{eff}} \tag{3.125}$$

where L_{eff} is the effective length. The estimated Raman threshold is about 500 mW for typical values of the parameters that appear in (3.125) (i.e., for $A_{eff} = 50\ \mu m^2$, $L_{eff} = 20\ km$, and $g_R(\Omega_R) = g_{Rmax} = 7 \times 10^{-13}$ m/W.)

Stimulated Raman scattering can be effectively used for optical signal amplification, since it can enhance the optical signal level by transferring the energy from the pump to the signal. Raman amplifiers can improve the performance of optical transmission systems by providing an additional optical power margin, as presented in Section 6.1.1. On the other hand, the SRS effect could be quite detrimental in the WDM transmission systems since the Raman gain spectrum is very broad, which

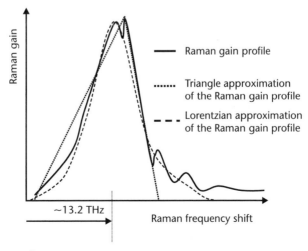

Figure 3.26 Raman gain spectrum.

enables the energy transfer from the lower to higher wavelength channels. In this situation the optical fiber acts as the Raman amplifier since longer wavelengths are amplified by using power carried by lower wavelengths.

Effective signal amplification by the SRS effect can occur for WDM channels spaced up to 150 nm. The shortest wavelength channel within the WDM system acts as a pump for several long wavelength channels, while undergoing the most intense depletion. The other shorter wavelengths will also be depleted, but the depletion rate decreases with the wavelength increase, as illustrated in Figure 3.27. On the other hand, longer wavelengths will gain some power in proportion to their position, while the gain will be the highest for the longest wavelength.

The energy transfer between two channels is bit-pattern dependent and occurs only if both wavelengths are synchronously loaded with 1 bits, as illustrated in Figure 3.28 for wavelengths λ_A and λ_B. The energy transfer will be reduced if chromatic dispersion is higher since higher chromatic dispersion will cause a greater difference in the group velocities associated with different channels. The difference in velocities will reduce the time of pulse overlapping, thus reducing the duration of an effective energy exchange. This is known as the walk-off phenomenon, which also applies to XPM and four-wave mixing effects. The evaluation of the impact that the SRS effect has on the system transmission characteristics will be done in Section 4.5.

3.3.10 Stimulated Brillouin Scattering

Brillouin scattering is a physical process that occurs when an optical signal interacts with acoustical phonons, rather than with the glass molecules. During this process an incident optical signal reflects backward from the grating formed by acoustic vibrations and downshifts in frequency, as illustrated in Figure 3.29.

The acoustic vibrations originate from the thermal effect if the power of an incident optical signal is relatively small. In this case the amount of backward scattered Brillouin signal is also small [31, 33]. If the power of the incident optical signal goes up, it increases the material density through the electrostrictive effect [2, 13]. The change in density enhances acoustic vibrations and forces Brillouin scattering to move from a spontaneous to a stimulated regime, which means that SBS is initiated.

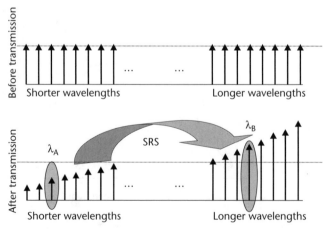

Figure 3.27 Stimulated Raman scattering and power transfer between WDM channels.

3.3 Signal Impairments

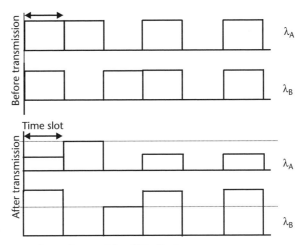

Figure 3.28 Bit-pattern dependence of the SRS effect.

The SBS process can also be explained as a positive feedback mechanism, in which the incident optical signal (or the pump) interacts with the Stokes signal and creates a beat frequency $\omega_{beat} = \omega_B = \omega_P - \omega_S$. Notice that this scattering process is the same in nature as one mentioned in Section 2.6.3, which refers to the acousto-optical filters. However, there is no external electric field applied in this case, since the electric field at the beat frequency ω_{beat} is created from inside rather than by applying an external microwave transducer.

The parametric interaction between pump, Stokes signal, and acoustical waves requires both the energy and momentum conservation [2, 33]. The energy is effectively conserved through the frequency downshift, while the momentum conservation occurs through the backward direction of the Stokes signal. The frequency downshift is expressed by the Brillouin shift Ω_B, which is given as [29]

$$\Omega_B = \frac{2n\omega_P V_A}{c} \tag{3.126}$$

where n is the refractive index of the fiber material, V_A is the velocity of the acoustic wave, c is the light speed in vacuum, and ω_P is the optical pump frequency. Equation (3.126) can also be rewritten in the form

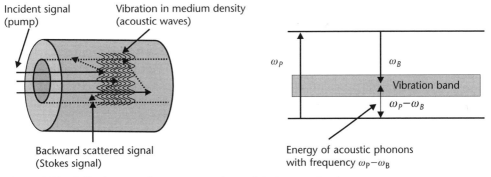

Figure 3.29 Brillouin scattering process and associated energy levels.

$$f_B = \frac{\Omega_B}{2\pi} = \frac{2nV_A}{\lambda_p} \quad (3.127)$$

where the standard relation between frequency and wavelength ($\omega_p = 2\pi c/\lambda_p$) was used. By inserting typical values of parameters (V_A = 5.96 km/s, λ_p = 1.550 nm, n = 1.45) into (3.127), the frequency shift becomes f_B = 11.5 GHz. It is worth mentioning that this frequency shift is fiber material dependent, and can vary from 10.5 to almost 12 GHz for different fiber materials [33].

The SBS process is governed by the following set of coupled equations [31]:

$$\frac{d\Pi_p}{dz} = -g_B \Pi_p \Pi_S - \alpha_p \Pi_p \quad (3.128)$$

$$-\frac{d\Pi_S}{dz} = g_B \Pi_p \Pi_S - \alpha_S \Pi_S \quad (3.129)$$

where $\Pi_p = P_p/A_{eff}$ and $\Pi_S = P_S/A_{eff}$ are the intensities of the pump and the Stokes signal, respectively, z is the axial coordinate, g_B is the Brillouin amplification coefficient (gain), and α_B and α_S are the fiber attenuation coefficients of the pump and the Stokes signal, respectively. The minus sign in (3.129) points to the fact that the scattered signal has a backward direction.

The scattered Stokes photons will not have equal frequencies, but will be dispersed within a certain frequency band. The number of photons corresponding to any specified frequency within the band determines the value of the Brillouin gain with respect to that specific frequency. The spectrum of the Brillouin gain is related to the lifetime of acoustic phonons, which can be characterized by the time constant T_B. The gain spectrum can be approximated by a Lorentzian spectral profile given as

$$g_B(\omega_B) = \frac{g_B(\Omega_B)}{1 + (\omega_B - \Omega_B)^2 T_B^2} \quad (3.130)$$

where Ω_B is the Brillouin frequency shift calculated by (3.126). It is known that not just Ω_B, but also the width of the function given by (3.130), will depend on characteristics of the fiber material. In addition, the Brillouin gain is dependent on the fiber waveguide characteristics. The SBS gain bandwidth will be about 17 MHz at λ_p = 1.520 nm for pure silica, while it can be almost 100 MHz in doped ilica fibers. The typical value of the SBS gain bandwidth is about 50 MHz.

The maximum value of the SBS gain $g_B(\Omega_B) = g_{Bmax}$ is also dependent on type of fiber material. It is between 10^{-10} and 10^{-11} m/W for silica-based optical fibers, and for wavelengths above 1 μm. The threshold value of the pump power, above which Brillouin scattering takes a stimulated character, can be estimated by following the same method as in the SRS case. The threshold power is defined as the incident power at which the half of the pump power is eventually converted to the backward Stokes signal. The incident power that corresponds to the SBS threshold P_{Bth} is given as [31]

$$P_{Bth} \approx \frac{21 A_{eff}}{g_{B\,max} L_{eff}} \quad (3.131)$$

where L_{eff} is the effective length. The estimated Brillouin threshold is about 7 mW for typical values of parameters from (3.131) (i.e., for A_{eff} = 50 μm^2, L_{eff} = 20 km, and g_{Bmax} = 5×10^{-11} m/W).

Although the SBS effect can potentially be used for optical signal amplification, its narrow gain bandwidth will limit the application area. On the other hand, the SBS effect can be quite detrimental in optical transmission systems, since the transfer of the signal energy to the Stokes signal has the same effect as the signal attenuation. In addition, the back-reflected light contributes to the optical noise and could even enter into the resonant cavity of the transmitting laser. Fortunately enough, there are some methods that can minimize the SBS effect. The impact of the SBS effect on the transmission system characteristics, as well as methods for its suppression, is covered in Section 4.5.

3.4 Summary

A number of system parameters have been introduced. They can be classified into three general categories:

1. Parameters related to the optical signal;
2. Parameters that define the noise in optical transmission system;
3. Parameters that degrade the system performance, but do not have a noisy nature.

The reader is advised to pay more attention to fundamental mathematical relations introduced in this chapter. This are given by the following equations: (3.3), (3.10), (3.25), (3.30), (3.32), (3.42), (3.43), (3.82), and (3.90). The reader is also advised to refer to data related to optical fiber types from Table 3.2, which can be used as basic reference points.

References

[1] Desurvire, E., *Erbium Doped Fiber Amplifiers*, New York: Wiley, 1994.

[2] Saleh, B. E., and A. M. Teich, *Fundamentals of Photonics*, New York: Wiley, 1991.

[3] Okano, Y., et al., "Laser Mode Partition Evaluation for Optical Fiber Transmission," *IEEE Trans. on Communications*, Vol. COM-28, 1980, pp. 238–243.

[4] Ogawa, K., "Analysis of Mode Partition Noise in Laser Transmission Systems," *IEEE J. Quantum Electron.*, Vol. QE-18, 1982, pp. 849–855.

[5] Agrawal, G. P., and N. K. Duta, *Semiconductor Lasers*, 2nd ed., New York: Van Nostrand Reinhold, 1993.

[6] Cvijetic, M., *Coherent and Nonlinear Lightwave Communications*, Norwood, MA: Artech House, 1996.

[7] Lachs, G., *Fiber Optic Communications: System Analysis and Enhancements*, New York: McGraw-Hill, 1998.

[8] Personic, S. D., *Optical Fiber Transmission Systems*, New York: Plenum, 1981.

[9] Papoulis, A., *Probability, Random Variables and Stochastic Processes*, New York: McGraw-Hill, 1984.

[10] Gower, J., *Optical Communication Systems*, 2nd ed., Upper Saddle River, NJ: Prentice Hall, 1993.

[11] Agrawal, G. P., *Fiber Optic Communication Systems*, 3rd ed., New York: Wiley, 2002.

[12] Robinson, F. N. H., *Noise and Fluctuations in Electronic Devices and Circuits*, Oxford, England: Oxford University Press, 1974.

[13] Born, M. E. Wolf, *Principles of Optics*, 7th ed., New York: Cambridge University Press, 1999.

[14] Henry, C. H., "Theory of the Linewidth of the Semiconductor Lasers," *IEEE J. Quantum Electron*, Vol. QE-18, 1982, pp. 259–264.

[15] Lee, T. P., "Recent Advances in Long-Wavelength Lasers for Optical Fiber Communications," *IRE Proc.*, Vol. 19, 1991, pp. 253–276.

[16] Keiser, G. E., *Optical Fiber Communications*, 3rd ed., New York: McGraw-Hill, 2000.

[17] Marcuse, D., "Interdependence of Material and Waveguide Dispersion," *Applied Optics*, Vol. 18, 1979, pp. 2930–2932.

[18] ITU-T Rec. G.652, "Characteristics of Single-Mode Optical Fiber Cable," ITU-T (04/97), 1997.

[19] ITU-T Rec. G.653, "Characteristics of Dispersion-Shifted Single Mode Optical Fiber Cable," ITU-T (04/97), 1997.

[20] ITU-T Rec. G.655, "Characteristics of Non-Zero Dispersion Shifted Single-Mode Optical Fiber Cable," ITU-T (10/00), 2000.

[21] Heidemann, R., "Investigations of the Dominant Dispersion Penalties Occurring in Multigigabit Direct Detection Systems," *IEEE J. Lightwave Techn.*, Vol. LT-6, 1988, pp. 1693–1697.

[22] Kogelnik, H., L. E. Nelson, and R. M. Jobson, "Polarization Mode Dispersion," in *Optical Fiber Communications*, I. P. Kaminov and T. Li, (eds), San Diego, CA: Academic Press, 2002.

[23] Staif, M., A. Mecozzi, and J. Nagel, "Mean Square Magnitude of All Orders of PMD and the Relation with the Bandwidth of the Principal States," *IEEE Photonics Techn. Lett.*, Vol. 12, 2000, pp. 53–55.

[24] Ciprut, P., et al., "Second-Order PMD: Impact on Analog and Digital Transmissions," *IEEE J. Lightwave Techn.*, Vol. LT-16, 1998, pp. 757–771.

[25] Galtarossa, A., et al., "Statistical Description of Optical System Performances Due to Random Coupling on the Principal States of Polarization," *IEEE Photon. Techn. Lett.*, Vol. 14, 2002, pp. 1307–1309.

[26] Foschini, G. J., and C. D. Pole, "Statistical Theory of Polarization Mode Dispersion in Single-Mode Fibers," *IEEE J. Lightwave Techn.*, Vol. LT-9, 1991, pp. 1439–1456.

[27] Karlsson, M., et al., "A Comparison of Different PMD Compensation Techniques," *Proceedings of European Conference on Optical Communications-ECOC*, Munich, 2002, Vol. II, pp. 33–35.

[28] Ramaswami R., and K. N. Sivarajan, *Optical Networks*, San Francisco, CA: Morgan Kaufmann Publishers, 1998.

[29] Agrawal, G. P., *Nonlinear Fiber Optics*, Third Edition, San Diego, CA: Academic Press, 2001.

[30] Shibata, N., et al., "Experimental Verification of Efficiency of Wave Generation Through Four-Wave Mixing in Low-Loss Dispersion Shifted Single-Mode Optical Fibers," *Electron. Letters*, Vol. 24, 1988, pp. 1528–1530.

[31] Smith, R. G., "Optical Power Handling Capacity of Low Loss Optical Fibers as Determined by Stimulated Raman and Brillouin Scattering," *Applied Optics*, Vol. 11, 1972, pp. 2489–2494.

[32] Stolen, R. G., and E. P. Ippen, "Raman Gain in Glass Optical Waveguides," *Applied Phys. Letters*, Vol. 22, 1973, pp. 276–278.

[33] Buck, J., *Fundamentals of Optical Fibers*, New York: Wiley, 1995.

CHAPTER 4
Assessment of the Optical Transmission Limitations and Penalties

This chapter evaluates the impact of key parameters introduced in Chapter 3 to transmission system performance. The impact is quantified through the power penalties associated with the signal transmission along the lightwave path. The reader should be able to understand the importance and impact of each specific parameter to the system performance, as well as their values when the impact does not create transmission issues.

An optical signal that has traveled along a lightwave path is converted to photocurrent by the photodiode at the receiving side. All impairments accompanying the optical signal will be also transferred to the electrical level, which means that the total photocurrent will contain both the signal component and corruptive additives. In addition, new impairments will be generated in the optical receiver, either during the photodetection process or through the amplification stage (refer to Figure 5.1).

Both the distorted signal and corruptive additives will eventually come together to the decision circuit that recognizes logical levels at defined clock intervals. The signal level at the decision point should be as high as possible to keep a required distance from the noise level, and to compensate for penalties due to impact of other impairments (crosstalk, timing jitter, and so forth).

In this chapter we will first evaluate the signal level and level of different noise components that are mixed with the signal at the decision point. This evaluation will be used later in Chapter 5 to calculate both the signal-to-noise ratio and the receiver sensitivity. Next, the impacts of other impairments will be quantified through the receiver sensitivity degradation with respect to a reference case. The receiver sensitivity degradation can also be considered as the signal power penalty that needs to be compensated for. It can be done by allocating some power margin, which is defined as an increase in the signal power that is needed to compensate for the impact of each specific impairment. By increasing the signal power, the signal-to-noise ratio is kept at the same level that would exist in a case where no impairment was involved.

This power penalty has a real meaning in some situations, such as those related to the impact of chromatic dispersion or nonideal extinction ratio, for example, in which an increase in signal power would certainly compensate for the penalty associated with the impairment in question. On the other hand, it has a different meaning if it is related to other specific cases, such as the impact of nonlinear effects. That is because an increase in signal power would not be beneficial since it would also increase the impairment impact. However, the power penalty serves as an engineering parameter that helps to optimize system characteristics.

The impact of different impairments is illustrated in Figure 4.1(a). They are observed through pulse level decrease, pulse shape distortion, or as noise additives. The receiver sensitivity degradation due to pulse level decrease can be evaluated directly, while the evaluation of the pulse shape distortion has a more complex character. Finally, the impact of the noise is evaluated through the averaged power of the stochastic process.

4.1 Attenuation Impact

Optical fiber attenuates the optical signal level through inserted fiber loss. Output optical power P_2 from the optical fiber can be calculated from input power P_1 and the optical attenuation coefficient α, and it is given as

$$P_2(L) = P_1 \exp(-\alpha L) \tag{4.1}$$

where L is the transmission length. Please recall that coefficient α in (4.1) is expressed in km^{-1}. If parameters P_2, P_1, and α are all expressed in decibels, then (4.1) becomes $P_2 = P_1 - \alpha L$. Since coefficient α is wavelength dependent, as shown in Figure 2.13, the output power will also depend on the wavelength of transmitted optical signal. That wavelength dependence is one of contributors to unequal signal-to noise ratio associated with individual optical channels. A dynamic adjustment of optical signal levels at different points along a lightwave path will eventually correct the impact of wavelength-dependent optical fiber loss.

The fiber loss coefficient is about 0.2 dB/km for single-mode optical fibers based on doped silica. It can be even lower for pure silica core fibers (PSCFs), in which case the loss of $\alpha = 0.1484$ dB/km was recently reported [1]. The loss increases in value up to about 20% for other wavelengths covered by C and L wavelength bands. The attenuation in the S band is even higher and can be more than 50% higher than the attenuation around 1,550 nm.

An optical signal is additionally attenuated at the optical splices and connectors along the lightwave path. Further on, the cabling process inserts some additional loses. Optical signal attenuation inserted due to fiber splicing and cabling is often distributed over the entire transmission length and added to the coefficient α. Such a distributive approach is very useful in the system engineering process. It is common

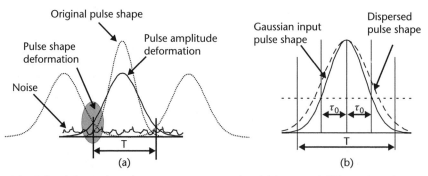

Figure 4.1 Pulse deformations that cause power penalty: (a) impact of different impairments, and (b) Gaussian pulse as a referent input shape.

practice to add an additional fraction of about 10% to the coefficient α to account for the impact of fiber splicing and cabling. This is, of course, case when no measurements data are available.

An optical signal is also attenuated by different optical elements placed along the lightwave path. Such elements include optical multiplexers, optical filters, and optical couplers. The impact of these elements to the signal level is characterized by the signal insertion losses that are equivalent to one time decrease in the optical signal power. If both the optical power and different losses are expressed in decibels, (4.1) takes a more general form and becomes

$$P_2 = P_1 - \alpha L - \alpha_c - \alpha_{others} \qquad (4.2)$$

where P_2 is the optical power of the signal that passed through different elements (fiber, splices, filters, and so forth), P_1 is the power at the beginning of the lightwave path, α_c refers to splicing and connector losses, and α_{others} represents all other insertion losses that might occur along the lightwave path. The power P_2 will eventually determine the signal-to-noise ratio, which means that any impairment related to the change in optical power translates directly to the power penalty.

4.2 Noise Impact

The impact of different noise components is measured by their noise power. Any noise leads to sensitivity degradation of an optical receiver, which is expressed through the power penalty. It is common to calculate the power of basic noise components (thermal noise, shot noise) first, and then to evaluate the impact of the other individual noise components that have practical meaning in each specific case. Some noise components can often be neglected in engineering considerations since their practical impact is minimal (refer to Table 4.1 for the importance of individual noise components).

It was mentioned in the previous chapter that the photocurrent I_p is a sum of the signal and noise contributions. Accordingly, it can be expressed as

$$I_p = I + i_{noise} = I + \left(i_{sn} + i_{dcn} + i_{the} + i_{beat} + i_{cross} + i_{int} + i_{jitt}\right) \qquad (4.3)$$

where I is the signal current calculated as a product of incoming optical signal power P and photodiode responsivity R (recall that R is expressed in amperes per watt.) Noise components are expressed by the fluctuating portion of the total photocurrent around value I. This fluctuating part includes currents i_{sn}, i_{dcn}, i_{the}, i_{beat}, i_{cross}, i_{int}, and i_{jitt}, which correspond to quantum shot noise, dark current noise, thermal noise, beat noise components (including signal to spontaneous emission beating noise, and spontaneous emission to spontaneous emission beating noise), crosstalk noise (including all inband and out-of-band components), intensity noise (including laser intensity noise, modal noise, and mode partition noise), and timing jitter noise, respectively.

The total noise power for this general case can be expressed by the following practical equation:

$$\langle i^2 \rangle = \langle i^2 \rangle_{sn} + \langle i^2 \rangle_{dcn} + \langle i^2 \rangle_{the} + \langle i^2 \rangle_{beat} + \langle i^2 \rangle_{others} \qquad (4.4)$$

The powers of the first four terms on the right side of (4.4) will be evaluated numerically, and the impact of all other components (intensity noise components, crosstalk components, and timing jitter noise) will be evaluated in Chapter 5 through the associated power penalties and receiver sensitivity degradation. We will evaluate the noise powers by calculating the noise spectral densities. Please recall that noise power equals the product of spectral density and the noise filter bandwidth.

The spectral density of dark current noise can be calculated by using (3.28) as

$$S_{dcn}(f) = 2q\langle M \rangle^2 F(M) I_d \approx 2q \langle M \rangle^{2+x} I_d \qquad (4.5)$$

This equation can be applied to both PIN photodiode and APD. We will assume the following reference values for photodiode parameters:

- I_d = 5 nA, M = 1, and x = 0 for PIN photodiode, which will produce $S_{dcn} \sim 1.6 \times 10^{-27}$ A^2/Hz;
- I_d = 5 nA, M = 10, and x = 0.7 for APD photodiode, which will produce the spectral density $S_{dcn} \sim 8 \times 10^{-25}$ A^2/Hz.

The spectral density of the thermal noise in can be obtained from (3.31), and expressed as

$$S_{the}(f) = \frac{4k\Theta F_{ne}}{R_L} = I_{the}^2 F_{ne} \qquad (4.6)$$

where I_{the}, which is expressed in A/Hz$^{1/2}$, is parameter equivalent to the standard deviation of the thermal current. This parameter is usually up to several pA/Hz$^{1/2}$. Parameter q in the previous equation represents the electron charge ($q = 1.6 \times 10^{-19}$ C), k is the Boltzmann's constant ($k = 1.38 \times 10^{-23}$ J/K), Θ is the absolute temperature in Kelvin, R_L is the load resistance expressed in ohms (Ω), and F_{ne} is the noise figure of the front-end amplifier (refer to Figure 3.7.) Assuming that I_{the} = 3 pA/Hz$^{1/2}$, and F_{ne} = 2, one obtains $S_{the} \sim 2 \times 10^{-23}$ A^2/Hz. This value for the spectral density of the thermal noise is at least about two orders of magnitude higher than the one related to the dark current noise.

The spectral density of the shot noise can be obtained from (3.23) and (3.25), and has a form

$$S_{sn}(f) = 2q \langle M \rangle^{2+x} I \qquad (4.7)$$

Reference values of the short noise power density can be calculated by assuming that:

- Optical power of P = –20 dBm falls to PIN photodiode with responsivity R = 0.8, internal gain M = 1, and noise parameter x = 0, which produces $I = 8\mu$A and $S_{sn/PIN} \sim 2.6 \times 10^{-24}$ A^2/Hz.

- Optical power of $P = -20$ dBm falls to APD with responsivity $R = 0.8$, internal gain $M = 10$, and noise parameter $x = 0.7$, which produces $I = 8\ \mu A$ and $S_{sn/APD} \sim 1.28 \times 10^{-21}\ A^2/Hz$.

The shot noise component produced in the PIN photodiode will be at the same level as the thermal noise component if optical power of -10.65 dBm (86 μW) comes to the photodiode that has responsively $R = 0.8$ A/W.

In a case where an optical amplifier precedes the photodiode, the major noise contributions come from the beat-noise components (signal to spontaneous emission noise, and spontaneous emission noise to spontaneous emission noise). The spectral density of the beat-noise components can be obtained from (3.42) and (3.43) as

$$S_{beat}(f) = S_{sig-sp}(f) + S_{sp-sp}(f) \tag{4.8}$$

where

$$S_{sig-sp} = 4R^2 G P S_{sp} \tag{4.9}$$

$$S_{sp-sp} = 2R^2 S^2_{sp}\left[2B_{op} - \Delta f\right] \tag{4.10}$$

Recall that G is the gain coefficient of optical amplifier, while B_{op} and Δf are the bandwidths of the optical and electrical filters, respectively. It is also important to recall that the shot noise increases if the optical signal is preamplified. The spectral density of the shot noise in this case can be obtained from (3.41), and it becomes

$$S_{sn,Amp} = 2qR\left[GP + S_{sp}B_{op}\right] \tag{4.11}$$

The spectral densities from (4.8) to (4.10) become $S_{sig-sp} \sim 1.05 \times 10^{-19}$ A^2/Hz, $S_{sp-sp} \sim 0.53 \times 10^{-22}$ A^2/Hz, and $S_{sn,Amp} \sim 2.5 \times 10^{-22}$ A^2/Hz, respectively. These results are obtained for the following typical values of parameters related to the amplifier and photodiode: $P = -20$ dBm, $G = 100$, $F_{no} = 3.2$ (which is 5 dB), $R = 0.8$ A/W, $B_{op} = 0.1\ nm$, and $\Delta f = 0.5 B_{op}$. As we can see, the beat-noise components are larger than the shot noise component in PIN photodiodes.

When considering the impact of individual noise components, one should differentiate two cases, which are related to the detection of 1 and 0 bits, respectively. Such a distinction is important to properly evaluate the levels of the shot noise, the beat-noise components, and the intensity related noise components. The extinction ratio can be used as a measure of the difference in receiving power associated with 1 and 0 bits. The comparison of different noise components, which are related to 1 and 0 bits, is shown in Table 4.1 for both PIN photodiode and APD. It was assumed that the extinction ratio is 10.

Table 4.1 will be used as an engineering reference when considering the impact of the noise components in different detection scenarios, as presented in Chapter 5. The power of individual noise components will be found by multiplying the spectral density with the optical receiver bandwidth.

Table 4.1 Typical Values of Spectral Densities Associated with Noise Components

Noise Spectral Density, [A²/Hz]	PIN		APD	
	1 Bit	0 Bit	1 Bit	0 Bit
Dark current	1.6×10^{-27}	1.6×10^{-27}	0.8×10^{-24}	0.8×10^{-24}
Thermal noise	2×10^{-23}	2×10^{-23}	2×10^{-23}	2×10^{-23}
Shot noise without preamp	2.6×10^{-24}	2.6×10^{-25}	1.28×10^{-21} (M=10)	1.28×10^{-22} (M=10)
Shot noise with preamp	2.6×10^{-22}	2.6×10^{-23}	1.97×10^{-20} (M=5)	1.97×10^{-21} (M=5)
Signal-ASE beat noise	1.04×10^{-19}	1.04×10^{-20}	0.52×10^{-18}	0.52×10^{-19}
ASE-ASE beat noise	0.54×10^{-22}	0.54×10^{-22}	2.7×10^{-20}	2.7×10^{-20}

4.3 Modal Dispersion Impact

The impact of chromatic dispersion on the system performance is a major factor that causes the degradation of the pulse waveform in single-mode optical fibers. As for multimode fibers, the impact of chromatic dispersion is generally smaller than the impact of intermodal dispersion. However, it can contribute significantly to the total pulse broadening if transmission is done far away from the zero-dispersion region.

It was mentioned in Section 3.3.4 that the impact of the intermodal dispersion is expressed through the fiber bandwidth given by (3.51). The fiber bandwidth is a distance-dependent parameter that can be calculated from the available fiber data for each specified transmission length. It is useful from the system perspective to convert the fiber bandwidth to the so-called pulse rise time in multimode optical fibers and consider it as a part of a total system rise time [2] (refer to Section 5.4.2). The fiber bandwidth can be converted to the pulse rise time in optical fiber by using the following relation:

$$t_{fib,L} = \frac{U}{B_{fib,L}} = \frac{UL^{\mu}}{B_{fib}} \qquad (4.12)$$

where $t_{fib,L}$ is the pulse rise time with respect to distance L, B_{fib} is optical fiber bandwidth of 1-km long optical fiber, $B_{fib,L}$ is the bandwidth of the specified fiber length, and μ is a coefficient that can take values in the range from 0.5 to 1. Parameter U in (4.12) represents the fact that the rise time is also related to the modulation format. It is 0.35 for nonreturn to zero (NRZ) modulation format, and 0.7 for return-to-zero (RZ) modulation formats. The pulse rise time for 1-km long optical fiber, which is calculated by (4.12), can vary anywhere from 0.5 ns (for graded-index fibers) to 100 ns (for step-index optical fibers) [3].

As for single-mode optical fibers, chromatic dispersion is the only dispersion component. The total impact of chromatic dispersion in single-mode optical fibers is considered in conjunction with several parameters that are related not just to characteristics of the optical fiber, but also to properties of the transmission signal and to characteristics of the light source [4]. The transmission system should be engineered to minimize the impact of chromatic dispersion. This impact can be evaluated through the signal power penalty, which is an increase in signal power that is needed to keep the SNR unchanged. The power penalty occurs due to intersymbol interference imposed by chromatic dispersion.

The impact of chromatic dispersion can be evaluated by assuming that the pulse spreading due to dispersion should not exceed a critical limit. That limit can be defined either by fraction $\delta_{s,chrom}$ of the signal that leaks outside of the bit period, or by the broadening ratio $\delta_{b,chrom}$ of widths associated to the input and output pulse shapes. These parameters do not translate directly to the power penalty since they are related to the pulse shape distortion rather than to the amplitude decrease. The exact evaluation of the power penalty could be rather complicated since it is related to a specific signal pulse shape. Instead, a reasonable approximate evaluation can be done by assuming that the pulse takes a Gaussian shape (introduced in Section 3.3.4) and by using (3.81) as a starting point.

It is often more convenient to use broadening factor $\delta_{b,chrom}$ as a measure of the pulse spreading since it is already calculated for most relevant transmission scenarios (refer to Section 3.3.4). The broadening factor can be expressed as

$$\sigma_{b,chrom} = \frac{\sigma_{chrom}}{\sigma_0(B)} \quad (4.13)$$

where σ_{chrom} is the pulse RMS width at the fiber end, and σ_0 is the pulse RMS at the fiber input. Please note that σ_0 is related to the signal bit rate $B = 1/T$, where T defines the length of the bit interval.

If the input has the Gaussian pulse shape, it will be confined within the time slot T if it satisfies the following relation:

$$\tau_0 = \sigma_0 \sqrt{2} \leq \frac{T}{4} = \frac{1}{4B} \quad (4.14)$$

where σ_0 now relates to RMS of the Gaussian pulse, while τ_0 represents the half-width at the 1/e intensity point of the Gaussian pulse, as shown in Figure 4.1(b). In fact, as it was shown in [5, 6], (4.14) guarantees that almost 100% of the pulse energy is contained within the pulse interval. Any increase of the parameter σ_{chrom} above the value σ_0 will indicate that there is some power penalty associated with that specific case. On the other hand, there is no penalty unless the RMS of the output pulse exceeds the RMS of the input pulse. The power penalty could even be negative if an initial pulse compression is observed (refer to Figures 4.2 to 4.4).

The power penalty ΔP_{chrom} due to chromatic dispersion can be evaluated by using the following formula [5]:

$$\Delta P_{chrom} \approx 10 \log(\delta_{b,chrom}) \quad (4.15)$$

The above formula is relatively simple, but the main task is related to calculating the broadening factor $\delta_{b,chrom}$. There are also some other more complex formulas that calculate the power penalty due to chromatic dispersion, such as one presented in [6]. However, it is important to mention that the above formula is quite effective from the systems engineering perspective.

The dispersion penalty from (4.15) can be calculated by using (3.81) to determine the broadening factor $\delta_{b,chrom}$. It is often appropriate to apply a more specific equation from Table 3.1, which is related to the transmission scenario in question. In a general case, the power penalty is a function of the initial pulse width, source linewidth, initial chirp parameter, chromatic dispersion parameter, and the transmission length. Let us consider the most common scenario when high-speed transmission is conducted at wavelengths out of the zero-dispersion region and when the light source spectrum is much smaller than the signal spectrum. The pulse broadening parameter can be found by using (3.85).

The results are obtained for $\sigma_0 = 70.71$ ps, $\sigma_0 = 17.68$ ps, and $\sigma_0 = 4.41$ ps, which correspond to the three representative high-speed bit rates (i.e., to 2.5, 10, and 40 Gbps, respectively). It is assumed that a single-mode optical fiber with chromatic dispersion coefficient $D = 17$ ps/(nm-km) is used. The results are shown in Figures 4.2 to 4.4.

We can see from Figures 4.2 to 4.4 that the dispersion penalty is highly dependent on the initial value C_0 of the chirp parameter. The dispersion penalty can even be negative, as mentioned above, which is related to the fact that the pulse undergoes an initial compression [4, 5]. Such a case can occur if negative initial chirp is combined with positive chromatic dispersion, and vice-versa. A dispersion penalty limit can be established by defining the power penalty ceiling. For example, it can be 0.5 or 1 dB, as shown by the dotted lines in Figures 4.2 to 4.4. In some cases, such as in nonamplified point-to-point transmission, an even larger power penalty, such as 2 dB, can be tolerated.

Dispersion should be periodically compensated in high-speed transmission systems that employ optical amplifiers along the transmission line. In that situation, the dispersion penalty refers to the amount of dispersion that can be tolerated. Accordingly, the amount of chromatic dispersion that can be tolerated is known as the *dispersion tolerance*. The dispersion tolerance with respect to 1-dB limit is summarized in Table 4.2 for different optical fibers in high bit-rate transmission systems. The

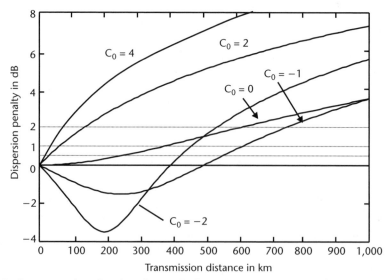

Figure 4.2 Power penalty related to chromatic dispersion impact on 2.5-Gbps signal bit rates.

4.3 Modal Dispersion Impact

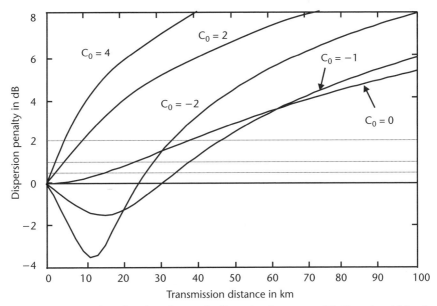

Figure 4.3 Power penalty related to chromatic dispersion impact on 10-Gbps signal bit rates.

lengths of different fiber types, which are associated with the dispersion tolerance, are also shown in Table 4.2.

The evaluation of dispersion penalty can be simplified in two extreme situations related to cases when the optical source spectrum $\Delta\lambda$ is relatively wide, and when a chirpless Gaussian pulse propagates through the optical fiber. We can assume that the input Gaussian pulse has an optimum width $2\tau_0 = (\beta_2 L)^{1/2}$, where β_2 is the GVD parameter, and $2\tau_0$ represents the full width at the 1/e intensity point (FWEM). These two extreme cases can be characterized by the following equations [7, 8]:

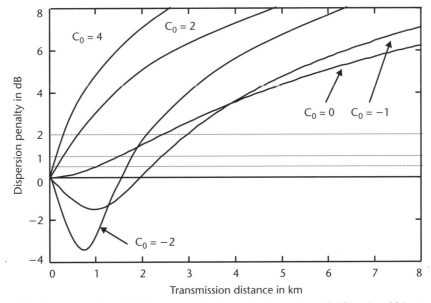

Figure 4.4 Power penalty related to chromatic dispersion impact on 40-Gbps signal bit rates.

Table 4.2 Dispersion Tolerance and Fiber Lengths Associated with Dispersion Tolerance

Bit Rate, Gbps	Dispersion Tolerance, ps/nm	Fiber Length Associated, for $D = 17$ ps/km-nm	Fiber Length Associated, for $D = 8$ ps/km-nm	Fiber Length Associated, for $D = 4$ ps/km-nm
2.5	11,000	650 km	1,380 km	2,770 km
10	720	42 km	90 km	180 km
40	45	2.6 km	5.6 km	11 km

$$\Delta\lambda |D| LB < \delta_{s,chrom} \quad \text{for the source with large spectral linewidth} \quad (4.16)$$

$$B\lambda[|D|L / 2\pi c]^{1/2} < \delta_{s,chrom} \quad \text{for an external modulation of CW laser} \quad (4.17)$$

where $B = 1/T$ is the signal bit rate, $\Delta\lambda$ is the source spectral linewidth, L is the transmission distance, and $\delta_{s,chrom}$ is the fraction of the signal that leaks outside of the bit period T. It is worth mentioning that the limit of dispersion penalty is recommended in some standard documents, such as [7], which states that the spreading factor should be up to $\delta_{s,chrom} = 0.306$ for the chromatic dispersion penalty to be below 1 dB. At the same time, it should be lower than $\delta_{s,chrom} = 0.491$ for the dispersion penalty to be lower than 2 dB.

The impact of chromatic dispersion can be rather severe for higher bit rates and longer distances. It could either be suppressed by proper selection of the parameters related to the pulse shape and to the optical modulator, or be compensated by using different compensation methods (see Section 6.1). In most cases, chromatic dispersion will have to be compensated and suppressed to a level that will cause a minimal power penalty, which is often lower than 0.5 dB. In addition, the effect of chromatic dispersion, which is observed through the ISI, can be partially canceled by using special filtering methods that require a heavy signal procession at the electrical level (refer to Section 6.1). The dispersion tolerance could be increased well above the limits from Table 4.1 if an appropriate combination of different methods is applied. Therefore, the numbers from Table 4.1 should be read as a reference, rather than the maximum achievable values.

4.4 Polarization Mode Dispersion Impact

PMD, which was discussed in Section 3.3.5, could be a serious problem in high-speed optical transmission systems. From an engineering perspective, some power margin should be allocated to account for the PMD impact. It is important to mention that the polarization mode effect can be suppressed by using different compensation methods [9]. The total pulse broadening due to PMD can be calculated by using (3.90), since the second-order PMD term can be neglected in most practical cases where the bit rate does not exceed 10 Gbps.

The impact of chromatic dispersion can be evaluated by calculating the fraction $\delta_{s,PMD}$ of the signal that leaks outside of the bit period due to the PMD impact, which is

$$\delta_{s,PMD} = \frac{\sigma_{PMD}}{T} = \sigma_{PMD} B \qquad (4.18)$$

where $B = 1/T$ is signal bit rate, and σ_{PMD} is the RMS of the output pulse. The fraction $\delta_{s,PMD}$ should be less than a specified value. The power penalty related to the PMD effect can be evaluated by using formula presented in [6], which relates the pulse spreading with the power penalties.

The power penalty due to pulse shape deformation is approximated in [6] by equation

$$\Delta P_p \approx -10 \log(1 - d_P) \qquad (4.19)$$

where

$$d_P \approx erfc(\xi) + 2 \sum_{i=1}^{\infty} \exp(-i^2 \xi^2) \{ erf[(i+1)\xi] - erf[(i-1)\xi] \} \qquad (4.20)$$

and

$$\xi = \frac{T}{2\tau_0} \frac{\sigma_0}{\sigma} \qquad (4.21)$$

Parameters σ and σ_0 are related to the RMS of the output and input pulses, respectively. If we now use (4.14), (4.18), (4.20), and (4.21) for the Gaussian shaped pulses, we can obtain that $\xi = 1/2\delta$. Consequently, the parameter d_{PMD} can be defined through the fraction parameter $\delta_{s,PMD}$ as

$$d_{PMD} \approx erfc\left(\frac{1}{2\delta_{s,PMD}}\right) + 2 \sum_{i=1}^{\infty} \exp\left(\frac{-i^2}{4\delta_{s,PMD}^2}\right) \left[erf\left(\frac{i+1}{2\delta_{s,PMD}}\right) - erf\left(\frac{i-1}{2\delta_{s,PMD}}\right) \right] \qquad (4.22)$$

Functions erf and $erfc = 1 - erf$ are well known as the "error function" and the "error function complement," respectively [10].

We can again assume that the input pulse has a Gaussian shape and that (4.14), which connects the RMS widths and broadening factor, can be applied. In addition, we can assume that the first-order PMD is a dominant effect. The fraction parameter $\delta_{s,PMD}$ can be found from (3.90), and it becomes

$$\delta_{s,PMD} = \frac{1}{4}\left[1 + 2\frac{\Delta\tau_{P1}^2}{\sigma_{in}^2}\zeta(1-\zeta)\right]^{1/2} = \frac{1}{4}\left[1 + 64\frac{\Delta\tau_{P1}^2}{T^2}\zeta(1-\zeta)\right]^{1/2} \qquad (4.23)$$

where $\Delta\tau_{P1}$ defines the differential delay between two principal polarization states over the fiber length L, and ζ represents the power splitting of the signal between two principal polarization states (recall Section 3.3.5).

The power penalty due to PMD can be calculated as

$$\Delta P_{PMD} \approx -10 \log(1 - d_{PMD}) \qquad (4.24)$$

The results obtained by using (4.22) to (4.24) for several practical cases are plotted in Figure 4.4. The summation in (4.22) was done for index $i = 1, 2, \ldots, 10$. We can again establish a tolerable penalty limit by drawing 1- and 2-dB penalty lines, as shown in Figure 4.5.

A more practical approach would be to follow the recommendations for maximum allowable pulse spreading due to PMD effect. Such a recommendation was issued by ITU-T [11] which states that the pulse spreading factor due to PMD should be up to $\delta_{s,PMD} = 0.30$ for the power penalty to be below 1 dB. If we recall that differential delay oscillates around its average value $<D_{P1}>$, and that the probability that an instantaneous value will exceed $3<D_{P1}>$ is rather small, the following relation can be established:

$$3\langle D_{P1}\rangle\sqrt{L} < 0.3T \tag{4.25}$$

which leads to formula

$$\langle D_{P1}\rangle\sqrt{L} < 0.1T \tag{4.26}$$

The requirements for the accumulated average first-order PMD, and the actual differential group delay, expressed as three times the average delay, is summarized in Table 4.3 for high-speed bit rates.

We can compare the PMD effect impact against the impact of chromatic dispersion by plotting the functional dependence $B(L)$ corresponding to 1-dB power penalty for both cases. For that purpose, we will use (4.15) and (4.24). Several functional curves showing bit-rate-distance dependence for some cases of practical interest are shown in Figure 4.6. This figure can serve as a reference, which points to distances after which the dispersion compensation is needed. Chromatic dispersion compensation is done periodically along the lightwave path, while

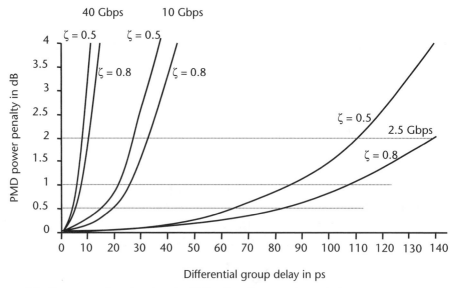

Figure 4.5 Power penalty related to the PMD effect for several practical cases.

4.4 Polarization Mode Dispersion Impact

Table 4.3 Requirements for the Average Value and the Actual Value of the First-Order PMD

Bit Rate, Gbps	The Average PMD Tolerated, ps	Actual PMD Tolerated, ps
2.5	40	120
10	10	30
40	2.5	7.5

polarization mode dispersion compensation is usually done at the end of the lightwave path.

It is obvious that the PMD effect becomes a critical factor for bit rates of 10 Gbps if the first-order PMD is higher than 0.5 ps/(km)$^{1/2}$, and PMD becomes dominant for 40 Gbps if the fibers have the first-order PMD that exceeds 0.05 ps/(km)$^{1/2}$. These numbers can serve as basic criteria to determine when PMD compensation might be necessary.

Finally, in addition to transmission optical fibers, the PMD effect can also occur in other optical elements along the lightwave path. Such elements are optical amplifiers, dispersion compensating modules (DCM), and optical switches. It is important to include the impact of the pulse spreading in these elements, especially if systems engineering is related to 40-Gbps bit rate, or to 10-Gbps bit rate if transmitted over distances more than 1,000 km. The contribution of optical different elements to the pulse spreading can be expressed as

$$\sigma_{PMD,addit} = \left(\sum_{i}^{J} \sigma_{i,PMD}^{2} \right)^{1/2} \tag{4.27}$$

where $\sigma_{i,PMD}$ are contributions to PMD effect from optical elements (amplifiers, DCM, optical switches). These values can be found in the associated data sheets. As

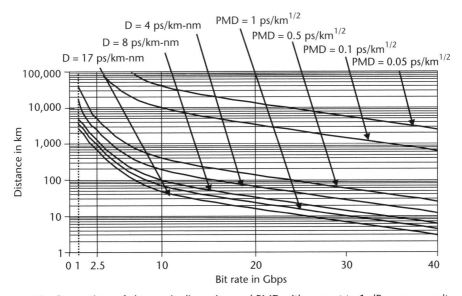

Figure 4.6 Comparison of chromatic dispersion and PMD with respect to 1-dB power penalty.

an example, typical PMD values of optical modules and functional elements deployed within a single span are listed in Table 4.4

It is easy to prove that (4.27) returns values in excess of 0.5 ps per optical fiber span, assuming that there are 3 to 5 optical elements that contribute significantly to the total PMD effect. The generalized formula that includes the entire PMD acquired in transmission fibers and optical modules can be derived from (4.25) as

$$\sqrt{\left[3\langle D_{P1}\rangle\sqrt{L}\right]^2 + \sigma_{PMD,addit}^2} < 0.3T \tag{4.28}$$

4.5 Impact of Nonlinear Effects

The nonlinear effects that can occur in optical fibers were discussed in Section 3.3. Some of them, such as self-phase modulation or cross-phase modulation, can degrade the system performance through the signal spectrum broadening and pulse shape distortion. The other effects degrade the system performance either through nonlinear crosstalk (as in the case of four-wave mixing and stimulated Raman scattering) or by signal power depletion (as in the case of stimulated Brillouin scattering). The power penalty associated with nonlinear effects does not have the same meaning as in cases associated with dispersion or attenuation, since the impact of nonlinearities cannot be compensated by an eventual increase of the optical signal power. That is because any increase in signal power will also increase the impact of nonlinearities. Therefore, the power penalty can be considered as a measure of the impacts of nonlinearities. Regardless of this fact, it is useful to establish some engineering criteria with respect to the impact of nonlinear effects. In the next section, such criteria will be established for all relevant nonlinear effects.

In stimulated Brillouin scattering the acoustic phonons are involved in interaction with light photons, and that occurs over a very narrow spectral linewidth $\Delta v_{SBS} \sim$ [50–100] MHz [recall (3.130)]. The interaction is rather weak and almost negligible if channel spacing is larger than 100 MHz. The SBS process depletes the propagating optical signal by transferring the power to the backward scattered light (i.e., to the Stokes wave). The process becomes very intensive if the incident power per channel is higher than some threshold value P_{Bth}. The threshold value, which is expressed by (3.131), is estimated to be about 7 mW.

The SBS penalty can be reduced by either keeping the power per channel below the SBS threshold, or by broadening the linewidth of the light source. The most practical way of the linewidth broadening is through the signal dithering. By applying the spectral linewidth broadening, (3.131) for the SBS threshold becomes [12, 13]

Table 4.4 Typical PMD Values of Optical Modules and Functional Elements in a Single Span

Module	Actual PMD, ps
Optical amplifier	0.15 to 0.3
Dispersion compensating module	0.25 to 0.7
Optical switch	0.2
Optical isolator	Up to 0.02

$$P_{Bth} = \frac{21bA_{eff}}{g_{B\max}L_{eff}}\left(1 + \frac{\Delta v_{laser}}{\Delta v_{SBS}}\right) \quad (4.29)$$

where Δv_{laser} is a broadened value of the source linewidth. As an example, the SBS threshold rises to $P_{Bth} \sim 16$ mW = 12 dBm if the source linewidth is broadened to be Δv_{laser} = 250 MHz. This value is good enough to prevent any serious impact of the SBS effect.

The impacts of both the FWM and SRS effects can be evaluated through the power penalty associated with the nonlinear crosstalk. Equation (3.44) can be utilized for this purpose since both FWM and SBS can be treated as out-of-band nonlinear crosstalk effects. However, let us repeat again that the negative effect due to either FWM or SRS cannot be compensated by the optical signal power increase.

The out-of-band crosstalk can be measured through the ratio

$$\delta = \frac{\sum_{i=1; i \neq n}^{M} P_i}{P_n} \quad (4.30)$$

where the nominator contains portions of the intruding powers, which originate from all channels except the channel in question, and the denominator refers to optical power of the channel in question.

In the four-wave mixing process, crosstalk components are generated at optical frequencies $v_{ijk} = v_i + v_j - v_k$. It occurs in situations when three wavelengths with frequencies v_i, v_j, and v_k propagate through the fiber and a phase matching condition among them is satisfied. The phase matching is defined through the relation between the propagation constants of the optical waves involved in process, which is expressed by (3.115). However, (3.155) commonly takes a more practical form given by (3.116), which is referred to as the degenerate case (such as the case associated with the WDM transmission).

The optical power of a resultant new optical wave can be calculated as in [14]

$$P_{ijk} = \frac{\alpha^2}{\alpha^2 + \Delta\beta^2}\left[1 + \frac{4\exp(-\alpha l)\sin^2(\Delta\beta l/2)}{[1-\exp(-\alpha l)]^2}\right]\left(\frac{2\pi v_{ijk} n_2 d_{ijk}}{3cA_{eff}}\right)^2 P_i P_j P_k L_{eff}^2 \quad (4.31)$$

where P_{ijk} ($i,j,k = 1...N$) is the power of the newly generated wave, n_2 is the nonlinear refractive index, and d_{ijk} is the degeneracy factor (d_{ijk} = 3 for degenerate case, while d_{ijk} = 6 for nondegenerate case—refer to Figure 3.23). Equation (4.31) can be simplified by assuming that powers of all channels are equal, which is generally true in advanced WDM systems that employ dynamic optical power equalization along the lightwave paths. In addition, we can insert phase condition given by (3.116) into (4.31) whenever considering WDM systems with equal channel spacing. Therefore, the power of a newly generated optical wave through the FWM process in a WDM system can be expressed as

$$P_{ijk} = \frac{\alpha^2}{\alpha^2 + \left[\beta_2 \left(\Delta v_{ijk}/2\pi\right)^2\right]^2} \times$$

$$\left[1 + \frac{4\exp(-\alpha l)\sin^2\left\{l\beta_2\left[\beta_2\left(\Delta v_{ijk}/2\pi\right)^2\right]/2\right\}}{[1-\exp(-\alpha l)]^2}\right] \times \quad (4.32)$$

$$\left(\frac{2\pi v_{ijk} n_2 d_{ijk}}{3cA_{eff}}\right)^2 P^3 L_{eff}^{\;2}$$

where P is the optical power per WDM channel, and β_2 is the GVD coefficient introduced by (3.61) and (3.62). Please recall that $\beta_2 = -D\lambda^2/2\pi c$, where D is chromatic dispersion coefficient and λ is the optical wavelength.

The power penalty related to the FWM impact can be calculated as

$$\Delta P_{FWM} = -10\log[1 - \delta_{FWM}] \quad (4.33)$$

where the crosstalk factor δ_{FWM} is calculated as

$$\delta_{FWM} = \frac{\sum_{i,j,k=1(\neq n)}^{M} P_{ijk}}{P_n} \quad (4.34)$$

If we want to keep the power penalty below 1 dB, δ_{FWM} factor should be lower than $\delta_{FWM} = 0.2$. Equation (4.32) gives us a good idea how to accommodate targeted power penalty by playing with the GVD parameter, channel spacing, and the optical power per channel. As an illustration, we plotted a family of curves related to three different channel spacing schemes (i.e., for 50-, 100-, and 200-GHz channel spacing), which is shown in Figure 4.7. The calculation was aimed to the WDM system with 80 channels. However, only several neighboring channels surrounding the channel in question were effectively contributing to the level of the crosstalk since the weight coefficient, which is associated with the intensity of interaction between different WDM channels, decreases rapidly with an increase of the channels spacing, as shown in Figure 4.8. For example, the impact of the neighboring channels spaced up and down by 50 GHz from the channel in question will be two times stronger than the impact of two channels that are spaced up and down by 300 GHz.

The curves in Figure 4.7 were obtained by calculating only the impact of channels that are placed within 600 GHz, either up or down from the channel in question. It means that there was the total number of 24 interacting channels for 50-GHz channel spacing, 12 interacting channels for 100-GHz channel spacing, and 6 interacting channels for 200-GHz channel spacing. It was assumed that $A_{eff} = 80\,\mu m$, and $D = 8$ ps/nm-km ($\beta_2 \sim -10$ ps^2/km).

From a practical perspective, it is important to make sure that chromatic dispersion in transmission fiber lies above a critical value, which is 2.5 to 3.5 ps/nm-km [15, 16]. This is a precondition that might be followed by the optimization procedure that involves the channel spacing and optical power per channel. Accordingly,

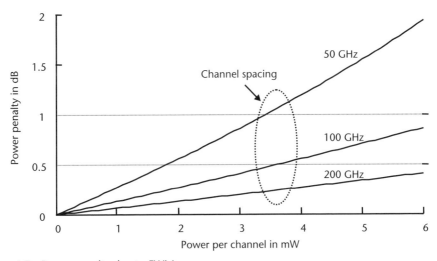

Figure 4.7 Power penalty due to FWM.

the zero-dispersion region does not provide favorable conditions for WDM transmission since the FWM effect will severely degrade the system performance. This fact explains the reason why the NZDSFs were introduced.

In addition to optimizing parameters in (4.32), the FWM effect can be reduced by imposing unequal optical channel spacing. This technique does not reduce the level of crosstalk, but rather prevents it from coinciding with any active WDM channel. Therefore, the crosstalk noise will not be captured during the photodetection process since it stays out of the photodiode bandwidth. This approach, although effective, is not quite practical since it introduces nonstandardized solutions and makes interoperability and optical networking more difficult. In addition, this is generally more expensive than a solution related to the equal channel spacing.

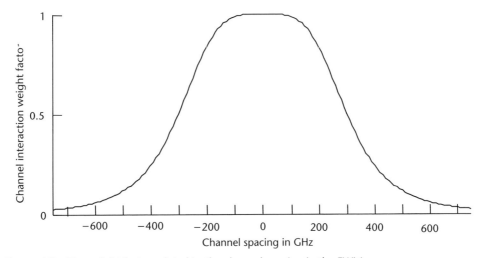

Figure 4.8 The weight factor related to the channel spacing in the FWM process.

Stimulated Raman scattering is broadband in nature and its gain coefficient covers a much wider wavelength region than the SBS gain. It was explained in Section 3.3.9 that optical channels spaced up to 125 nm from each other could be effectively coupled by the SRS process, possibly in both directions. This is the main reason why the SRS effect can be used to design an efficient optical amplifier, in spite of the fact that the gain peak $g_R \sim 7 \times 10^{-13}$ m/W is much smaller than the gain peak associated with the SBS process (refer to Section 6.1.1).

The SRS coupling occurs only if both channels carry 1 bits at any specific moment, as shown in Figure 3.28. The power penalty occurs only in the originating channel and can be expressed is

$$\Delta P_{SRS} = -10\log(1 - \delta_{Raman}) \qquad (4.35)$$

where δ_{Raman} represents the coefficient that is proportional to the power portion leaking out of the channel in question. The coefficient δ_{Raman} can be calculated as [7, 15]

$$\delta_{Raman} = \sum_{i=1}^{M-1} g_R \frac{i\Delta\lambda_{ch}}{\Delta\lambda_R} \frac{PL_{eff}}{2A_{eff}} = \frac{g_R \Delta\lambda_{ch} PL_{eff} M(M-1)}{4\Delta\lambda_R A_{eff}} \qquad (4.36)$$

where $\Delta\lambda_R \sim 125$ nm is the maximum spacing between the channels involved in the SRS process, $\Delta\lambda_{ch}$ is the channel spacing, P is the power per channel, and M is the number of channels. Equation (4.36) is obtained under the assumption that powers of all optical channels were equal before the SRS effect took place.

In order to keep the penalty below 0.5 dB, it should be $\delta_{Raman} < 0.1$, or

$$PM(M-1)\Delta\lambda_{ch} L_{eff} = P_{tot}\Delta\lambda_{total} L_{eff} < 40 Wnmkm \qquad (4.37)$$

where $\Delta\lambda_{total}$ is the total optical bandwidth occupied by all channels. Equations (4.36) and (4.37) were derived in [7] by using results obtained in [13] and under the assumption that there was no chromatic dispersion involved. However, The SRS effect is reduced if there is chromatic dispersion present, since different channels travel with different velocities, while the probability of an overlapping between pulses decreases. The coefficient δ_{Raman} expressed by (4.36) looses almost half of its value if chromatic dispersion exceeds a critical limit (which is again 2.5 to 3.5 ps/nm-km).

As an illustration of the above conditions, we calculated the power penalty associated with the case when chromatic dispersion is above the critical limit, which is shown in Figure 4.9. The power penalty was calculated for several representative cases by using (4.35) and (4.36). However, the coefficient δ_{Raman} was decreased by half to reflect the positive impact of chromatic dispersion. As we can see from Figure 4.9, the power penalty can be decreased by placing optical channels closer to each other, or by decreasing the power per channel. Since the power decrease will reduce the SNR, the power reduction will be beneficial only if power penalty decreases faster than the power itself, which may or may not be the case (refer to the individual case from Figure 4.9). The SRS effect in a multichannel system can be neutralized by equalization of the powers associated with individual WDM channels, which is usually done through the dynamic gain equalization applied periodically along the lightwave path.

4.5 Impact of Nonlinear Effects

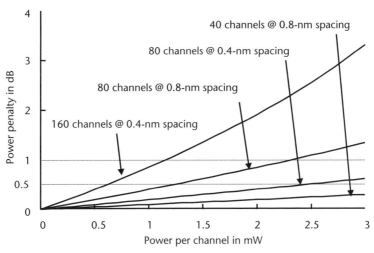

Figure 4.9 Power penalty due to SRS.

The self-phase modulation effect is closely connected with the chromatic dispersion impact. SPM does not cause any crosstalk or power depletion. However, it induces the broadening of the pulse spectrum, which then interacts with chromatic dispersion and enhances the pulse spreading. The hypothetical power penalty due to SPM can be estimated by using the same approach that was used for chromatic dispersion; that is,

$$\Delta P_{SPM} \approx 10\log(\delta_{b,SPM}) \tag{4.38}$$

where $\delta_{b,SPM}$ is the broadening factor due to SPM effect, which can be calculated by using (3.111) as

$$\delta_{b,SPM} = \frac{\sigma_{SPM}}{\sigma_0} = \left[1 + \frac{\sqrt{2}L_{eff}L\beta_2}{2L_{nel}\sigma_0^2} + \left(1 + \frac{4}{3\sqrt{3}}\frac{L_{eff}^2}{L_{nell}^2}\right)\frac{L^2\beta_2^2}{4\sigma_0^4}\right]^{1/2} \tag{4.39}$$

Equation (4.39) is plotted in Figure 4.10 for the 10-Gbps bit rate, and for several different values of the input power and chromatic dispersion parameter. It was again assumed that the input pulse has a Gaussian shape, and that (4.14) can be applied. Therefore, the parameter σ_0 from (4.39) takes the value σ_0 = 17.68 ps. As we can see, even a modest value of input optical power can cause considerable power penalties if transmission is done in the normal dispersion region (negative chromatic dispersion coefficient). On the other hand, the SPM effect can help to suppress the impact of chromatic dispersion in the anomalous dispersion region. This possibility is further explored through two special techniques known as chirped RZ coding and soliton transmission, which are often used in high-speed transmission systems. These methods will be explained in Section 6.1.3.

Cross-phase modulation is another nonlinear effect that causes changes in the optical signal phase. This phase shift is pulse-pattern dependent, and it is converted to the power fluctuations in the presence of chromatic dispersion. Therefore, the SNR will be diminished through the intensity-like noise, which is also pulse-pattern

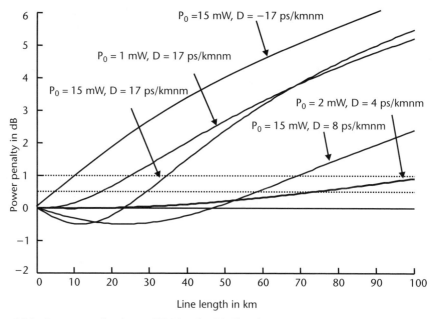

Figure 4.10 Power penalty due to SPM for the 10-Gbps bit rate.

dependent. The XPM effect can be reduced by optical power reduction since the RMS of these intensity fluctuations is dependent on the optical power. A rough estimate of XPM effect was done in [12] by assuming that the total phase shift Φ_m due to XPM should be lower than one radian. Therefore, by calculating the total phase shift from (3.112) and (3.113), and by assuming that $\Phi_m < 1$, the following restriction for optical channel power is obtained:

$$P_{ch,XPM} \leq \frac{\alpha}{\gamma(2M-1)} \qquad (4.40)$$

where α is the fiber attenuation, M is the total number of channels, and γ is the non-linear coefficient introduced by (3.105). Equation (4.40) returns the value of 0.25 mW if the number of channels is 40, and the value of 0.125 mW if the number of channels is 80.

Equation (4.40) helps us to establish a reference line, but it does not account for channel walk-off, which occurs because different channels travel with different speeds. The XPM effect can occur only if two pulses overlap in the time domain, and that means that maximum power per channel will be higher than the value expressed by (4.40). The phase shift due to XPM is eventually converted to amplitude variations in the same fashion as in the SPM case. In addition, the bit pattern and power variations between different channels will cause an asymmetric pulse overlapping that results in the net frequency shift between different channels, which is observed as a timing jitter in the time domain. Therefore, the system performance degradation due to the XPM effect occurs not just because there are noisy variations in amplitude, but because of the timing jitter that has been introduced.

The XPM effect is one of the most serious impairments that can severely degrade performance of multichannel optical transmission systems. A more accurate way to evaluate the impact of the XPM effect is by solving nonlinear wave equations numerically for different bit patterns [refer to (7.32) to (7.34)].

4.6 Summary

Optical fiber attenuation, thermal noise, quantum shot noise, and chromatic dispersion should be considered as fundamental parameters in any systems engineering considerations. On the other hand, polarization mode dispersion and different types on nonlinear effects are becoming more important if the system operates in a regime involving high bit rates and high optical powers. We can assume that such a high bit rate exceeds 10 Gbps, while high optical power is loosely defined as power exceeding a few milliwatts. The reader should benefit from data presented in Tables 4.1 to 4.3, and diagrams presented in Figures 4.2 to 4.10, which are produced for the several cases of practical interest.

References

[1] Nagayama, K., "Ultra Low Loss Fiber with Low Nonlinearity and Extension of Submarine Transmissions," *IEEE LEOS Newsletter*, Vol. 16, 2002, pp. 3–4.

[2] Keiser, G. E., *Optical Fiber Communications*, 3rd ed., New York: McGraw-Hill, 2000.

[3] *Lightwave Optical Engineering Sourcebook, 2003 Worldwide Directory*, Lightwave 2003 Edition, Nashua, NH: PennWell, 2003.

[4] Cvijetic, M., "Performance Evaluation of Externally Modulated High-Bit-Rate Lightwave Systems," *IEEE Photonics Techn. Letters*, Vol. 9, 1997, pp. 687–689.

[5] Agrawal, G. P., *Fiber Optic Communication Systems*, 3rd ed., New York: Wiley, 2002.

[6] Kazovski, L., S. Benedetto, and A. Willner, *Optical Fiber Communication Systems*, Norwood, MA: Artech House, 1996.

[7] Ramaswami, R., and K. N. Sivarajan, *Optical Networks*, San Francisco, CA: Morgan Kaufmann Publishers, 1998.

[8] Bellcore document GR-253 CORE, SONET Transport Systems: Common Generic Criteria, 1995.

[9] Karlsson, M., et al., "A Comparison of Different PMD Compensation Techniques," *Europ. Conf. Opt. Commun., ECOC 2000*, Vol II, pp. 33–35.

[10] Abramovitz, M., and I. A. Stegun, *Handbook of Mathematical Functions*, New York: Dover, 1970.

[11] ITU-T Rec. G.691, "Optical Interfaces for Single-Channel STM-64, STM-256 and Other SDH Systems with Optical Amplifiers," ITU-T (10/00), 2000.

[12] Agrawal, G. P., *Nonlinear Fiber Optics*, 3rd ed., San Diego, CA: Academic Press, 2001.

[13] Stolen, R., "Nonlinearity in Fiber Transmission," *IRE Proc.*, Vol. 68, 1980, pp. 1232–1236.

[14] Shibata, N., R. P. Braun, and R. G. Waarts, "Phase-Mismatch Dependence of Efficiency of Wave Generation through Four-Wave Mixing in a Single-Mode Optical Fiber," *IEEE Journal of Quantum Electronics*, Vol. QE-23, 1987, pp. 1205–1210.

[15] Chraplivy, A. R., "Limitations in Lightwave Communications Imposed by Optical Fibers Nonlinearities," *IEEE J. Lightwave Techn.*, Vol. LT-8, 1990, pp. 1548–1557.

[16] Buck, J. A., *Fundamentals of Optical Fibers*, New York: Wiley, 1995.

CHAPTER 5
Optical Transmission Systems Engineering

This chapter deals with basic principles of optical transmission systems engineering. It introduces the bit error rate as an ultimate measure of the transmission system performance. It also introduces fundamental mathematical formulas that can be used to evaluate the system performance in commonly occurring scenarios. The chapter also contains a description of simulation tools and computer-aided methods that can be used in the systems engineering process.

The optical receiver terminates the lightwave path and separates the signal from all corruptive additives. The incoming optical signal is first converted in the photocurrent by the photodiode and then is processed electronically in subsequent functional blocks (refer to Figure 5.1). Optical noise, such as the intensity noise or amplified spontaneous emission, is also converted to the electrical current. In addition to incoming optical noise, some receiver-specific noise components, such as quantum shot noise and thermal noise, have also been generated during the photodetection process. The assessment of different impairments that degrade the system performance has been done in Chapter 4.

5.1 Transmission Quality Definition

The functional scheme of an optical receiver is shown in Figure 5.1. An incoming optical signal may be amplified by an optical preamplifier before it comes to the photodiode. In addition, it can be processed optically at the receiving side by some other elements, such as optical demultiplexers or optical filters. The optical signal is first converted to photocurrent and then processed by several electronic stages. The electronic processing includes amplification in the front-end amplifier and main amplifier, noise filtering, and pulse shape equalization. The total current is eventually converted to electrical voltage that contains signal and different corruptive additives, which come together with the signal to the decision point. The decision process is based on comparison of the electrical voltage at the sampling moments with specified threshold value. The 0 bits will be recovered if the total voltage is lower than the threshold level, while the 1 bits will be recovered if the voltage at the sampling moments is higher than the threshold level. However, the result of the decision process may be deceiving since each voltage sample is a mixture of the signal and corruptive ingredients (noise, nonlinearities, and so forth) that vary in time.

The transmission quality is measured by the received SNR, which is defined as the ratio of the signal level to the noise level at the sampling points. The parameter

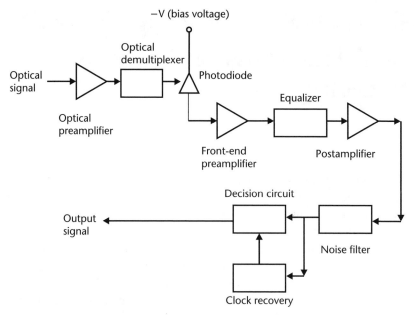

Figure 5.1 Functional scheme of an optical receiver.

that is most often used to measure the transmission performance is the BER. The value of BER is interrelated with SNR and defines the probability that a digital signal space (or 0 bits) will be mistaken for a digital signal mark (or 1 bits), and vice-versa. The fluctuating signal levels that correspond to 1 or 0 bits can be characterized by corresponding probability density functions, as shown in Figure 5.2. These levels fluctuate around their average values I_1 and I_0, which are associated with 1 and 0 bits, respectively.

These two currents can be expressed as $I_0 = RP_0$ and $I_1 = RP_1$ where P_0 is the optical power during 0 bit, P_1 is the incoming optical output power during 1 bit, and R is the responsivity of the photodiode. Any current fluctuations around the average

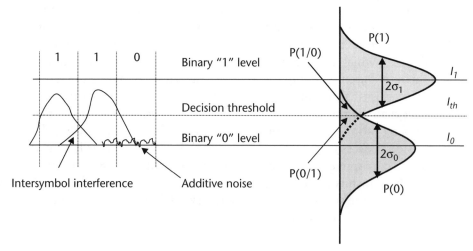

Figure 5.2 Probability density functions related to 1 and 0 bits.

value are associated with the noise. The noise intensity can be characterized by standard deviations σ_1 and σ_0, which are related to 1 and 0 bits, respectively. At the same time, the noise power associated with 1 and 0 bits can be characterized by variances $(\sigma_1)^2$ and $(\sigma_0)^2$.

The current levels are sampled at moments corresponding to the recovered signal clock at the decision circuit, and compared with the threshold value I_{th}. Bit 1 is recovered if the sampled value is higher than the threshold, while 0 bit is recovered if the sampled value is lower than the threshold. The decision is correct if it coincides with the situation at the transmitting side. However, an error occurs if a 1 bit is recovered when a 0 bit was sent. The reason for this error lies in the fact that fluctuating current at the decision instant was high enough to cross the threshold value and was recognized as the level associated with the 1 bit. Another false decision occurs if bit 0 was recovered when bit 1 was sent. Such a decision was done since the current fluctuation around the average value I_1 was relatively high in negative direction. It eventually went below the threshold level, and when compared to threshold, a false decision was made.

The BER accounts for both cases of false decision. The total probability of a false decision can be expressed as

$$BER = p(1)P(0/1) + p(0)P(1/0) = 0.5[P(0/1) + P(1/0)] \tag{5.1}$$

where $p(0)$ and $p(1)$ are probabilities that 0 and 1 bits were received. We can safely assume that $p(0) = p(1) = 0.5$, which applies for a longer data bit stream. The probabilities $P(0/1)$ and $P(1/0)$ are respectively related to cases when 0 was recovered while 1 was sent, and when 1 was recovered while 0 was sent. Probability $P(0/1)$ is represented by the area under the $P(1)$ function that is placed below the threshold level, as shown in Figure 5.2. At the same time the probability $P(1/0)$ can be identified as the area under function $P(0)$ that lies above the threshold line.

The calculation of BER by (5.1) involves both the signal and noise parameters. Signal is characterized by average values I_1 and I_0, while the total noise is characterized by standard deviations σ_1 and σ_0, which depend on the intensity of different noise components that might contribute to fluctuation of the total current. Therefore, it is

$$\sigma_1 = \sqrt{\langle i_1^2 \rangle_{total}} \tag{5.2}$$

$$\sigma_0 = \sqrt{\langle i_0^2 \rangle_{total}} \tag{5.3}$$

where $i_{1,total}$ and $i_{0,total}$ are the fluctuating currents that are related to 1 and 0 bits, respectively.

The statistics of the current fluctuations at the sampling points are rather complex, and an exact calculation of BER is rather tedious. However, there are several fairly good approximations that are used so far to evaluate the BER in optical receivers [1–3]. The simplest yet effective method is based on the assumption that both probability functions related to noise are the Gaussian distributions, which are characterized by the mean and standard deviation (refer to Figure 5.2).

The Gaussian model for noise functions leads to the following expressions for $P(0/1)$ and $P(1/0)$ [1, 2]:

$$P(0/1) = \frac{1}{\sigma_1 \sqrt{2\pi}} \int_{-\infty}^{I_{th}} \exp\left[-\frac{(I - I_1)^2}{2\sigma_1^2}\right] dI = \frac{1}{2} erfc\left(\frac{I_1 - I_{th}}{\sigma_1 \sqrt{2}}\right) \quad (5.4)$$

$$P(1/0) = \frac{1}{\sigma_0 \sqrt{2\pi}} \int_{I_{th}}^{\infty} \exp\left[-\frac{(I - I_0)^2}{2\sigma_0^2}\right] dI = \frac{1}{2} erfc\left(\frac{I_{th} - I_0}{\sigma_0 \sqrt{2}}\right) \quad (5.5)$$

where $erfc(x)$ is the complementary error function, which is defined as [4]

$$erfc(x) = \frac{2}{\sqrt{\pi}} \int_x^{\infty} e^{-y^2/2} dy \quad (5.6)$$

Both probabilities from (5.4) and (5.5) depend on the threshold value I_{th}, which means that the threshold value can be adjusted in order to reduce the probability of false detection. The threshold adjustment can be done by equalizing arguments in (5.4) and (5.5), which leads to

$$\frac{(I_1 - I_{th})^2}{2\sigma_1^2} = \frac{(I_{th} - I_0)^2}{2\sigma_0^2} \quad (5.7)$$

An optimum threshold value obtained from (5.7) is

$$I_{th} = \frac{\sigma_1 I_0 + \sigma_0 I_1}{\sigma_1 + \sigma_0} \quad (5.8)$$

Equations (5.4) to (5.8) can be used to transform (5.1) into form

$$BER(Q) = \frac{erfc(Q/\sqrt{2})}{2} \approx \frac{\exp(Q^2/2)}{Q\sqrt{2\pi}} \quad (5.9)$$

The approximate expression on the right side of above equation is reasonably accurate for values $Q > 4$.

The so-called Q-factor from (5.9) is defined as

$$Q = \frac{I_1 - I_0}{\sigma_1 + \sigma_0} \quad (5.10)$$

This parameter is often taken as a direct measure of the SNR, while the exact relationship between Q-factor and SNR depends on the detection scheme that is used. The Q-factor can be evaluated experimentally through a so-called eye diagram presented at the oscilloscope screen. The eye diagram is obtained by superposition of several sequences of the received signal waveform on top of each other. Each sequence is usually several bits long. The eye diagram should be as open as possible and as clear as possible, and it can be very useful for estimating the impact of different impairments (refer to Section 7.9 and Figure 7.4).

Functional dependence $BER(Q)$ is shown in Figure 5.3 and serves as one of the most important tools used in transmission systems engineering. This function returns several useful reference points, and they are:

- $BER = 10^{-9}$, which corresponds to $Q = 6$. If expressed in decibels, this becomes $20 \log(6) = 15.65$ dB.
- $BER = 10^{-12}$, which corresponds to $Q = 7$, or 16.90 dB.
- $BER = 10^{-15}$, which corresponds to $Q = 8$, or 18.06 dB.

Please notice that factor 20, rather than 10 was used to calculate the decibels value. It is because Q-factor is related to the electrical level, while $20\log(Q)$ measures the power level related to electrical parameters (voltage, current). It should be also mentioned that Q^2 is sometimes used in association with the optical domain, which is based on the fact that the value $20\log(Q)$ is equal to $10\log(Q^2)$.

The performance of the optical receiver can be evaluated by the parameter known as receiver sensitivity. The receiver sensitivity P_R is defined as the average power needed to achieve BER that is lower or equal to a specified value; that is,

$$P_R = \frac{P_0 + P_1}{2} = \frac{N_{photons} h\nu}{2T} = \langle N_{photons} \rangle B h\nu \qquad (5.11)$$

where P_1 and P_0 are power levels related to 1 and 0 bits, respectively, T is the bit duration, B is signal bit rate, $N_{photons}$ is the average number of photons carried by each 1 bit, $<N_{photons}> = N_{photons}/2$ is the average number of photons extended over streams of 1 and 0 bits.

Receiver sensitivity is very often specified with respect to the three values of BER written above. A minimum number of photons that is needed to achieve specified BER could be evaluated through (3.17), which can be rewritten as

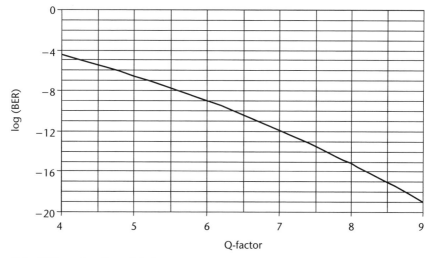

Figure 5.3 BER as a function of Q-factor.

$$P(n) = \frac{N_{photons}^n e^{-N_{photons}}}{n!} \quad (5.12)$$

The above equation returns the probability of generating n electron-hole pairs for the average number $N_{photons}$ of photons carried by each 1 bit. The BER of an ideal optical receiver can be calculated by using (5.1) and (5.12). The probability P(1/0) will be zero, since it is not possible to generate any electrons if there are no incoming photons. On the other hand, there is some probability that zero level will be recognized even if there are incoming photons. This probability can be obtained from (5.12) for n = 0. Therefore, the receiver sensitivity of an ideal optical receiver can be calculated as

$$BER = 0.5[P(0/1) + P(1/0)] = \frac{e^{-N_{photons}}}{2} \quad (5.13)$$

Equations (5.11) to (5.13) can be used to establish initial reference points related to the receiver sensitivity of an ideal optical receiver, and to specified bit rates. It is done below with respect to three exemplary high-speed bit rates, and for the wavelength of 1,550 nm. The calculated of the receiver sensitivity are listed with respect to specified BER, and they are:

- For BER = 10^{-9}, $N_{photons}$ = 20 or <$N_{photons}$> = 10 is needed, which translates to the following receiver sensitivities:
  - P_R = −54.94 dBm, for signal bit rate B = 1/T = 2.5 Gbps;
  - P_R = −48.92 dBm, for signal bit rate B = 10 Gbps;
  - P_R = −42.90 dBm, for signal bit rate B = 40 Gbps.
- For BER = 10^{-12}, $N_{photons}$ = 27 or <$N_{photons}$> = 14 is needed, which translates to the following receiver sensitivities:
  - P_R = −53.64 dBm, for signal bit rate B = 2.5 Gbps;
  - P_R = −47.62 dBm, for signal bit rate B = 10 Gbps;
  - P_R = −41.59 dBm, for signal bit rate B = 40 Gbps.
- For BER = 10^{-15}, $N_{photons}$ = 34 or <$N_{photons}$> = 17 is needed, which translates to the following receiver sensitivities:
  - P_R = −52.63 dBm, for signal bit rate B = 2.5 Gbps;
  - P_R = −46.61 dBm, for signal bit rate B = 10 Gbps;
  - P_R = −40.59 dBm, for signal bit rate B = 40 Gbps.

As we can see, the number of photons that is needed to achieve specified BER could be acquired easier within longer bit intervals. That is the reason why receiver sensitivity is better for lower bit rates (better receiver sensitivity is associated with the lower power P_R).

5.2 Receiver Sensitivity Handling

Any specific practical case can be also characterized with associated receiver sensitivity, which will be different than one related to an ideal optical receiver. That is because the sensitivity of an ideal optical receiver is determined only by the quantum limit of photodetection, while the impact of any other signal impairments is not included. However, such impact should be included when considering any practical case. Different impairments will degrade receiver sensitivity, which is observed through an increase in the power P_R required to keep the specified BER. The difference between the receiver sensitivities, which are related to an ideal optical receiver and to the practical version, can be considered as the optical power penalty associated to that specific case.

The biggest contribution to the overall receiver sensitivity degradation comes from the noise side. Since the noise accompanies the signal in all practical cases, it is convenient to evaluate the receiver sensitivity degradation due to noise impact and establish new reference points that are more relevant from the engineering perspective than the ones related to an ideal receiver case.

5.2.1 Receiver Sensitivity Defined by Shot Noise and Thermal Noise

The impacts of the thermal and quantum shot noise components should be evaluated first since they are present in any photodetection case. This can be done by considering a direct detection scheme with no optical preamplification involved. We can assume that the signal has an indefinite extinction ratio, which means that the power P_0 carried by "0" bits can be neglected. In such a case, the receiver sensitivity from (5.11) becomes

$$P_R = \frac{P_1}{2} = \frac{RI_1}{2} \tag{5.14}$$

where R is photodiode responsivity given in A/W.

We can use values from Table 4.1 to recognize the noise components that are dominant in different detection scenarios. The other noise components that have a smaller contribution can be neglected. Accordingly, one can recognize that the thermal noise dominates for 0 bits in the direct detection scenario, while both the thermal noise and quantum shot noise contribute to the total noise for 1 bits. In addition, the impact of the dark current noise can be neglected. Therefore, the parameters that define Q-factor will have the following values:

$$I_1 = R\langle M \rangle P_1 = 2\langle M \rangle R P_R \tag{5.15}$$

$$\sigma_1^2 = \langle i_1^2 \rangle_{total} = \langle i_1^2 \rangle_{sn} + \langle i^2 \rangle_{the} = 2q\langle M \rangle^2 F(M) I_1 \Delta f + \frac{4k\Theta F_{ne} \Delta f}{R_L} \tag{5.16}$$

$$\sigma_0^2 = \langle i_0^2 \rangle_{total} = \langle i^2 \rangle_{the} = \frac{4k\Theta F_{ne} \Delta f}{R_L} \tag{5.17}$$

where <M> is the average value of photodiode amplification parameter, while $F(M)$ is the noise factor. The amplification parameter is always larger than 1 if APD is used, while it equals 1 if the PIN photodiode is employed. The noise factor $F(M)$ has unity if it is related to the PIN photodiodes, and it can be evaluated by using (3.26) and (3.27) if it is related to APD.

The Q-factor in this case can be expressed as

$$Q = \frac{I_1}{\sigma_1 + \sigma_0} = \frac{2\langle M\rangle RP_R}{\left[2q\langle M\rangle^2 F(M)(2\langle M\rangle RP_R)\Delta f + \dfrac{4k\Theta F_{ne}\Delta f}{R_L}\right]^{1/2} + \left[\dfrac{4k\Theta F_{ne}\Delta f}{R_L}\right]^{1/2}} \quad (5.18)$$

It is possible to solve the above equation with respect to P_R, and it becomes [5]

$$P_R = \frac{Q\sqrt{\langle i^2\rangle_{the}}}{\langle M\rangle R} + \frac{qQ^2 F(M)\Delta f}{R} = \frac{1}{\langle M\rangle}\left[\frac{Q(4k\Theta F_{ne}\Delta f)^{1/2}}{R\sqrt{R_L}}\right] + \frac{qQ^2 F(M)\Delta f}{R} \quad (5.19)$$

The thermal noise term dominates in (5.19) if PIN photodiode is used, since $\langle M\rangle = F(M) = 1$, and such a detection scenario is recognized as the thermal-noise limited case [5, 6]. The receiver sensitivity in the thermal noise limited case can be calculated by neglecting the shot noise contribution in (5.19), which leads to

$$P_{R,PIN} \approx \frac{\sigma_{the}Q}{R} = \frac{Q(4k\Theta F_{ne}\Delta f)^{1/2}}{R\sqrt{R_L}} \quad (5.20)$$

Therefore, the receiver sensitivity is determined by the receiver bandwidth Δf, the load resistor R_L, and the noise figure of the front-end amplifier F_{ne}. We can compare the thermal noise limited case with an ideal optical receiver case by assuming that $\Delta f = B$, although the receiver bandwidth depends on the modulation format and can range from $0.5B$ to B (refer to Section 5.4.1).

The receiver sensitivity for the thermal noise limited case and sensitivity degradation from the value associated with an ideal receiver is shown in Figure 5.4 for several bit rates of practical interest (i.e., for $B = 0.1, 1, 2.5, 10,$ and 40 Gbps) and BER = 10^{-12}. It was assumed that $R = 0.8$ A/W and $R_L = 50\Omega$. Please notice that the sensitivity degradation is smaller for higher bit rates. That is because in this case the

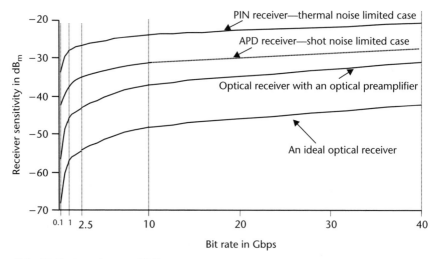

Figure 5.4 Optical receiver sensitivity.

noise power increases in proportion to square root of the signal bandwidth (bit rate), rather that in proportion to the bit rate.

The SNR in the thermal noise limited case can be calculated as

$$SNR = \frac{I_1^2}{\sigma_1^2} = 4Q^2 \qquad (5.21)$$

Accordingly, the following SNR values can be associated with Q-factors:

- SNR = 144 or 21.58 dB for Q = 6, which provides BER = 10^{-9};
- SNR = 196 or 22.92 dB for Q = 7, which provides BER = 10^{-12};
- SNR = 256 or 24.08 dB for Q = 8, which provides BER = 10^{-15}.

If APD are used, the avalanche amplification will enhance the signal by factor $<M>$. At the same time, the additional shot noise, which is proportional to noise factor $F(M)$, will be generated. In this case the shot noise contribution may become comparable or even larger than the thermal noise contribution. It is possible to find an optimum amplification factor $<M>_{opt}$ that will optimize the receiver sensitivity, while minimizing the function $P_R(<M>)$ given by (5.19). For that purpose, it is useful to employ (3.26) and express $F(M)$ as

$$F(M) = k_N \langle M \rangle + (1 - k_N)\left[2 - \frac{1}{\langle M \rangle}\right] \qquad (5.22)$$

Please recall that parameter k_N is referred to as the ionization coefficient that measures the ability of a carrier to generate other carriers in the avalanche amplification process. The ionization coefficient takes the values in the range from 0 to 1 (refer to Table 3.1). The optimization procedure applied to (5.19) leads to the following value of the APD receiver sensitivity [5, 6]:

$$P_{R,APD} = \frac{qQ^2\left[2k_N \langle M \rangle_{opt} + 2(1 - k_N)\right]\Delta f}{R} \qquad (5.23)$$

where an optimum value of the gain coefficient is given as

$$\langle M \rangle_{opt} = \left(\frac{\sqrt{\langle i^2 \rangle_{the}}}{k_N q Q \Delta f} + \frac{k_N - 1}{k_N}\right)^{1/2} \approx \left(\frac{\sqrt{\langle i^2 \rangle_{the}}}{k_N q Q \Delta f}\right)^{1/2} \qquad (5.24)$$

We can use the noise parameters from Table 3.1 to evaluate sensitivity of the APD optical receivers. It is easy to show that the APD receiver sensitivity is typically enhanced by at least 5 to 10 dB as compared to sensitivity of the PIN-based optical receivers. This benefit is limited to bit rates up to 10 Gbps, since the APD frequency range prevents them to be deployed in optical receivers operating efficiently at bit rates higher than 10 Gbps. The receiver sensitivity of an optimized APD-based optical receiver is shown in Figure 5.4. The application area is limited to 10 Gbps, and that is the reason why a dashed line is drawn in Figure 5.4 when referring to the APD case.

5.2.2 Receiver Sensitivity Defined by Optical Preamplifier

Receiver sensitivity can be greatly enhanced in cases where an optical amplifier is employed in front of the photodiode. This method, also known as optical preamplification, is the most efficient if it is used in combination with PIN photodiodes, since combination with APD could introduce relatively high shot noise and diminish the benefit of preamplification. However, if APD is used in combination with optical preamplifiers, the amplification coefficient <M> should be adjusted to relatively low values. For example, it might be a good choice to keep <M> to be lower than 5.

Receiver sensitivity can be evaluated by using the same approach that was used in the previous section. We can again assume that there is an indefinite extinction ratio, which means that the power P_0 carried by 0 bits can be neglected. In addition, Table 4.1 can be used to estimate the significance of the individual noise components involved in this case. As we see, the beat-noise components will be the strongest ones. This applies even if APD is used since the avalanche gain should be adjusted to lower values.

For a detection scheme with optical preamplification we have again that $P_R = P_1/2$, while the noise parameters can be obtained from (4.8) and (4.9). Therefore, it is

$$I_1 = RGP_1 = 2GRP_R \tag{5.25}$$

$$\sigma_1^2 = \langle i_1^2 \rangle_{total} = \langle i^2 \rangle_{sig-sp} + \langle i^2 \rangle_{sig-sp} = 8R^2 GP_R S_{sp}\Delta f + 2R^2 S^2{}_{sp}[2B_{op} - \Delta f]\Delta f \tag{5.26}$$

$$\sigma_0^2 = \langle i_0^2 \rangle_{total} = \langle i^2 \rangle_{sp-sp} = 2R^2 S^2{}_{sp}[2B_{op} - \Delta f]\Delta f \tag{5.27}$$

$$Q = \frac{I_1}{\sigma_1 + \sigma_0} = \frac{2GRP_R}{\left[\langle i^2 \rangle_{sig-sp} + \langle i^2 \rangle_{sp-sp}\right]^{1/2} + \left[\langle i^2 \rangle_{sp-sp}\right]^{1/2}} \tag{5.28}$$

Please notice that noise related to 0 bits is determined just by spontaneous-spontaneous beat component, since we neglected the power P_0 carried by 0 bits. The receiver sensitivity extracted from (5.25) to (5.28) is

$$P_R = \frac{2S_{sp}\Delta f\left[Q^2 - Q(B_{op}/\Delta f)^{1/2}\right]}{G-1} = F_{no}hf\Delta f\left[Q^2 - Q(B_{op}/\Delta f)^{1/2}\right] \tag{5.29}$$

where the following relationship between the power density of spontaneous emission noise and the optical amplifier noise figure was used:

$$S_{sp} = \frac{F_{no}hf(G-1)}{2} \tag{5.30}$$

It is clear from (5.29) that optical amplifiers with low noise figure should be used in order to fully utilize benefits from optical amplification. In addition, it is necessary to adjust the optical filter bandwidth to be as close as possible to the signal bandwidth.

The receiver sensitivity related to the receiver with optical preamplification has been calculated for five exemplary bit rates, and for BER = 10^{-12}, and shown in Figure 5.4 in parallel with results related to other cases. All results were obtained by assuming that optical filter has a bandwidth $B_{op} = 2\Delta f$. The results shown in Figure 5.4 prove that optical amplification is very beneficial since receiver sensitivity associated with this detection scheme is the one closest to sensitivity of an ideal optical receiver.

5.2.3 Optical Signal-to-Noise Ratio

The optical amplification process is accompanied by the generation of ASE noise that accumulates along the transmission line. The power of the ASE noise is calculated with respect to a specified optical bandwidth B_{op}. This noise can be measured as well, which is usually done by the optical spectrum analyzer (OSA) that registers the power of the ASE noise with respect to assigned optical bandwidth region. In addition, it is possible to calculate and measure the ratio of the optical signal power and the ASE noise at any specific point along the lightwave path.

The optical signal-to-noise ratio (OSNR) measured along the lightwave path can be defined as

$$\text{OSNR} = \frac{P_S}{P_{ASE}} = \frac{P_S}{S_{sp} B_{op}} = \frac{P_S}{2n_{sp} hf(G-1)B_{op}} \quad (5.31)$$

where P_s is the optical signal power at any specific point, n_{sp} is the spontaneous emission factor, G is the optical amplifier gain, B_{op} is the bandwidth of the optical filter, h is the Planck's constant, and f is frequency of the optical signal. The factor 2 in the denominator at the right side of (5.31) accounts for two polarization modes of ASE, where each of them carries optical power equal to $n_{sp} hf(G-1)B_{op}$. Optical filter bandwidth, which is related to bandwidth of the OSA, is usually declared during the measurement process. For example, there are a number of measurements in the optical domain that are done within the optical bandwidth equal 0.1 nm, which is approximately 12.5 GHz if applied to the 1,550-nm wavelength region.

OSNR can be also measured at the receiver entrance point just before photodetection takes place. In such a case, the power of the optical signal can be related to the receiver sensitivity by using relation $P_S \sim 2P_R$, while the OSNR can be expressed by using (5.29) and (5.31):

$$\text{OSNR} \approx \frac{P_R}{n_{sp} hf(G-1)B_{op}} = \frac{2\Delta f \left[Q^2 - Q\left(B_{op} / \Delta f \right)^{1/2} \right]}{(G-1)B_{op}} \quad (5.32)$$

The OSNR value will be eventually converted to an electrical equivalent that defines both the Q-factor and BER. The Q-factor in this case will be mainly determined by the ASE noise that is eventually converted to the beat-noise components. The following relationship between the Q-factor and OSNR can be established if we neglect all other noise contributions except the beat-noise components [7]:

$$Q = \frac{2OSNR\sqrt{B_{op}/\Delta f}}{1+\sqrt{1+4OSNR}} \qquad (5.33)$$

Equation (5.33) can be further simplified if we assume that there is just one dominant noise term. If we refer to (5.28), the dominant term would be signal-spontaneous beat-noise, which leads to following simplified formula:

$$Q = \sqrt{\frac{OSNR}{2}\frac{B_{op}}{\Delta f}} \qquad (5.34)$$

We can conclude that the Q-factor is the most important parameter since it is directly correlated with BER, and can provide information about system performance. From a systems engineering and performance monitoring perspective, it became necessary to have a good correlation between Q-factor and both SNR and OSNR in different detection scenarios. In some situations it can be done by using approximate empiric formulas, such as those given by (5.21) and (5.34), but in many other scenarios it is useful to establish more precise relationship between Q-factor and OSNR. The relation between Q-factor and OSNR is transmission case specific, which means that it should be calculated for input parameters related to the transmission case in question. For example, (5.19), (5.20), (5.23), and (5.29) can be used to identify the receiver sensitivity related to a specific performance requirement defined through BER or Q-factor. Such receiver sensitivity can be used afterwards as a reference to identify the SNR and OSNR.

More precise correlation between Q-factor, SNR, and OSNR can be established by using numerical calculations, possibly with some measured data inputs. This process can be also performed on a dynamic base. Namely, OSNR can be measured along the lightwave path, while the values for Q parameter and BER can be calculated for that specific scenario. The calculated values of the Q-factor can be compared with established reference and used in the system performance monitoring process (refer to Section 5.4.4). This approach is obviously more sophisticated, but it might be an inevitable part of any high-speed transmission systems engineering, especially in an optical networking environment.

5.3 Power Penalty Handling

In the previous section, the receiver sensitivity was defined with respect to the receiver noise for several basic detection scenarios. The highest sensitivity (the lowest value of the received optical power that is needed) corresponds to an ideal optical receiver, and it is determined by the quantum limit of photodetection. That sensitivity is degraded in real optical receivers due to the impact of two principal noise contributions: the thermal noise (in PIN photodiodes) and quantum shot noise (in APD). Optical amplifiers can considerably enhance the receiver sensitivity since the

5.3 Power Penalty Handling

benefit due to optical signal amplification is still larger than the penalty due to additional beat-noise components.

By including the impact of the noise components in three basic real case detection scenarios, new reference points have been established. The next step in engineering considerations is to include the impact of other impairments that degrade the receiver sensitivity. Some of them, such as a finite extinction ratio and chromatic dispersion, are very important and should be considered in most practical situations, while the others, such as fiber modal noise and timing jitter, might play an important role only in some specific cases.

All impairments degrade the receiver sensitivity by increasing the value of the optical power P_R needed to achieve a specified BER, which was evaluated in the previous section. It means that higher optical power is needed to achieve the same transmission quality. Accordingly, each individual impairment will cause receiver sensitivity degradation, which is evaluated by the power penalty. The power penalty is equal to the increase in signal power that is needed to keep the Q-factor and BER at the same level that would exist if no impairments were present [8].

This is illustrated in Figure 5.5, which connects the receiver sensitivity and sensitivity degradation with Q-factor and BER. As an example, if BER = 10^{-12} is specified, it would need for the Q-factor to be equal or larger than 7. This value of Q-factor can be translated to the receiver sensitivity, or to the OSNR required to satisfy the system BER specification. Such a correlation, which leads to identification of the required OSNR, is shown in Figure 5.5. Any impairment will degrade receiver sensitivity by some extent, which will lead to a new lower value of Q-factor and to higher BER. Figure 5.5 also shows several impairments acting together to

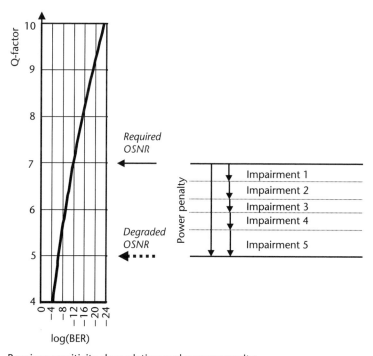

Figure 5.5 Receiver sensitivity degradation and power penalty.

force the Q-factor to take a lower value (Q = 5), while BER becomes much smaller (BER ~ 10^{-4}).

Receiver sensitivity degradation can be caused by different impairments that are related either to the signal transmission or to nonideal conditions during modulation and photodetection processes. Receiver sensitivity degradation related to transmission effects can be expressed through the associated power penalties. These power penalties were evaluated in Chapter 4 for all transmission related impairments. In the next section we will estimate the power penalties related to other relevant nontransmission-related effects, and explain the process of the power margin allocation that is needed to account for different power penalties.

5.3.1 Power Penalty Due to Extinction Ratio

The extinction ratio was introduced with (3.3). It is defined by parameter $R_{ex} = P_1/P_0$, where P_1 is the power of 1 bit, while P_0 is the power of 0 bit. In essence, the extinction ratio is determined by the energy carried by 0 bits, and that is the reason why its inverse value $r_{ex} = 1/R_{ex} = P_0/P_1$ is commonly used. The energy carried by 0 bits is relatively high in direct modulation schemes if the laser is biased above or close to the threshold value. On the other hand, the extinction ratio in external modulator schemes is determined by the bias voltage and structure of the modulator.

The Q-factor associated to a finite extinction ratio can be calculated by replacing P_0 with $r_{ex}P_1$ in (5.10) and (5.11), which leads to

$$Q(r_{ex})\big|_{PIN} = \left[\frac{1-r_{ex}}{1+r_{ex}}\right]\frac{2RP_R}{\sigma_1 + \sigma_0} \quad (5.35)$$

This value can be related to the receiver sensitivity by using methods presented in Section 5.2. As an example, if detection is done by PIN photodiode we can use (5.20) to evaluate the receiver sensitivity through the Q-factor. In such a case, the receiver sensitivity is given as

$$P_{R,PIN}(r_{ex}) = \left[\frac{1+r_{ex}}{1-r_{ex}}\right]\frac{\sigma_{the}Q(0)}{R} = \left[\frac{1+r_{ex}}{1-r_{ex}}\right]P_R(0) \quad (5.36)$$

while the power penalty due to nonideal extinction ratio becomes

$$\Delta P_{ex}(r_{ex})\big|_{PIN} = 10\log\left[\frac{1+r_{ex}}{1-r_{ex}}\right] \quad (5.37)$$

The situation is more complex if APD is used for photodetection since the excitation ratio has an impact on the optimum value of the APD gain. It was first shown in [9] that such an optimum value of the APD gain decreases if the extinction ratio is not zero, which leads to the receiver sensitivity degradation. The impact of nonideal extinction ratio to receiver sensitivity of APD-based optical receivers can be approximately evaluated if we assume that the quantum shot noise is the dominant noise factor, and that $\sigma_0 \sim r_{ex}\sigma_1$. With this assumption the Q-factor can be expressed as

5.3 Power Penalty Handling

$$Q(r_{ex})\big|_{APD} \approx Q_{\inf}\left[\frac{1-r_{ex}}{1+r_{ex}}\right] \quad (5.38)$$

where $Q_{\inf}$ corresponds to an ideal case with infinite extinction ratio, which refers to (5.23). The approximate value given by (5.38) can be now inserted in (5.23). The power penalty can be calculated as a difference in receiver sensitivities that correspond to $Q_{\inf}$ and $Q(r_{ex})$.

The power penalty due to nonideal extinction ratio for the PIN- and APD-based optical receivers is shown in Figure 5.6. The inverse value of the extinction ratio for PIN-based optical receivers should be smaller than 0.06 to keep the power penalty below 0.5 dB, while it should be smaller than 0.12 to keep the power penalty below 1 dB.

5.3.2 Power Penalty Due to Intensity Noise

The intensity noise is related to intensity fluctuations of the incoming optical signal. The biggest contribution to these fluctuations comes from the light source. However, the intensity noise can be enhanced by some other effects, such as multiple light reflections along the lightwave path, and conversion of the phase noise to intensity noise. Any light intensity fluctuations are converted to the electrical noise in the photodiode and added to the noise components that already exist (i.e., to thermal noise, quantum shot noise, and beat noise).

Both the SNR and the receiver sensitivity will be degraded due to the impact of the intensity noise. An exact evaluation of the power penalty that is related to the intensity noise is rather complex. However, a simplified approach presented in [6] provides a fairly good estimate of the impact of the overall intensity noise. It was assumed in [6] that the power of the intensity noise can be simply added to the

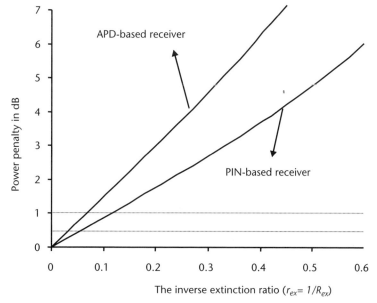

Figure 5.6 Power penalty due to nonideal extinction ratio.

powers of the thermal and shot noise, so that values of noise powers, which are related to 1 bits and 0 bits, become

$$\sigma_1^2 = \langle i_1^2 \rangle_{total}$$
$$= \langle i_1^2 \rangle_{sn} + \langle i^2 \rangle_{the} + \langle i^2 \rangle_{int} \qquad (5.39)$$
$$= 2q\langle M \rangle^2 F(M) I_1 \Delta f + \frac{4k\Theta F_{ne} \Delta f}{R_L} + (RP_1 r_{int})^2$$

$$\sigma_0^2 = \langle i_0^2 \rangle_{total} = \langle i^2 \rangle_{the} = \frac{4k\Theta F_{ne} \Delta f}{R_L} \qquad (5.40)$$

The power of the noise caused by intensity fluctuations is expressed as

$$\langle i \rangle_{int}^2 = (RP_1)^2 r_{int}^2 = \frac{(RP_1)^2}{2\pi} \int_{-\infty}^{\infty} RIN(\omega) d\omega = 2(RP_1)^2 RIN_{laser} \Delta f \qquad (5.41)$$

where $RIN(\omega)$ is the so-called relative intensity noise spectrum defined by (3.14), and r_{int} is the parameter that measures the intensity fluctuations, which was introduced by (3.15). The RIN_{laser} parameter in (5.41), which is related to the average value of the RIN spectrum, is smaller than –160 dB/Hz for high-quality lasers (which translates to $r_{int} \sim 0.004$).

Receiver sensitivity degradation due to the impact of the intensity noise can be evaluated by assuming that receiver sensitivity is $P_R = P_1/2$, and by inserting (5.39) to (5.41) into (5.10) afterwards. The following equation can easily be obtained:

$$P_R(r_{int}) = \frac{P_R(0)}{1 - r_{int}^2 Q^2} \qquad (5.42)$$

It is worth mentioning that (5.42) was obtained under the assumption that the impact of the extinction ratio can be neglected. The power penalty due to the impact of the intensity noise can be calculated as

$$\Delta P_{int} = 10 \log \left[\frac{P_R(r_{int})}{P_R(0)} \right] = -10 \log \left[1 - r_{int}^2 Q^2 \right] \qquad (5.43)$$

The power penalty calculated by (5.43) is shown in Figure 5.7 for three values of the Q-factor (i.e., for Q = 6,7,8). The power penalty is smaller than 0.5 dB for r_{int} values ranging from 0 to 0.42 if Q = 8, or for r_{int} values ranging from 0 to 0.55 for Q = 6. There is a very sharp increase in power penalty if r_{int} goes above some critical value, which is about $r_{int} = 0.12$ for Q = 8, and $r_{int} = 0.15$ for Q = 6. The power penalty can become very high and totally degrade the receiver sensitivity if parameter r_{int} lies above these critical values.

There are the other factors, such as reflections, phase noise to intensity noise conversion, the mode partition noise, and the reflection noise, which also contribute to the total intensity noise. The impact of these noise ingredients can be estimated through an effective intensity noise parameter r_{eff} defined as

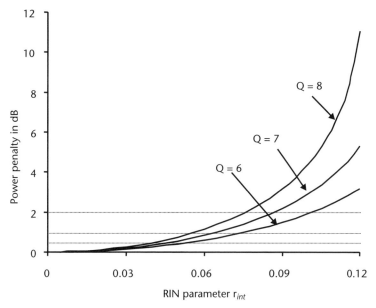

Figure 5.7 Power penalty due to intensity noise.

$$r_{eff} = \sqrt{\left(r_{int}^2 + r_{ref}^2 + r_{part}^2 + r_{phase}^2\right)} \tag{5.44}$$

where the individual components on the right side of (5.44) are related to the intensity noise, reflection noise, mode partition noise, and converted phase noise, respectively. Therefore, it is necessary to evaluate each of these individual components first, which might not be an easy task.

The problem is simplified if the DFB lasers are used in combination with single-mode fibers, since the intensity noise contribution is basically determined by the first term in brackets on the right side. That is because we assume that the reflection noise is reduced by optical isolators inserted along the lightwave path. However, in the case when either multimode lasers or VCSEL are used, the contributions from other terms from (5.44) can become dominant.

The parameter related to the reflection induced intensity noise can be expressed as

$$r_{ref} \approx \frac{(r_1 r_2)^{1/2}}{\alpha_{iso}} \tag{5.45}$$

where r_1 and r_2 are reflection coefficients of two disjoints introduced by (3.16), and α_{iso} is the attenuation coefficient of the optical isolator. This refection coefficient becomes $r_{ref} \sim 3.6\%$ or -14.4 dB if it is related to connectors that include the air/glass interface and if there is no optical isolator employed. However, it becomes comparable or less than -55 dB if there is the optical isolator in place, since the attenuation of an optical isolator is usually higher than 40 dB.

The parameter due to intensity noise introduced by the mode partition effect can be expressed as [6]

$$r_{part} \approx (k/\sqrt{2})\{1-\exp[-(\pi BLD\sigma_\lambda)^2]\} \quad (5.46)$$

where k is a coefficient that varies in the range from 0.6 to 0.8, B is signal bit rate, L is the transmission distance, D is the chromatic dispersion coefficient, and σ_λ represents the spectral linewidth of the light source. Please recall from Section 3.2.1 that the impact of the mode partition noise can be almost entirely suppressed if the product $BLD\sigma_\lambda$ is less than 0.075, which corresponds to the value $r_{part} \sim 0.15$.

5.3.3 Power Penalty Due to Timing Jitter

Timing jitter is a well-known cause of transmission system performance degradation [10]. It arises due to fluctuations of the clock recovery instants, since the sampling time also fluctuates around some reference positions. Any fluctuations in time around specified instants are directly related to the noisy nature of the incoming signal. This will cause the sampling intervals to fluctuate around the bit center, which means that the sampled value of the signal will not always be aligned with the signal maximum. The timing jitter is measured by a random time variable Δt that measures the fluctuation of any specific sampling instant from the bit center.

The impact of the timing jitter at the decision point is similar to the impact of the intensity noise since any variations in sampling times are eventually converted to intensity variations of samples. Such a negative impact can be suppressed by an increase in the signal power, which is equal to the induced power penalty. The impact of the jitter to receiver sensitivity can be analyzed by using rather complex numerical methods, such as one presented in [11]. There is also an approximate approach proposed in [6] that offers a good understanding of the statistics involved in the timing jitter process. It was assumed in [6] that the noise parameters from (5.10) can be expressed as

$$\sigma_1^2 = \langle i_1^2 \rangle_{total} = \langle i^2 \rangle_{the} + \langle i^2 \rangle_{jitt} = \frac{4k\Theta F_{ne}\Delta f}{R_L} + \langle \Delta i^2{}_{jitt} \rangle \quad (5.47)$$

$$\sigma_0^2 = \langle i_0^2 \rangle_{total} = \langle i^2 \rangle_{the} = \frac{4k\Theta F_{ne}\Delta f}{R_L} \quad (5.48)$$

where $\langle i^2 \rangle_{jitt}$ is the power of the noise component due to timing jitter, and $\langle \Delta i_{jitt} \rangle$ is the standard deviation of current fluctuations induced by time variations Δt. It is clear that the pulse shape will have a big impact on the timing jitter that is generated. The current fluctuations can be evaluated more precisely if we know the function $h_{out}(t)$ that governs the shape of the signal current. In the general case, we can assume that function $h_{out}(t)$ takes the raised cosine shape [refer to Section 7.9 and to (7.63)].

In order to evaluate the noise power, it is necessary to find the variance $\langle \Delta i^2_{jitt} \rangle$ of the stochastic variable Δi_{jitt}. It was assumed in [6] that Δi_{jitt} follows the Gaussian distribution and that its variance can be calculated as

$$\langle \Delta i^2_{jitt} \rangle = 8I_1^2\left[(B\sigma_{\Delta t})^2(\pi^2/3-2)\right]^2 \quad (5.49)$$

where $\sigma_{\Delta t}$ is the standard deviation of time fluctuations Δt.

The receiver sensitivity can be evaluated by inserting (5.47) to (5.49) in expressions given by (5.14), (5.18), and (5.20), which eventually leads to

$$Q = \frac{I_1 - \langle \Delta i_{jitt} \rangle}{\sigma_1 + \sigma_0}$$

$$= \frac{I_1\left[1 - (B\sigma_{\Delta t})^2 (2\pi^2/3 - 4)\right]}{\left[8I_1^2\left[(B\sigma_{\Delta t})^2 (\pi^2/3 - 2)\right]^2 + \frac{4k\Theta F_{ne} \Delta f}{R_L}\right]^{1/2} + \left[\frac{4k\Theta F_{ne} \Delta f}{R_L}\right]^{1/2}} \quad (5.50)$$

and

$$P_R(\sigma_{\Delta t}) = P_R(0) \frac{1 - (B\sigma_{\Delta t})^2 (2\pi^2/3 - 4)}{\left[1 - (B\sigma_{\Delta t})^2 (2\pi^2/3 - 4)\right]^2 - 8Q^2\left[(B\sigma_{\Delta t})^2 (\pi^2/3 - 2)\right]^2} \quad (5.51)$$

The power penalty due to the impact of the timing jitter can be calculated from (5.51) as

$$\Delta P_{jitt} = 10\log\left[\frac{P_R(\sigma_{\Delta t})}{P_R(0)}\right] \quad (5.52)$$

The power penalty calculated by (5.51) and (5.52) is below 0.5 dB if the RMS value of the jitter is lower than 10% of the bit interval (i.e., if $\sigma_{\Delta t} B < 0.1$). However, if the RMS value of the jitter is higher than 20% of the bit interval (i.e., if $\sigma_{\Delta t} B > 0.2$), it would produce indefinite power penalties.

Equations (5.51) and (5.52) are good for an approximate evaluation of the power penalty. More precise calculations, such as one presented in [11], shows that the power penalty is even higher than the one estimated by (5.52). However, the criteria $\sigma_{\Delta t} B < 0.1$ is quite valid to be used as a guidance in systems engineering considerations. This criterion also implies that a power margin of about 0.5 dB should be allocated to account for the impact of timing jitter.

5.3.4 Power Penalty Due to Signal Crosstalk

The crosstalk noise is related to multichannel systems and can be either "out-of-band" or "inband" in nature, as discussed in Section 3.2.9. The out-of-band crosstalk occurs when the fraction of the power of an optical channel spreads outside of the channel bandwidth and mixes with the signals of neighboring channels. Therefore, an optical receiver of any specific optical channel can capture the interfering optical power and convert it to the electrical current, which can be expressed as

$$i_{cross,out} = \sum_{n \neq m}^{M} RP_n X_n = RP \sum_{n \neq m}^{M} X_n \quad (5.53)$$

where R is photodiode responsivity, P is optical power per channel, and X_n is the portion of the nth channel power that has been captured by the optical receiver of the mth optical channel. For convenience sake, the parameter X_n can be identified as "the crosstalk ratio." The crosstalk current (or better to say the crosstalk noise) can be treated as an intensity noise. Consequently, the impact of the out-of-band crosstalk noise can be evaluated by applying (5.43). Accordingly, the following set of equations can be established:

$$r^2_{cross,out} = \left[\sum_{n\neq m}^{M} X_n\right]^2 \tag{5.54}$$

$$\langle i^2 \rangle_{cross,out} = (RP_1)^2 \, r^2_{cross,out} = (RP_1)^2 \left[\sum_{n\neq m}^{M} X_n\right]^2 \tag{5.55}$$

$$\Delta P_{cross,out} = -10\log\left[1 - r^2_{cross,out} Q^2\right] \tag{5.56}$$

The same approach can be used when considering the impact of inband crosstalk noise to the receiver sensitivity. Please recall that the inband crosstalk noise occurs when interfering optical power has the same wavelength as the wavelength of the optical channel in question—please refer to Section 3.2.9. The following set of equations can be established by applying (5.43) to this specific case:

$$i_{cross,in} = 2RP \sum_{n \neq m}^{M} \sqrt{X_n} \tag{5.57}$$

$$\langle i^2 \rangle_{cross,in} = (RP_1)^2 \, r^2_{cross,in} = 2(RP_1)^2 \left[\sum_{n\neq m}^{M} \sqrt{X_n}\right]^2 \tag{5.58}$$

$$\Delta P_{cross,in} = -10\log\left[1 - r^2_{cross,in} Q^2\right] \tag{5.59}$$

Since both out-of band and inband crosstalk can be identified as the intensity noise-like impairments, the total impact of the crosstalk can be evaluated by using an equivalent noise parameter

$$r_{cross} = \sqrt{r^2_{cross,out} + r^2_{cross,in}} \tag{5.60}$$

This parameter can be inserted in (5.59) in order to evaluate the receiver sensitivity degradation due to the total crosstalk effect, so that it becomes

$$\Delta P_{cross} = -10\log\left[1 - r^2_{cross} Q^2\right] \tag{5.61}$$

It is convenient to evaluate the total impact of the crosstalk noise by assuming that there is a single crosstalk contribution that dominates in the sums in (5.55) and (5.58), while the impact of other terms can be neglected. This assumption basically means that this single crosstalk acts as an equivalent crosstalk ratio X_{eq}. The concept of an equivalent crosstalk ratio was used in (5.56), (5.59), and (5.61) to calculate the crosstalk-related power penalties. The results, which are calculated by assuming that the Q-factor is 7, are plotted in Figure 5.8.

As we can see, the power penalty due to out-of-band crosstalk goes over the 1-dB line if the crosstalk ratio becomes larger than 6.7%. This ratio also means that the power penalty will be lower than 1 dB if the interfering optical power is at least 11.7 dB below the power level associated with the channel in question. At the same time, the power penalty crosses the 1-dB line if an inband crosstalk ratio becomes higher than 0.85% (which corresponds to an interfering optical power of 20.65 dB below the signal level). These numbers and diagrams from Figure 5.8 can be used as a reference in the component selection process. From a system point of view, the power penalty should be kept below 0.5 dB, which means that out-of-band crosstalk should be at least 13.4 dB below the signal in question, while inband crosstalk should be at least 23.5 dB below the signal in question.

5.3.5 Comparative Review of Power Penalties

Let us summarize the impact of different impairments to the optical receiver sensitivity by comparing the associated power penalties. For this purpose, it is useful to produce a joint chart of power penalties imposed by different impairments that were discussed in Chapters 4 and 5. The chart can be produced by applying the following methodology. First, the reference point for all curves from the chart will be the power penalty of 0.5 dB. Second, we can recognize a value of the defining parameter, which is associated with this power penalty, for each individual impairment. Such a value can be simply identified as the σ-value. For example, the σ-value

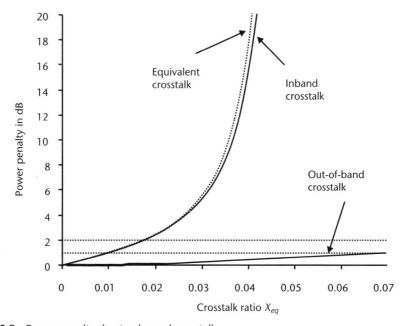

Figure 5.8 Power penalty due to channel crosstalk.

is associated with the inverse extinction ratio of 0.06, or with the RIN parameter of 0.042 (refer to Figures 5.6 and 5.7).

Finally, each specific parameter related to the 0.5-dB power penalty was increased up to three times with respect to its σ-value, and corresponding power penalties were plotted, as shown in Figure 5.9. As an illustration of the process, we can mention that the power penalty was calculated for the inverse extinction ratio of 0.06, 0.12, and 0.18, while the same was done for the RIN parameter of 0.042, 0.084, and 0.168. The plotted curves from Figure 5.9 do not have a reference meaning. However, they can be used to observe how sensitive the power penalty is with respect to the relative change of each individual impairment.

As we see, the power penalty, as a function of σ-value, is the most sensitive to changes that occur with respect to four-wave mixing, out-of-band crosstalk, the intensity noise, and the timing jitter. On the other hand, it is less sensitive to changes with respect to the extinction ratio, inband crosstalk, and chromatic dispersion (which was calculated for the chirpless input signal). Finally, it is moderately sensitive to changes in polarization mode dispersion and self-phase modulation.

The family of curves from Figure 5.9 helps us to recognize that different types of impairments should be treated differently. The parameters that induce bigger changes to the power penalty function should be kept under tighter control. On the other hand, the impact of parameters that induce smaller changes to the power penalty function in Figure 5.9 can be accounted for by a proper power margin allocation.

It is also important to notice that the evaluation process of different impairments presented in this chapter can be treated as the worst case scenario if we assume that associated power penalties will simply add to each other. However, several impairments acting together will rarely produce an overall impact that resembles the worst case scenario. Accordingly, a statistical approach, such as one using the Monte-Carlo simulation, can be applied to evaluate the overall power penalty as a function of the number of parameters.

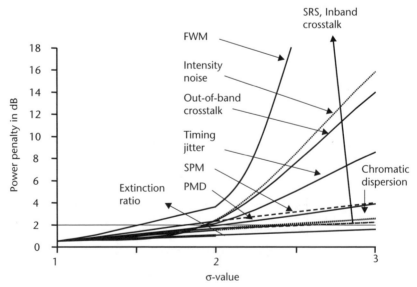

Figure 5.9 The impact of deviation of impairments from values associated with a 0.5-dB power penalty.

An engineering summary of different impairments is shown in Table 5.1. It outlines the significance of each impairment and the engineering approach with respect to it. It is important to notice that the margin allocation plays one of the most important roles in the overall engineering process. A reasonably accurate process of margin allocation includes the following steps:

- Allocate margin for the more predicable impacts. Allocate up to 0.5 dB to compensate for nonideal extinction ratio, up to 1 dB to compensate for PMD effect, and up to 0.5 dB to compensate for chromatic dispersion.
- Use statistical approach to allocate the power margin that covers the effects of the crosstalk noise, stimulated Raman scattering, and the intensity noise. It is more reasonable to allocate 1 dB to cover all three effects than to allocate 0.5 dB per each of them.

Table 5.1 Summary of Different Impairments

Impairment	Significance	Power Penalty That Can Be Tolerated	Will an Increase in Signal Power Help?	Recommended Engineering Approach
Chromatic dispersion	High	0.5 dB	Yes	Allocate 0.5 dB and compensate the rest
PMD	High for bit rates above 10 Gbps	1 dB	Yes	Allocate 1 dB and compensate the rest
Attenuation	High	Difference between the incoming power level and receiver sensitivity	Yes	Compensate by amplification
Thermal noise	Moderate	N/A (serves as a reference)	Yes	Use better front-end amplifiers
Shot noise	High	N/A (serves as a reference)	Yes	Optimize avalanche gain coefficient
Beat noise	High	N/A (serves as a reference)	No	Decrease optical amplifier noise figure
Intensity noise	Moderate	0.5 dB 1 dB in some cases	Yes	Use lasers with RIN 150 dB
Extinction ratio	Moderate	0.5 dB	Yes	Allocate 0.5-dB power margin
Crosstalk	High	0.5 dB	No	Select better elements, allocate margin
SPM	Moderate	0.5 dB	It may	Allocate 0.5-dB margin and use special modulation methods if needed
FWM	High	N/A	No, it is just the opposite	Operate outside critical wavelength region
SRS	Moderate	0.5 dB	No	Keep power low
SBS	Low	No	No	Use dithering
XPM	Highest in multichannel systems	No	No	Decrease power

- Account for component aging and the polarization-dependent losses, which might occur along the lightwave path, by allocating up to 3 dB for each of them.
- Assign the system margin of about 3 to 4 dB to account for all other effects that might happen during the system operation, such as temperature changes, unexpected fiber cuts, and the discrepancy that might occur during the module replacement.
- Allocate an additional power margin of about 2 dB in some special cases, such as use of VCELS over multimode optical fibers.

5.3.6 Handling of Accumulation Effects

In general, there are a number of cascaded optical amplifiers along the lightpath. We can assume that they are spaced l km apart, where parameter l defines the span length, as shown in Figure 5.10. The span loss between two amplifiers is $\Gamma_{span} = \exp(-\alpha l)$, assuming that the fiber attenuation coefficient is α. The purpose of each in-line optical amplifier is to amplify an incoming optical signal just enough to compensate for the loss at the previous span.

However, each amplifier also generates some spontaneous emission noise in parallel with the optical signal amplification. Both the signal and the spontaneous emission noise continue to propagate together to be amplified by the following optical amplifiers. The buildup of amplifier noise is the most critical factor in achieving specified systems performance when dealing with longer distances. The ASE noise will be accumulated and increased after each subsequent span, thus contributing to the total noise and to SNR deterioration. In addition, the ASE power will contribute to the saturation of optical amplifiers, thus reducing the amplifier gain and the amplified signal level.

This process is illustrated in Figure 5.11. As we see, the optical signal that starts from some launched level will experience optical attenuation over the span length before it is enhanced by an optical amplifier. The optical amplifier helps to restore the output power level, but generates the spontaneous emission noise by itself, and amplifies the spontaneous emission noise coming from the previous amplifier. The ASE noise level will increase after each amplifier and will eventually force optical amplifiers into saturation regime. Since the output power stays on the same level, the increase in ASE will diminish the optical signal level. The end result will be a decrease in the OSNR, as illustrated in Figure 5.11.

Described picture can be transferred to a mathematical relation that connects the signal and noise parameters with respect to the transmission length. We can assume that the optical amplifier gain G is adjusted just to compensate for the span

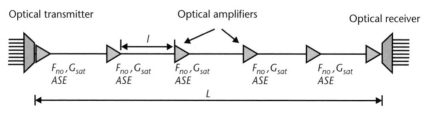

Figure 5.10 Transmission system employing optical amplifiers.

5.3 Power Penalty Handling

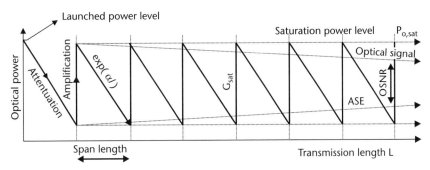

Figure 5.11 Signal amplification and noise accumulation in an amplified transmission system.

loss Γ_{span}. Otherwise, if the gain was larger than the span loss, the signal power would increase gradually throughout the amplifier chain forcing the amplifier saturation regime. Recall that the saturation means that the amplifier gain drops as input power increases. At the end of the settlement process, the amplifiers will enter into the saturation regime, while the total gain will drop from its initial value $G_0 = G_{max}$ to a saturated value G_{sat} [refer to (3.10)].

The output power from the optical amplifier is determined by the saturated value $P_{o,sat}$, which is given by (3.9). From a practical point of view, it is important to consider just a spatial steady-state condition that was reached in long-haul transmission systems, in which both the saturated output power $P_{o,sat}$ and the gain G_{sat} remain the same, thus governing the signal amplification process. The output power per channel in this case will be $P_{out} = P_{o,sat}/M$, where M is the number of optical channels that are multiplexed and amplified together.

As already mentioned, OSNR gradually decreases along the lightwave path, since the accumulated ASE noise gradually makes up a more significant portion of the limited total power from an optical amplifier. The steady-state gain, or saturated gain, will be slightly smaller than the span signal loss, due to the added noise at each amplifier point. Therefore, the best engineering approach is to choose a saturated gain that is very close to the span loss. In such a case, the following balance between the signal and accumulated noise can be established:

$$P_{o,sat} \exp(-\alpha l) G_{sat} + P_{sp} = P_{o,sat} \qquad (5.62)$$

where P_{sp} is the ASE power defined by (3.36). In this case, the ASE power can be expressed as

$$P_{sp}(v) = 2 S_{sp}(v) B_{op} = (G_{sat} - 1) F_{no} h v B_{op} \qquad (5.63)$$

where $S_{sp}(v)$ is the spectral density of the spontaneous emission that is given by (3.32), v is the optical frequency, B_{op} is the optical filter bandwidth, $F_{no} = 2n_{sp}$ is the noise figure of the optical amplifier, and n_{sp} is the spontaneous emission factor defined by (3.34). It is important to recall that the saturated gain also satisfies (3.10), which can be rewritten as

$$G_{sat} = 1 + \frac{P_{sat}}{P_{in}} \ln \frac{G_{max}}{G_{sat}} = 1 + \frac{P_{sat}}{P_{o,sat} \exp(-\alpha l)} \ln \frac{G_{max}}{G_{sat}} \qquad (5.64)$$

Equation (5.63) can be now used to evaluate the total noise power that is associated with a steady state reached at the end of the transmission line, and it becomes

$$NP_{sp}(v) = N(G_{sat} - 1)F_{no} h v B_{op} = (L/l)(e^{\alpha l} - 1)F_{no} h v B_{op} \qquad (5.65)$$

where $N = L/l$ represents the number of spans at the transmission line that is equal to the number of optical amplifiers employed along the line. Please note that the saturated gain value was chosen just to compensate for the signal loss at the preceding span.

The OSNR, calculated per channel basis at the end of the transmission line, can be expressed as

$$OSNR = \frac{P_{s,sat}/M - F_{no} h v B_{op}(e^{\alpha l} - 1)L/l}{F_{no} h v B_{op}(e^{\alpha l} - 1)L/l} = \frac{P_{ch} - F_{no} h v B_{op}(e^{\alpha l} - 1)N}{F_{no} h v B_{op}(e^{\alpha l} - 1)N} \qquad (5.66)$$

where $P_{o,sat}$ is the total launched optical power at the optical amplifier output, and P_{ch} is the launched power within an individual optical channel.

In order to fulfill the specified OSNR, the launched channel power should satisfy the equation

$$P_{ch} \geq (OSNR + 1)\left[F_{no} h v B_{op}(e^{\alpha l} - 1)N\right] \approx OSNR\left[F_{no} h v B_{op}(e^{\alpha l} - 1)N\right] \qquad (5.67)$$

If we did not have to worry about nonlinearities, we would maximize the power per channel to easily achieve to requirement given by (5.67). However, the real story is different since any power increase will boost the nonlinear effects.

Equation (5.67) can be converted to decibel units if the operator $10\log(\cdot)$ is applied to both sides, in which case it becomes

$$P_{ch} \geq OSNR + F_{no} + \alpha l + 10\log(N) + 10\log(h v B_{op}) \qquad (5.68)$$

where $OSNR$, P_{ch}, and F_{no} are expressed in decibels, l is expressed in kilometers, and α is expressed in dB/km. This equation can also be written as

$$OSNR \approx \frac{2Q^2 \Delta f}{B_{op}} \geq P_{ch} - F_{no} - \alpha l - 10\log(N) - 10\log(h v B_{op}) \qquad (5.69)$$

The OSNR value can be now used as an input parameter to calculate the total length L than can be achieved. That value can be obtained by using the relation between OSNR and Q-factor, given by (5.32) to (5.34). As an example, (5.34) was used to express the connection between OSNR and Q-factor in the above equation. Therefore, we can first calculate OSNR for a specified Q-factor, and then use (5.69) to evaluate the other parameters that can help to achieve the transmission goal. These parameters include launched optical power per channel, noise figure of optical amplifiers, number of fiber spans, optical fiber bandwidth, and optical loss per fiber span.

There are a few important conclusions based on (5.69). First, the OSNR for a specified link length can be increased by increasing the output optical power. Second, this can be also done by decreasing the number of optical amplifiers. In addition, the OSNR can be improved by decreasing the amplifier noise figure and optical loss per fiber span. However, the number of amplifiers can be decreased only if the span length is increased, which means that the total signal attenuation will be also increased. Therefore, the overall picture becomes more complex, and some trade-offs may be needed.

We can use (5.63) to plot the number of fiber spans as a function of signal and noise parameters, as shown in Figure 5.12. Note that Figure 5.12 contains four curves that were produced by using (5.69) and two curves that are related to the case when some system margin was included. This was done for comparison sake to show the difference between a basic case and more realistic situation.

It is important to notice the following facts when considering curves from Figure 5.12. First, the power increase that could be beneficial according to Figure 5.12 should also be judged through the impact of nonlinearities. Second, some power margin ΔP needs to be allocated in advance to compensate for power penalties due to receiver sensitivity degradation. The situation changes if some power margin is allocated in advance, and it can be observed through a decrease in the number of the fiber spans. Please notice the difference in Figure 5.12 between two cases corresponding to the zero power margin allocation and to $\Delta P = 4.5$-dB power margin allocation. The process of the power margin allocation is discussed in the next section.

5.4 Systems Engineering and Margin Allocation

Transmission systems engineering is based on a specified transmission quality that is expressed through BER and considers the overall conditions related to the optical

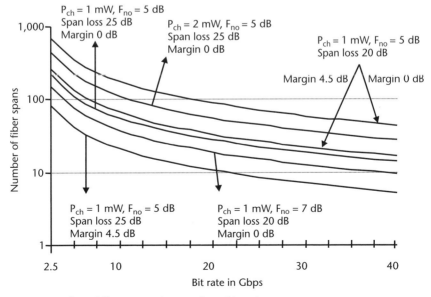

Figure 5.12 Number of fiber spans that can be achieved.

signal and different impairments. There are several basic steps that are involved in the engineering process:

- Understand systems requirements in terms of distance, bit rate, and BER.
- Check for systems viability, in terms of elements that are available (i.e., fibers, sources, photodetectors, amplifiers).
- Set up the major system parameters (i.e., output signal power, amplifier spacing) that satisfy the initial requirements.
- Allocate margins to account for penalties that can be compensated.
- Go through second iteration by repeating the process again in order to perform a fine-tuning of parameters and to identify possible trade-offs.

Understanding system requirements and checking for system viability is a prerequisite of a thorough engineering process. The analysis, which refers to the transmission line power budget and the system transmission speed, is usually carried out while checking for the system viability. This analysis is closely related to the fiber type that will be used for signal transmission, and it recognizes if there is need to use any other elements, such as optical amplifiers, which should be placed between optical transmitters and optical receivers.

The simplest procedure refers to the case where no other elements—except for optical fibers and fiber splices—are placed between optical transmitter and optical receiver. This case is known as a point-to-point transmission. In the general case, the point-to-point transmission is done for distances that are shorter than 120 to 150 km and for bit rates that are lower than 1 Gbps, although there might be some situations that break this rule.

The first step in checking for transmission viability is to verify if the link power budget is high enough to enable transmission. That power budget is determined as the power difference between the output power from optical transmitter and the receiver sensitivity associated with specified BER. The power budget should be high enough to cover different losses (fiber attenuation, splice losses) and to include a certain system power margin. The system power margin is needed to account for component degradation due to aging or for environmental (temperature) effects.

With a clear picture of the link power budget, checking for transmission viability should continue with an analysis related to system bandwidth to make sure that the specified bit rate can be transmitted over a distance that looks viable from the power budget point of view. The system bandwidth can be limited due to slow components (such as light source, photodiode, and electronics associated with them) or due to limited transmission bandwidth of the optical fiber. If system requirements cannot be met with components that are available, they could be replaced. Such an option is more realistic for transmitters and receivers than for optical fiber, since the fiber is already in place in most situations.

In the general case, it is the optical fiber that imposes the transmission limit in point-to-point transmission systems. There are several types of the point-to-point systems. If using the transmission length as a criterion, they can be classified as very short reach (VSR) with lengths measured in hundreds of meters, short reach (SR) with lengths measured in kilometers, long reach (LR) with lengths measured in tens and hundreds of kilometers, and ultralong reach (ULR) with lengths measured in thousands of kilometers. If using the bit rate as a criterion, the point-to-point systems can be low speed with bit rates measured in tens of megabits per second, medium speed with bit rates measured in hundreds of megabits per second, and high speed with bit rates measured in gigabits per second. Finally, from an application perspective, all point-to-point systems are either power budget limited (or loss limited), or transmission speed limited (or bandwidth limited).

5.4.1 Systems Engineering of Power-Budget Limited Point-to Point Lightwave Systems

As was mentioned above, the engineering process starts with the power budget considerations. In cases where no optical amplifiers are employed, the power budget is expressed through the following relation:

$$P_{out} - \alpha L - \alpha_c - \Delta P_M \geq P_R(Q, \Delta f) \tag{5.70}$$

where P_{out} is the output optical power from the light source pigtail, α is attenuation coefficient of the optical fiber, α_c is the signal loss related to optical splices and connectors, and P_R is the receiver sensitivity related to specified BER. Parameter ΔP_M is the system margin that is needed to account for different effects such as aging or temperature change. The structure of the system margin will be discussed shortly. Note that P_R is expressed as a function of Q-factor and the receiver bandwidth (or the signal bit rate). All parameters in (5.70) are expressed in decibels, except the attenuation coefficient, which is expressed in dB/km.

We can differentiate several cases here, based on what parameters are specified in advance. The common case occurs if both transmission distance L and the bit rate B are specified, in which situation (5.70) should help to select cost-effective components that satisfy system requirements. The selection includes sources (LED versus laser), operating wavelength (the wavelength window), and the optical receiver (PIN photodiode versus APD-based optical receiver).

Generally speaking, all components are cheaper if they operate at shorter wavelengths. The lowest price is for components operating around 850 nm, and increases when shifting to wavelengths around 1,310 and 1,550 nm. As for light sources, LEDs are much cheaper then laser photodiodes, and Fabry-Perot lasers are cheaper than single-mode DFB lasers. On the receiver side, the APD-based receivers offer higher sensitivity than PIN-based receivers, however they need a high voltage supply that should be carefully controlled to avoid avalanche breakdown. In addition, APDs are more expensive than PINs, and that applies even if the APD is compared with an integrated combination of PIN- and FET-based front-end amplifiers. Therefore, the selection process should go from LED to lasers, and from PIN photodiodes to APD, while checking if the selected component can satisfy system requirements. If there is a choice to select the optical fiber, the process should go from multimode to single-mode optical fibers.

As an example, let us consider the case when transmission of the signal with bit rate of 200 Mbps should be done over the distance that is not shorter than 15 km (refer to Table 1.1 to recognize that this bit rate corresponds to the ESCON data channel). We can also assume that there is a requirement for BER to be lower than

10^{-12}. Let us start with LED as a source candidate and with both PIN and APD as photodetector candidates. We will assume that P_{out} = –12 dBm and that receiver sensitivities related to a bit rate of 200 Mbps with respect to PIN photodiodes and APD are –37 and –47 dBm, respectively. In addition, we can start with a multimode optical fiber and assume that the optical fiber bandwidth of 1-km fiber length is 1.8 GHz-km (i.e., B_{fib} = 1.8 GHz-km) [see (3.51)]. Let us first consider the cheapest option (at wavelength 850 nm) and the second cheapest option (at wavelength 1,300 nm). We will assume that the system margin is 5 dB, since it is a common approach to allocate the system margin in the range 4 to 6 dB [12] to cover for component aging and temperature effects. Finally, we will assume that splicing losses are included in the fiber attenuation coefficient (which becomes α = 3.0 dB/km at 850 nm, and α = 0.5 dB/km for 1,300 nm), while connector losses are 2 dB. This situation is illustrated in Table 5.2.

We can clearly see that transmission with neither combination operating at 850 nm is viable (it shown in italic in Table 5.2). The next available option is to use transmission at 1,300 nm, where the LED/PIN combination is the cheapest one that satisfies the requirements. If there were a request to transmit the signal over 28 km, the LED/APD combination operating at 1,300 nm could be deployed. Any requirement for transmission over 44 km can be satisfied only by using the laser diodes.

In addition to the system margin, a power margin that compensates for the impact of both the modal noise and mode partition noise should also be allocated if transmission is performed over multimode optical fibers, while using laser as a light source. The margin of 1 to 2 dB can serve for such a purpose. Note that we allocated 1 dB in the example presented in Table 5.2. The analysis presented above is applicable to the LAN environment.

The selection of transmission wavelength in the case presented above may be related to the wavelength availability, since it might happen that the wavelength that was originally selected is already occupied. If such a scenario occurs, some other wavelength should be considered. Optical transmission in the LAN environment often includes signal broadcast by using optical power splitting. In such a case, each optical coupler/splitter employed along the lightwave path should be accounted for

Table 5.2 Power Budget for 200-Mbps (ESCON Data Channel) Transmission, $Q = 7$

Parameter	LED		Laser	
Wavelength	850 nm	1,300 nm	850 nm	1,300 nm
Receiver sensitivity—PIN	–33 dBm	–34 dBm	–33 dBm	–34 dBm
Receiver sensitivity—APD	–41 dBm	–42 dBm	–41 dBm	–42 dBm
Output power	–12 dBm	–13 dBm	0 dBm	0 dBm
Connector losses	2 dB	2 dB	2 dB	2 dB
System margin	5 dB	5 dB	5 dB	5 dB
Additional margin	0 dB	0 dB	1 dB	1 dB
Available loss for PIN	14 dB	14 dB	25 dB	26 dB
Available loss for APD	22 dB	22 dB	33 dB	34 dB
Fiber loss per km	3.5 dB/km	0.5 dB/km	3.5 dB/km	0.5 dB/km
Transmission length for PIN	*4.1 km*	28 km	*7.1 km*	52 km
Transmission length for APD	*6.3 km*	44 km	*9.4 km*	68 km

by allocating 3-dB power-splitting loss. As an example, if there are 5 optical couplers, additional 15 dB should be added to the connection losses in Table 5.2.

Another transmission type that might be power budget limited is related to higher bit rates and shorter distances. For example, it applies to bit rates of 2.5, 10, and 40 Gbps if they need to be transmitted over distances ranging from a couple hundred meters to several kilometers. The analysis related to this case is similar to the one presented in Table 5.2, but with different starting points since only lasers should be considered as a light source. Both the FP lasers and VCSEL are good candidates for bit rates up to 2.5 Gbps, while VCSEL and the DFB lasers should be considered for bit rates up to 10 Gbps. As for 40-Gbps bit rates, DFB lasers monolithically integrated with electroabsorption modulators should be considered as primary candidates.

As for the power margin allocation with respect to this transmission scenario, the power penalties related to the intensity noise and the extinction ratio impact should be covered by some power margin, which comes in addition to system power margin. The power margin of up to 1 dB for each of these impairments is needed. The situation that might apply to high bit rates is illustrated in Table 5.3. In most cases, however, that transmission distance is not power budget limited, but rather bandwidth limited, as we will see shortly. Therefore, the values presented in Table 5.3 may not be quite relevant from the engineering perspective.

5.4.2 Systems Engineering of Bandwidth-Limited Point-to Point Lightwave Systems

The performance of an optical transmission system can be limited due to limited frequency bandwidth of the some of the key components that are used (such as light source, photodetector, or optical fiber). The optical fiber is the most important from the system bandwidth perspective. Optical fiber bandwidth in multimode fibers is limited mainly due to the modal dispersion effect since it is usually much larger than the chromatic dispersion, while the chromatic dispersion is the only factor that defines the bandwidth in single-mode optical fibers.

Table 5.3 Power Budget for Short-Reach High-Speed Transmission, Q – 7

Parameter	2.5 Gbps		10 Gbps		40 Gbps	
Wavelength, nm	1,310	1,550	1,310	1,550	1,310	1,550
Receiver sensitivity for PIN, dBm	27	27	24	24	21	21
Receiver sensitivity for APD, dBm	35	35	32	32	29	29
Output power, dBm	3	3	3	3	0	0
Connector losses, dBm	2	2	2	2	2	2
System margin, dBm	5	5	5	5	5	5
Additional margin, in dBm	2	2	2	2	2	2
Available loss for PIN, dBm	15	15	12	12	6	6
Available loss for APD, dBm	23	23	20	20	14	14
Fiber loss, dB/km	0.5	0.22	0.5	0.22	0.5	0.22
Transmission length for PIN, km	30	67	24	54	12	27
Transmission length for APD, km	46	100	40	90	28	64

The bandwidth of multimode optical fibers is characterized through the bandwidth B_{fib} of the 1-km optical fiber length, which is somewhere around 150 MHz-km for step-index optical fibers, and around 2 GHz-km for graded-index optical fibers. As for single-mode optical fibers, the available bandwidth also depends on the spectral linewidth of the light source that is used. Generally, we can distinguish two separate cases, which are related to single-mode fibers and sources with narrow linewidth and with wide linewidth, respectively.

We can use (3.51), (3.83), (3.85), and (4.12) to evaluate the available fiber bandwidth of any given length. This can be summarized as

$$BL^\mu \leq B_{fib} \quad \text{for multimode optical fibers} \quad (5.71)$$

$$BL \leq (4D\sigma_\lambda)^{-1} \quad \text{for single-mode fibers and larger source linewidth} \quad (5.72)$$

$$B^2 L \leq (16|\beta_2|)^{-1} \quad \text{for single-mode fibers and narrow source linewidth} \quad (5.73)$$

where B is signal bit rate, L is the transmission length, B_{fib} is the bandwidth of 1-km length of multimode optical fibers, μ is parameter that ranges from 0.5 to 1, D is the chromatic dispersion coefficient, σ_λ is the light source linewidth, and β_2 is the group velocity dispersion coefficient for single-mode optical fibers.

Equations (5.71) to (5.73) can be used in an assessment if the optical transmission system is power budget limited or bandwidth limited. It is useful to plot (5.71) to (5.73) in parallel with the curves that are related to optical power budget limitations. System limitations related to the optical power budget can be expressed by the following functional dependence that follows from (5.70):

$$L(\lambda, B) \leq \frac{P_{out}(\lambda, B) - [P_R(\lambda, B) + \alpha_c + \Delta P_M]}{\alpha(\lambda)} \quad (5.74)$$

where P_{out} is output optical signal power, P_R is receiver sensitivity, α is attenuation coefficient of the optical fiber, α_c is the signal loss related to optical splices and connectors, and ΔP_M is the system margin. For this purpose we can assume that there is no system margin, and that connector losses can be neglected.

Functional curves expressed by (5.71) to (5.74) are shown in Figure 5.13. The transmission distance is shown as a function of the bit rate, while optical wavelength serves as a parameter. Figure 5.13 can be used to identify the reference points and to recognize what are the critical limitation factors. It was assumed that the GVD coefficient β_2 in single-mode fibers is $\beta_2 = -20$ ps²/km, while it is $\beta_2 = -4$ ps²/km for NZDSFs.

In general, there are several conclusions with respect to system limitations when using different fiber types. First, the systems with step-index multimode optical fibers are bandwidth limited for all bit rates of practical interest (ranging from 1 Mbps to several megabits per second), while the achievable distance is up to several kilometers. Second, the systems with graded-index multimode optical fibers are generally power budget limited if the bit rate is up to about 100 Mbps, while they become bandwidth limited for higher bit rates. It is possible to transfer 1-Gbps bit

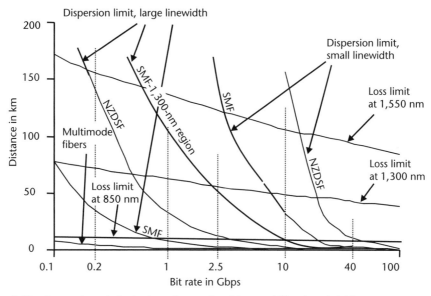

Figure 5.13 System length limitations due to signal loss and fiber bandwidth.

rate up to 1.5 to 2 km, or 10-Gbps bit rate up to 200m to 250m over graded-index multimode optical fibers.

Next, single-mode optical fibers are also power budget limited for bit rates up to several hundred megabits per second, if the optical source has a large linewidth. The power budget limit can be extended to bit rates of 2.5 Gbps and up if optical sources with a narrow linewidth are used. In addition, different types of single-mode optical fibers impose different kinds of limits with respect to a specific bit rate. As an example, standard single-mode optical fibers impose the power budget limit to signals with bit rates of 1 Gbps if operation is done at wavelengths around 1,300 nm, while NZDSFs impose the bandwidth limit for this specific bit rate and the operational wavelength.

The operation over single-mode fibers at wavelengths around 1,300 nm is the most advantageous for signals with bit rates up to 600 Mbps. It is important to mention that it is also possible to transmit a signal with bit rate of 2.5 Gbps over 50 to 60 km of SMF by applying a direct modulation scheme at the wavelength of 1,300 nm. The same arrangement can be used to transmit signals with 10-Gbps bit rate up to 10 to 15 km.

Single-mode optical fibers provide a bandwidth-limited transmission at wavelengths that belong to the 1,550-nm wavelength region. With this arrangement, it is possible to transmit signals with 10-Gbps bit rate up to 40 to 50 km, or signals with a bit rate of 40 Gbps up to 2 to 3 km. Please also recall the data from Table 4.2, which can also serve as a reference point with respect to transmission lengths that can be achieved for specified bit rate without using any dispersion compensation scheme. It can also be used to make an assessment of the amount of residual chromatic dispersion that can be tolerated in bandwidth-limited systems.

The transmission length of the bandwidth-limited systems is not determined just by the optical fiber bandwidth expressed by (5.71) to (5.73), but also by the frequency bandwidth (3-dB bandwidth) of the optical transmitter and optical receiver. The bandwidth limitation is related to the pulse rise times that occur in individual modules, and the following relation can be established:

$$T_r^2 \geq T_{tr}^2 + T_{fib}^2 + T_{rec}^2 \tag{5.75}$$

where T_r is the overall response time of the system, T_{tr} is the response time (i.e., the rise time) of the optical transmitter, T_{fib} is the response time of the optical fiber, and T_{rec} is the response time of the optical receiver. There is the following general relation between the response time and the 3-dB system bandwidth Δf [10–12]:

$$T_r = \frac{0.35}{\Delta f} \tag{5.76}$$

The above equation can be converted to a form that connects the response time with the signal bit rate. The relation between system bandwidth Δf and bit rate B is dependant on the digital modulation format that is used. It is $\Delta f = B$ if the RZ modulation format is used, and it is $\Delta f = 0.5B$ if the signal has a NRZ format.

Connection between the bandwidth and the rise time expressed by (5.76) can be applied to all key elements in the optical system (i.e., to optical transmitter, optical fiber, and optical receiver). With this, (5.75) becomes

$$\frac{1}{B^2} \geq \frac{1}{\Delta f_{tr}^2} + \frac{1}{B_{fib,L}^2} + \frac{1}{\Delta f_{rec}^2} \quad \text{for RZ format} \tag{5.77}$$

$$\frac{4}{B^2} \geq \frac{1}{\Delta f_{tr}^2} + \frac{1}{B_{fib,L}^2} + \frac{1}{\Delta f_{rec}^2} \quad \text{for NRZ format} \tag{5.78}$$

where $B_{fib,L}$ is the optical fiber bandwidth related to specified length L, which can be calculated from (5.71) to (5.73). Please notice that (5.71) to (5.73) should be solved first in order to obtain the value of parameter B, which is associated to specified transmission length L. The obtained value is then assigned to the fiber bandwidth $B_{fib,L}$.

Equations (5.77) and (5.78) can be used to verify whether the system is bandwidth limited or not. They should be solved to find the transmission length L, while the obtained result for parameter L should be compared with the value obtained from (5.70), which is related to the power budget limit scenario. Assuming that the transmitter and receiver satisfy the bandwidth requirement, the result of comparison will be determined by the optical fiber type and by operational wavelength. However, if it happens that transmitter and receiver do not satisfy requirements given by (5.77) and (5.78), faster alternatives should be considered.

5.4.3 Systems Engineering for High-Speed Optical Transmission Systems

Results related to point-to-point transmission can be generalized for an optically amplified system that employs some number of optical amplifiers. The power budget equation (5.70) can be now replaced by (5.69), which can be rewritten as

5.4 Systems Engineering and Margin Allocation

$$OSNR \approx \frac{2Q^2 \Delta f}{B_{op}} \geq P_{ch} - F_{no} - \alpha l - 10\log(N) - 10\log(hvB_{op}) - \Delta P \qquad (5.79)$$

Please notice that a power margin ΔP has been introduced to account for the impact of various impairments. The starting point when applying (5.79) will again be a specified value of BER, which is related to the Q-factor. The entrance parameters are the signal bit rate and the number of optical channels. It is worth mentioning that (5.77) and (5.78) should also be satisfied when considering high-speed transmission systems.

In-line optical amplifiers, which are essential network elements in high-speed transmission systems, are employed to compensate not only for attenuation losses but for chromatic dispersion as well, since they usually contain a dispersion-compensating module. Therefore, chromatic dispersion compensation is performed in parallel with optical amplification. Such compensation is not perfect for all channels in a mutichannel transmission system, and it is common that some residual dispersion will remain for some of the channels, thus causing the power penalty.

Equation (5.79) can also be written as

$$OSNR_{required} = OSNR + \Delta P \approx \frac{2Q^2 \Delta f}{B_{op}} + \Delta P \qquad (5.80)$$
$$\geq P_{ch} - F_{no} - \alpha l - 10\log(N) - 10\log(hvB_{op})$$

where $OSNR_{required}$ is a new optical signal-to-noise-ratio that should be achieved in order to accommodate power penalties imposed by various impairments. It is clear that this value should be higher than the original one by an amount equal to the power penalties, which are expected to occur. Equation (5.80) is the basic one that can be used in the engineering process related to high-speed long-haul transmission systems.

The parameters that are included in the engineering process are the output power per channel, span length, number of channels, channel spacing, and number of spans. The output power is a parameter that should be optimized by evaluating merits and demerits of its increase. That means that the power level can be increased only to the point where an increase in OSNR is higher than the power penalty imposed by nonlinear effects. The number of optical channels will have an impact on nonlinear effects and crosstalk noise. As for the fiber span length, it is commonly predetermined, and the system considerations are based on an evaluation of the maximum number of spans that can be accommodated under specified conditions.

Margin assignment plays a very important role in the systems engineering process. The assignment may be based on a conservative approach, in which case the total margin is a sum of individual contributions, as illustrated in Figure 5.5. Although the process of margin allocation may differ from case to case, it should follow basic facts that are summarized in Table 5.1. There are several impairments, such as chromatic dispersion and the extinction ratio, that do not have a stochastic character and thus require allocation of a quite specific margin.

On the other hand, there are some impairments, such as PMD and intensity noise, that are stochastic in nature and can combine differently with each other at any given moment. These impairments can be covered by a joint power margin, which is allocated for all of them. The value of the margin can be estimated by applying a statistical approach that evaluates a joint impact of several parameters. The simplest option is to consider these parameters to be random Gaussian variables and to assign the margin that is proportional to the total standard deviation of the summary stochastic process. The alternative approach is to use statistical modeling, such as one that employs the Monte-Carlo method, in order to get the most realistic outcome. In such a case, the margin allocation will be based on the results of the most realistic outcome. The main benefit from the statistical approach is that the allocated margin is not overvalued, as it might be if applying a conservative scenario. Finally, the margin allocation can be based on computer-aided engineering and the use of specialized simulation software. This will lead to more complex calculations and will provide a better estimate of some important effects that are otherwise rather difficult for evaluation (cross-phase modulation, for example).

One possible scenario that can be used with respect to margin allocation is shown in Table 5.4. There are two columns in Table 5.4, which are related to a conservative (or worst case) scenario and to a statistical approach. In the statistical approach scenario, the margin was assigned for the group of impairments by assuming that individual margins, otherwise related to the worst case scenario, can be treated as contributing variances in the multivariable Gaussian process. Such an approach is not accurate as one that includes a complete statistical treatment, but it can be considered as a middle way between the conservative scenario and the statistical approach. The margins in Table 5.4 were assigned for three values of Q-factor (i.e., for Q = 6, 7, 8), which correspond to BER of 10^{-9}, 10^{-12}, and 10^{-15}, respectively. Please notice that although some margins were allocated to account for nonlinear effects, it was assumed that the other steps, which would prevent severe system degradations, were already taken. This is particularly related to having an

Table 5.4 Margin Allocations for High-Speed Long-Haul Transmission Systems

Impairment	Margin Allocated for the Conservative Approach Scenario, Expressed in Decibels			Margin Allocated for the Statistical Approach Scenario, Expressed in Decibels		
	Q = 6	Q = 7	Q = 8	Q = 6	Q = 7	Q = 8
Chromatic dispersion	0.5	0.5	0.5	0.5	0.5	0.5
Extinction ratio	0.5	0.5	0.5	0.5	0.5	0.5
PMD	0.7	0.7	1			
Intensity noise	0.3	0.3	0.5	1.5	1.5	2
Crosstalk	0.5	0.5	1			
Polarization-related losses	1	1	1			
SPM	0.5	0.5	0.5			
FWM	0.5	0.5	0.5	1.5	1.5	1.5
SRS	0.5	0.5	0.5			
XPM	0.5	0.5	0.5			
Component aging	2	2	2	2	2	2
System margin	3	3	3	3	3	3
Total margin ΔP, in dB	10.5	10.5	11.5	9	9	9.5

operation outside of the zero-dispersion region in order to prevent an extreme rise of nonlinear effects. In addition, selecting the proper optical channel spacing serves this purpose as well.

It is necessary now to establish the required optical signal-to-noise ratio ($OSNR_{required}$), which is given by (5.80), for each specified case. For this purpose, an original value of the optical-signal-to-noise-ratio should be calculated first, and then added to the allocated power margin. The original value of OSNR can be calculated by using (5.32) to (5.34). As an example, (5.32) was used to calculate OSNR for several cases, which are specified by the bit rate B and Q-factor. The results have been calculated for both the NRZ modulation format (with $\Delta f = B/2$) and the RZ modulation format (with $\Delta f = B$), and they are shown in Table 5.5.

Now we can add the margin values from Table 5.4 to the OSNR values from Table 5.5 to obtain $OSNR_{required}$. The $OSNR_{required}$ values, which are obtained for the same values of Q-factor as those used in Table 5.5, are shown in Table 5.6. The values in Table 5.6 that are in italics correspond to the conservative scenario and are associated with inputs from the left part of the Table 5.4; while those in parentheses correspond to the statistical approach in the margin allocation and are associated with inputs from the right part of the Table 5.4. Newly calculated $OSNR_{required}$ values can now be used to evaluate the number of spans that can be achieved.

The process of margin assignment is illustrated in Figure 5.14. As we can see, the value of $OSNR_{required}$ corresponds to the Q-factor that is higher than the one that would be required if there were no power penalties. Therefore, since the power margin is allocated to account for the power penalties, new reference levels for OSNR and Q-factor have been established. Enough power margin should be available for the allocation. This means that either an output power increase, or some other means, should provide a relief against impairments.

Table 5.5 Original OSNR Required for Specific Bit Rate and Q Parameter

Parameter	Bit Rates, NRZ Coding			Bit Rates, RZ Coding		
	2.5 Gbps	10 Gbps	40 Gbps	2.5 Gbps	10 Gbps	40 Gbps
OSNR for Q = 6, dB	5.05	9.96	12.62	7.4	12.59	15.12
OSNR for Q = 7, dB	6.02	11.12	13.70	8.57	13.81	16.58
OSNR for Q = 8, dB	7.98	12.70	14.80	9.46	14.82	17.61

Table 5.6 Required $OSNR_{required}$ After Margin Is Included

Parameter	Bit Rates, NRZ Coding			Bit Rates, RZ Coding		
	2.5 Gbps	10 Gbps	40 Gbps	2.5 Gbps	10 Gbps	40 Gbps
$OSNR_{required}$ for Q = 6, dB	15.55	20.46	23.12	17.9	23.09	25.62
	(14.05)	(18.96)	(21.62)	(16.4)	(21.59)	(24.12)
$OSNR_{required}$ for Q = 7, dB	16.52	21.62	24.20	19.07	24.31	27.08
	(15.02)	(20.12)	(22.70)	(17.57)	(22.81)	(25.58)
$OSNR_{required}$ for Q = 8, dB	19.48	24.20	26.30	20.96	26.32	29.11
	(17.48)	(22.20)	(24.30)	(18.96)	(24.32)	(27.11)

In some situations, it is possible to increase the output channel power from an initial value P_{ch} to the value $P_{ch,required}$, which is enough to increase the OSNR in such a way that there will be enough power margin to compensate for penalties. The power can be increased even a little bit further and still be beneficial since the pace of the increase of OSNR will be larger than the penalties induced by nonlinear effects. However, if power increases above some level, which can be identified as an optimum power per channel $P_{ch,opt}$ in Figure 5.14, the total benefit will begin to shrink. That is because the impact of nonlinear effects on receiver sensitivity has become dominant and would eventually disappear if power continued to rise above the maximum value P_{max}.

The situation illustrated in Figure 5.14 is a favorable one, since it was possible to increase the power to compensate for power penalties. In some other cases, such as the one represented by dashed line in Figure 5.14, the power increase is not enough to provide the margin that is needed. However, there are some other methods to cope with the situation when a power increase does not provide enough power margin. That might include changes in the system configuration by increasing the optical channel spacing or using optical amplifiers with lower noise figure.

Some advanced methods, such as FEC, or advanced components, such as Raman amplifiers, can be used to provide the necessary power margin. These methods will be discussed in Chapter 6. Herewith we can only say that introduction of advanced methods and schemes, which provide an additional margin, can be understood as if there were a positive margin ($+\Delta P_{adv}$) on the right side of (5.79) that would offset the negative margin related to the power penalties. This concept will be explained in Chapter 6 (see Section 6.1).

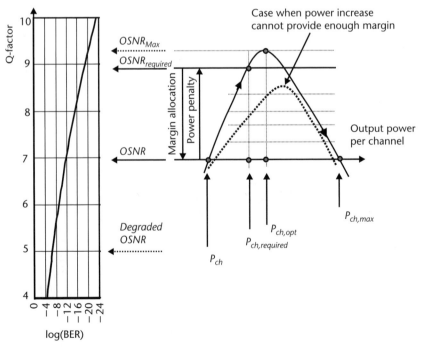

Figure 5.14 Power margin allocation.

5.4.4 Optical Performance Monitoring

Performance monitoring in optical transmission systems plays a special role for several reasons. First, it is extremely important for the network operator to control the overall status of the transmission lines, and the status of the network as a hole, in order to verify the fulfillment of the service layer agreement (SLA) established with customers. Second, the performance evaluation and detection of the signal degradation below a threshold level serves for an activation of the resilience mechanism. Next, the performance monitoring can also serve to anticipate the change in operation condition that could degrade the transmission quality.

Performance monitoring can be based either on digital or analog techniques. The digital technique, which is related to BER monitoring, can be properly evaluated only by using specially developed out-of service methods, in which a test signal is in place of the real signal. On the other side, BER can be approximately evaluated by measuring the block-error ratio on real signals, or estimated by using some analog methods.

All analog methods are approximate in nature and evaluate the analog parameters that are related to BER (mainly OSNR and signal crosstalk). The availability of a fast and accurate performance assessment can help the operator speed up the operations, thus saving on time and operational cost. From such a perspective, analog techniques can serve an important purpose, even if they are less accurate than the digital ones. However, it is the extremely important to translate the measured results to commonly used parameters, such as BER or Q-factor.

The three commonly used analog techniques are related to optical spectrum analysis, detection of the special pilot tone, and histogram method [13]. Optical spectrum analysis measures OSNR, optical frequency, and optical power. Optical channels within a composite WDM signals should be extracted from the composite signal by some filtering method (by using tunable filters, spatial filtering, and so forth) before the mentioned parameters are evaluated by using an optical spectrum analyzer (OSA). This method is the most comprehensive one, but it needs precise and reliable measurement equipment. On the other hand, the pilot tone technique, which adds a specific small sinusoidal tone to the laser bias at the transmission side, is a much simper method that can be used to detect both the intensity variations and signal crosstalk. The specific tone added to each optical channel can also serve as a tag for each channel, which is important for operation and maintenance purposes in an optical networking environment. The signal power of the specific optical channel is estimated by monitoring the amplitude of the pilot tone along the lightwave path. The amplitude of the pilot tone could even be used to extract the OSNR and ASE values if the total optical power is measured. It was shown experimentally that results obtained by using optical spectrum analysis and the pilot tone techniques are in good agreement if the OSNR is below 30 dB.

Neither optical spectrum analysis nor the pilot tone technique can be used to identify dispersion parameters or nonlinear distortion since they evaluate only average signal parameters. That was the reason to propose the monitoring technique that is based on a histogram of signal amplitudes. The histograms are extracted by asynchronous sampling of the signal and by using a very fast photodiode for signal detection. This method is bit-rate transparent and capable of capturing the amplitude variations. However, a relatively complex signal processing is needed to interpret the histogram data properly. The histogram method can be modified to be bit-rate specific by using synchronous measurement without any signal sampling. In such a case, the receiver design can be modified to include the second decision circuit with an adjustable threshold and digital counters.

The OSNR value measured by OSA, or by the pilot tone technique, can be correlated to Q-factor by using a modified (5.33), which is

$$Q = \frac{2OSNR_{band}\sqrt{B_{op}/\Delta f}}{1+\sqrt{1+4OSNR_{band}}} \tag{5.81}$$

where $OSNR_{band} = (B_{band}/B_{op})OSNR$ defines an optical signal-to-noise ratio applied to the measurement scenario, B_{op} is the optical filter bandwidth, Δf is the electrical filter bandwidth, and B_{band} is the resolution bandwidth of the instrument used (usually the OSA resolution bandwidth). It was shown that (5.81) gives a fair estimate of the Q-factor penalties, which are expressed as

$$\Delta Q = 10\log\left(\frac{Q}{Q_{ref}}\right) \tag{5.82}$$

where Q is the Q-factor value extracted from the measured OSNR in accordance with (5.81), and Q_{ref} is the reference value, which can be related to the value measured when equipment was provisioned. Therefore, the operation of the system can be kept under control if $OSNR$ is measured, while the Q-factor penalty is extracted from measured data. At the same time, the alarm thresholds are set to be associated with the maximum value of the Q-factor penalty that can be tolerated.

The impact of crosstalk noise can be estimated by using (5.57) to (5.61), while assuming that the $OSNR$ penalty can be evaluated from (5.61). Therefore, it is necessary to know the amount of crosstalk power to evaluate the degraded $OSNR$ value before Q-factor has been extracted, which can be done by the pilot tone monitoring technique. The implementation of this methodology, which includes both the $OSNR$ and the crosstalk monitoring, should start with an evaluation of the Q_{ref} during the channel provisioning process and continue by monitoring the value ΔQ based on measured data, and in accordance with (5.82).

The histogram technique helps to extract more detailed information from measured data with regards to signal degradation. The histogram data is usually divided in two groups, which are related to "1" and "0" bits. Each group is used to evaluate signal and noise parameters from (5.10) by assuming that they follow the Gaussian statistics. The histogram recording helps to evaluate BER as [13]

$$BER = \frac{1}{2p(1)}\sum_i H(I_{1,i})erfc\left(\frac{I_{1,i}-I_{th}}{\sigma_{1,i}\sqrt{2}}\right) + \frac{1}{2p(0)}\sum_i H(I_{0,i})erfc\left(\frac{I_{th}-I_{0,i}}{\sigma_{0,i}\sqrt{2}}\right) \tag{5.83}$$

where $H(I_{1,i})$ and $H(I_{0,i})$ are the occurrences of the signal amplitudes associated with 1 and 0 bits, respectively, which are imported from the histogram, and where $p(0)$ and $p(1)$ are probabilities that 0 and 1 bits were received [refer to (5.1) to (5.6)]. It is reasonable to assume that probabilities $p(0)$ and $p(1)$ are equal (i.e., that

$p(0) = p(1) = 0.5$). The accuracy of the BER evaluation by using (5.83) is within −20% to +40% if synchronous sampling is applied, and within −20% to 120% with asynchronous sampling applied [13]. The biggest asset of the histogram technique is that it can detect small signal degradations and in some cases can point to the causes of such degradation.

The digital techniques for performance evaluation of an optical channel are based on monitoring the error detection codes. Two well-known types of such codes are the cyclic-redundancy check (CRC) code and the bit-interleaved parity (BIP) code, which are commonly used in asynchronous and synchronous transmission systems, respectively. The digital techniques are mandatory to provide in-service performance monitoring and are commonly defined within different standard bodies [14, 15]. The data stream is monitored per block basis. The content of the data block is captured by the error-detection code bits that are also inserted in the data stream (usually at the at the end of the data frame, or in the overhead of the following frame). The imprinted error detection code can be calculated by using different methodologies. The simplest approach is in BIP error detection codes, where the BIP bit provides an even parity over all control bits. The error detection code is calculated again at the receiving side and compared with the transmitted code. The transmission error is registered if the bit-parity scheme is violated.

The digital techniques are very suitable for in-service performance monitoring, although results are highly dependent on the occurrence of error statistics. However, they are not quite suitable for fault localization and identification, or for the fine performance optimization during the operation process. This could be done by using a fast and simple analog technique that can be adjusted to specific application scenario. Therefore, the best approach may be to combine digital and analog methods in a way that will provide cost-effective, yet efficient, overall system performance monitoring.

5.4.5 Computer-Based Modeling and Systems Engineering

The transmission system engineering evaluates different parameters and validates the goal with respect to the established requirement, which is defied through specified BER. The engineering process may lead to the optimization of some parameters that are related either to the transmitter/receiver or to the transmission line. The approach presented so far introduced the engineering fundamentals and was related mostly to a conservative scenario. At the same time, a random character of some impairments was recognized and included in the process of power margin allocation. The presented engineering approach serves the following purposes:

- Provides understanding of parameters and processes involved in overall considerations;
- Establishes a conservative scenario that serves as a reference and reality check for any other calculation;
- Provides guidance and builds up the knowledge and skills needed for using numerical modeling and computer-based calculations.

However, the analytical engineering approach may not be enough to identify all possible optimization variants and to provide guidance for fine-tuning of different parameters that have an impact on the system performance. That is because there are a few dozen parameters that can be put in different combinations, and because the relation among them is not a simple one [16]. On the other hand, an analytical engineering approach is self-sustained and beneficial for systems that employ a moderate number of optical channels, and when transmission is done over moderate distances (up to several hundred kilometers).

The most complex case from the system engineering perspective involves long-haul transmission with a large number of WDM channels, in which optical routing may occur. In this situation, an analytical approach could be considered just as a conservative guidance and a reality check of the more precise engineering that can be done by using a computed-aided engineering approach. The real value of computer-aided engineering is in helping to optimize different parameters and helping to establish a more realistic system margin. It is important to mention again that considerable knowledge and skills are needed to apply software-modeling tools properly.

The computer-based programs and simulation tools are capable of optimizing a large number of parameters, which helps to achieve the engineering goal at a lower cost. There are several software packages available today, such as Virtual Photonics-VIP™, BroadNeD-BneD™, and OptiWave™, that are well written and relatively robust. They also offer the point-and-click interface and system composition. On the other hand, the user is responsible for efficiency and accuracy, which means that the user should be armed with knowledge to handle the situation properly.

There are also some software applications that are more open to the user's input. However, they require an even more sophisticated user to feed the model with data and establish a firmer connection with experimental results.

Simulation software can be written by using different building blocks and tools (i.e., C/C++, MathCad, MatLab) and can run on different platforms (Windows, Linux, Sun). The essential part of the software is the database, or a set of libraries that contain data related to optical modules, electronic modules, digital signal processing, protocols, and the mathematical fundamentals. As for the source code, software can be quite open for everything, open for module parts only, or entirely closed to the user input (i.e., just application oriented). One of the most important parameters of the software is the processing time, which should not be too long while enabling specified accuracy.

As already mentioned, some software packages are open for module modeling, which might be very advantageous in some situations. In such a case, proprietary equations and data can be used to obtain more reliable results. Modeling of components needs both the physical level and system level insights, which becomes quite clear if we recall the fundamentals of EDFA, Raman amplifiers, and external modulators. On the other hand, there is a black-box approach that follows a pre-established modeling pattern for given input parameters. In the optical amplifier case, for example, input parameters are the noise figure, dynamic gain curve, gain saturation curve, and the output power.

The computer-aided engineering process involves several steps that are illustrated in Figure 5.15. A random bit pattern is generated first to include a stream of

5.4 Systems Engineering and Margin Allocation

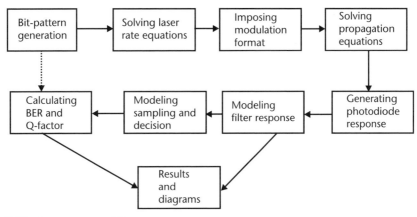

Figure 5.15 Computer modeling process.

"0" and "1" bits in a combination that is as close as possible to the case that can happen in reality. That means that the length of the pseudorandom sequence should be as long as possible. However, since the longer sequence increases the calculation time, it should be chosen carefully. The same issue happens when using the measurement equipment. From a practical viewpoint, it is advised to use the data stream length equal to $N = 2^7$ bits to get the initial results. It can be doubled in the second iteration to observe the impact of the sequence increase.

The next step involves modeling of the optical signal generation process. It may include either numerical solving of the laser rate equations, or just calculating the output signal parameters from the closed form equations. After that, the signal modulation process is modeled to account for the modulator transfer function and the signal code properties. It is very important to properly capture the frequency chirping process during the modulation process.

The following step in the modeling process includes solving of the wave equations. The coupled nonlinear Schrodinger equations are commonly used for this purpose in order to incorporate different propagation effects (chromatic dispersion, PMD, nonlinear effects) [see (7.29) to (7.34)]. The analysis of transmission characteristics is usually done in the frequency domain, since each frequency component evolves independently of each other. The Fourier transform of the electric field envelope is used to account for different effects, while the inverse Fourier transform is eventually used to convert the signal back to the time domain. This approach is good for evaluation of frequency-related effects, where time points do not evolve independently. The Fourier transform is not quite good enough for the evaluation of different nonlinear effects, since each time point evolves independently and the frequency calculation can be complicated. There are several approaches to account for nonlinear effects, and they are:

- Frequency decomposition and evaluation by assuming that each frequency propagates independently, which is good if the impact of nonlinear effects is not high.
- Split-step modeling, where derivatives are done in the frequency domain but multiplications are calculated in the time domain. In addition, a random noise

is added at in-line optical amplifiers. This approach is good for cross-coupling effects, but it can be very time consuming and may underestimate the quantity of the ASE noise.

- Multiple-length scale models, in which each process is related to the specific length associated with its generation (i.e., the micron-lengths for laser cavity and fiber core, the meter-length for pulse duration since the duration is converted to length by multiplying it with the light speed, the kilometer-lengths for attenuation and the nonlinear effects, the thousands of kilometers–length for dispersion map and soliton jitter). This method might be good from the system engineering perspective but only if it is correlated with experimental results and if it uses measured values of different key parameters as model inputs.
- Random process modeling, which may include raw data processing by using the Monte-Carlo method. It can also use some equations that deal with statistical parameters (i.e., with the mean and standard deviation). It usually includes the so-called linearization process, first to evaluate the waveforms without any noise, and next to linearize the evolution equations around the obtained solution. In the final phase, an evolution of the solution is found by assuming that input parameters are statistical variables. This approach may be good for some special cases but may also produce results that are different from experimental ones.
- Empirical methods that take known experimental results to extrapolate penalties associated with a new regime. This approach might be the most effective from a systems engineering perspective. On the other hand, it is not good in predicting what happens when conditions change.

The modeling of the photodetection process employs the transfer function of the photodetector to evaluate the signal transferal from optical to electrical levels. In addition, different noise components are produced by a random generator and added to the signal waveform after the photodetection process. The generation of random noise components can be done in accordance with different models, such as Poisson, Gaussian, and Maxwellian. It is common practice, though, to model both the thermal noise and the quantum shot noise through Gaussian statistics (refer to Sections 3.2 and 4.2). In addition to noise generation, the statistical variations attributed to polarization mode dispersion are also generated.

The Monte-Carlo method is commonly used to add both the noise components and PMD variations to numerical results that are related to the signal waveform after the photodetection process has been modeled. This part of the modeling is very important since all effects that occur in the optical receiver have a strong impact on the overall results and to a quantitative agreement with experimental data.

The next step includes the pulse shaping by the receiving filter. The filter characteristics, which include the transfer function and filter bandwidth, are very important for producing the pulse shape that minimizes intersymbol interference. The obtained pulse shape can be monitored through a simulated eye diagram (refer to Figure 7.4). The eye opening is directly correlated with the Q-factor, which is calculated before the decision process take place. The eye opening is commonly used for monitoring the Q-factor sensitivity to different parameters and for investigating the trade-offs that would lead to optimization of the system performance. Variations in the Q-factor can also be used to estimate the power penalty associated with different impairments and the power margin that should be allocated to compensate for them. In addition, it is also important to establish a very precise correlation between the Q-factor and SNR that will reflect the case in question. The relation between Q-factor and SNR can be used as an input for system performance monitoring.

The final step in the simulation process includes modeling of the decision circuit. In this step, the impacts of the decision threshold level and the timing jitter are accounted for. Both parameters are numerical variables that can be changed to estimate the associated power penalties. The output from the simulation module of the decision circuit is the signal bit-pattern, which can be compared with the input data bit stream in order to produce the BER curves.

Computer-aided models are generally very useful for more complex transmission setups, such as long-haul systems with optical amplifiers and multichannel transmission, which may include a large number of different elements. Each element is characterized by a set of parameters. If an empirical approach is used, just the nominal values of parameters would be entered into corresponding equations that produce a solution. However, there is a large number of different elements and pieces from the same category that are employed in practical systems, and they do not have exactly the same characteristics. (It is necessary to outline that each of these elements, such as optical fibers and cables, optical connectors, and optical multiplexers, satisfies the quality requirements.)

Therefore, we have a situation in practice where there are a large number of parameters, and each of them varies around its nominal value. A statistical approach is very effective in estimating the impact of parameter variations [17]. It is not likely that all parameters would simultaneously take such values that would produce the worst case scenario. Therefore, system requirements can be defined to meet requirements with some high probability. Usually, that probability can vary from 98% to 99.99%.

5.5 Summary

The system engineering process has been defined and has been demonstrated in several practical scenarios. The reader is advised to pay a special attention to key engineering formulas defined by the following equations: (5.9), (5.19), (5.29), (5.33), (5.34), (5.69), (5.74), (5.77) to (5.78), and (5.79). The reader should refer to Table 5.1 to get a quick and clear picture with respect to the significance of different engineering parameters. In addition, Tables 5.3 to 5.5 can be used as reference tools related to the most common cases. A thorough understanding of the system engineering process is essential for its practical implementation. By following the process presented in this chapter, the reader will know where the process will end up and have a solid analytical tool for a reality check in different scenarios. In addition, the reader will have insight into computer-aided engineering, with instructions of what can be expected if someone uses software modeling tools.

References

[1] Personic, S. D., et al., "A Detailed Comparison of Four Approaches for the Calculation of the Sensitivity of Optical Fiber System Receivers," *IEEE Trans. Commun.*, Vol. COM-25, 1977, pp. 541–548.

[2] Smith, D. R., and I. Garrett, "A Simplified Approach to Digital Optical Receiver Design," *Optical Quantum Electronics*, Vol. 10, 1978, pp. 211–221.

[3] Personic, S. D., *Optical Fiber Transmission Systems*, New York: Plenum, 1981.

[4] Abramovitz, M., and I. A. Stegun, *Handbook of Mathematical Functions*, New York: Dover, 1970.

[5] Personic, S. D., "Receiver Design for Digital Fiber-Optic Communication Systems," Bell Syst. Techn. J., Vol 52, 1973, pp. 843–886.

[6] Agrawal, G. P., *Fiber Optic Communication Systems*, 3rd ed., New York: Wiley, 2002.

[7] Marcuse, D., *Light Transmission Optics*, New York: Van Nostrand Reinhold, 1982.

[8] Ramaswami, R., and K. N. Sivarajan, *Optical Networks*, San Francisco, CA: Morgan Kaufmann Publishers, 1998.

[9] Muoi, T. V., "Receiver Design for High Speed Optical Fiber Systems," *IEEE J. Lightwave Techn.*, Vol. LT-2, 1984, pp. 243–267.

[10] Proakis, J. G., *Digital Communications*, 3rd ed., New York: McGraw-Hill, 1995.

[11] Kazovski, L., S. Benedetto, and A. Willner, *Optical Fiber Communication Systems*, Norwood, MA: Artech House, 1996.

[12] Keiser, G. E., *Optical Fiber Communications*, 3rd ed., New York: McGraw-Hill, 2000.

[13] Bendelli, G, et al., "Optical Performance Monitoring Techniques," *Proc. European Conference on Optical Communications ECOC*, Munich, 2000, Vol. 4, pp. 213–216.

[14] ITU-T Rec. G.751, "Digital Multiplex Equipments Operating at the Third Order Bit Rate of 34 368 kbit/s and the Fourth Order Bit Rate of 139 264 kbit/s and Using Positive Justification," ITU-T (11/88), 1988.

[15] ITU-T Rec. G.709/Y1331, "Interfaces for the Optical Transport Network (OTN)," ITU-T (02/01), 2001.

[16] Menyuk, C., "Modelling Nonlinear Lightwave Systems," *Tutorial at Optical Fiber Conference, OFC'98*, San Jose, CA, 1998.

[17] Jeruchim, M. C., P. Balaban, and K. S. Shamugan, *Simulation of Communication Systems*, New York: Plenum Press, 1992.

CHAPTER 6
Optical Transmission Enabling Technologies and Trade-offs

This chapter introduces advanced methods and technologies that could enhance the overall transmission performance. It also explains how to use benefits brought by the application of advanced methods and schemes to offset the impacts of impairments. Finally, it discusses the trade-offs related to the performance evaluation, as an integral part of the systems engineering process.

The systems engineering process presented in Chapter 5 was related to SNR calculations, evaluation of various impairments, and allocation of power margins. It was recognized that in some cases optimization of systems parameters and trade-offs between different options might be necessary in order to provide the required power margin.

The parameter optimization and trade-offs are very often related to advanced technologies and techniques that can improve the overall system performance, by enhancing the signal and suppressing the negative impact of different impairments. There are a number of technologies that can help to enhance the overall system performance. Some of them, such as using advanced optical amplifiers and advanced dispersion compensation schemes, help to suppress the noise level and signal waveform degradation. Others, such as forward error correction or advanced modulation schemes, can bring relief to the total SNR that is needed to achieve a specified bit error rate, or help to minimize the impact of nonlinear effects. In addition, there are some advanced transmission methods, such as soliton transmission, that can capitalize on the effect of mutual cancellation of different impairments. Finally, there are different detection schemes, such as coherent detection, that can bring a considerable improvement to the receiver sensitivity.

6.1 Enabling Technologies

Enabling technologies can be related either to components (such as amplifiers, compensators, and modulators) or to the methods (such as the forward error correction, modulation schemes, and detection schemes). We will examine the most important methods that might have a positive impact on the systems engineering process in years to come. Although the content discussed here can be considered to be just an illustration of the current state of art, it will outline the engineering benefits of using each of them and point in the direction of additional systems enhancement.

6.1.1 Optical Amplifiers

Optical amplifiers were discussed in Section 2.4 with an emphasis on EDFA. In this section we will examine the characteristics and common application schemes of Raman optical amplifiers since they can bring additional benefits and enhance the overall system performance. Raman amplifiers are based on the SRS effect, which was explained in Section 3.3. The point is that the transmission fiber itself becomes the amplification medium instead of using specialty-doped optical fiber for signal amplification. The amplification is not done through the process of stimulated emission, bur rather through an inelastic scattering, in which an incident primary photon generates the secondary photon with lower energy than itself, while the rest of the energy is transferred to the molecule vibrations.

While the SRS effect in optical fibers occurs among different optical channels, the SRS effect in Raman amplifiers occurs between a specialty pump and individual channels. The signal gain is created due to conversion of the pump photons to signal photons. If interaction occurs over distances (measured in tens of kilometers) of regular transmission fiber, we have the distributed amplification process and distributed Raman amplifiers (DRA). On the other hand, if the interaction occurs over limited fiber length, which is not part of the original transmission line, we have lumped (or discrete) Raman amplifiers. It is common to use DCFs to realize lumped Raman amplifiers. In such a case, lengths of a few kilometers of DCF allow for efficient use of the SRS effect. In this section we will pay more attention to distributed Raman amplifiers to explain the specifics of the amplification process and benefits from the application perspective. On the other hand, it is worth mentioning that lumped Raman amplifiers can be characterized by using the same methodology that was applied to other optical amplifiers (refer to Sections 2.4 and 3.3.1).

The pump is introduced in optical fiber through a broadband WDM coupler, which has the same structure as one used in EDFA, but here it can be applied for pumping in both the backward and forward directions. In addition, the Raman pumping scheme produces polarization-dependent gain, while the polarization dependence is minimal in EDFA. This is the main reason why special precautions are taken to equalize the impact of both polarizations in the pumping scheme, as shown in Figure 6.1. It could be done either by scrambling the input polarizations coming from the pump lasers, or by two orthogonal polarizations at the same wavelength that are combined together by a polarization beam combiner.

Backward pumping was often used in the early stage of Raman amplifier deployment, but recently both backward and forward pumping have been considered, while an optimization between backward versus forward portions has been performed [1]. The Raman amplifier characteristics can be studied by using (3.120) and (3.121), which can be rewritten in the following form related to the forward pumping:

$$\frac{dP_P}{dz} = -\left(\frac{g_R}{A_{eff}}\right)\left(\frac{\omega_P}{\omega_S}\right)P_P P_S - \alpha_P P_P \qquad (6.1)$$

$$\frac{dP_S}{dz} = \frac{g_R}{A_{eff}} P_P P_S - \alpha_S P_S \qquad (6.2)$$

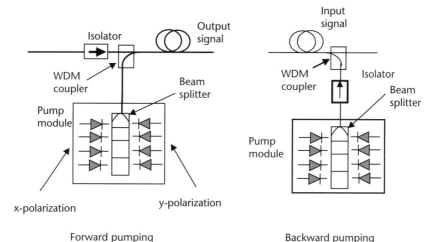

Figure 6.1 Schematics of Raman amplifiers.

where z is the axial coordinate, g_R is the Raman amplification coefficient (gain), α_R and α_s are the fiber attenuation coefficients for the pump and the signal, respectively, and A_{eff} is the effective cross-sectional area of the fiber core that is defined by (3.101). Note that the optical powers P_p and P_s, which refer to the pump and signal, respectively, were introduced in (6.1) and (6.2) to replace the associated signal intensities $\Pi_p = P_p/A_{eff}$, and $\Pi_s = P_s/A_{eff}$. It is also important to mention that in this case the scattered Stokes wave contributes and enhances the incoming signal.

The signal power at the output of the amplifier length L, which is found by solving (6.1) and (6.2), can be expressed as [2, 3]

$$P_s(L) = P_{S0} \exp(-\alpha_s L) \exp\left(\frac{g_R P_{P0} L_{eff}}{A_{eff}}\right) \approx P_{S0} \exp(-\alpha_s L) \exp\left(\frac{g_R P_{P0}}{\alpha_p A_{eff}}\right) \quad (6.3)$$

where $P_{S0} = P_s(0)$ is the launched signal power, P_{P0} is the input pump power, and L_{eff} is the effective length for the pump signal defined by (3.99). The approximation $L_{eff} \sim 1/\alpha_p$ was used in (6.3), which is quite valid if $L >> 1/\alpha_p$. The gain of the Raman amplifier can be obtained from (6.3) as

$$G_R = \frac{P_s(L)}{P_{S0}\exp(-\alpha_s L)} = \exp\left(\frac{g_R P_{P0} L_{eff}}{A_{eff}}\right) \approx \exp\left(\frac{g_R P_{P0}}{\alpha_p A_{eff}}\right) = \exp(g_0 L) \quad (6.4)$$

where the amplification coefficient g_0 from above equation can be expressed as

$$g_0 = g_R \left(\frac{P_{P0}}{A_{eff}}\right)\left(\frac{L_{eff}}{L}\right) \approx \frac{1}{L}\left[\frac{g_R(\omega) P_{P0}}{\alpha_p A_{eff}}\right] \quad (6.5)$$

There are several important conclusions that follow from (6.5). First, the Raman amplification coefficient keeps the same frequency dependence as the Raman gain spectrum [see Figure (3.25) as a reference]. Second, the Raman amplification coefficient is strongly dependent on fiber type through the effective cross-sectional area parameter. It is stronger for optical fibers having smaller cross-sectional area, and vice versa. The largest amplification coefficient is in DCFs, where it is six to eight times higher than in other fiber types. That is the reason why DCF fibers can be effectively used to construct the lumped Raman amplifiers.

The Raman gain increases exponentially with the pump power, while entering into saturation regime after the pump power exceeds the level of approximately 1W. This can be verified by solving (6.1) and (6.2) numerically. An approximate expression for saturated gain value, which was obtained in [3], is

$$G_{R,sat} = \frac{1+\rho_0}{\rho_0 + G_R^{-(1+r_0)}} \qquad (6.6)$$

where

$$\rho_0 = \frac{\omega_P}{\omega_S} \frac{P_{S0}}{P_{P0}} \qquad (6.7)$$

The amplifier gain is reduced by about 3 dB if the power of amplified signal becomes comparable with the input pump power P_{P0}. Since P_{P0} is relatively high, the amplifier will basically operate in a linear regime, while its gain will be $G_{R,sat} \sim G_R$.

In parallel with the signal amplification, the Raman amplifier also induces signal impairments. These impairments are related to the ASE noise, double Rayleigh backscattering (DRB) noise, and signal distortion due to nonlinear Kerr effect [1]. Two dominant noise sources (the ASE noise, and the DBR noise) are illustrated in Figure 6.2.

The power of ASE noise can be found by applying a general formula given by (7.43); that is,

$$P_{ASE} = 2n_{sp} h\nu_S \Delta\nu_R = \frac{2h\nu_S \Delta\nu_R}{1 - \exp\left(\frac{h\nu_S}{k\Theta}\right)} \qquad (6.8)$$

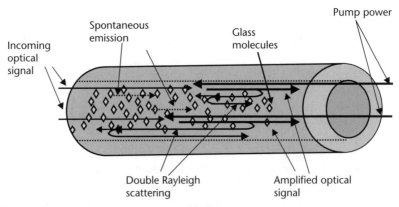

Figure 6.2 Signal and noise in the Raman amplification process.

where $v_s = \omega_s/2\pi$ is the optical frequency of the signal, n_{sp} is the spontaneous emission factor that is determined by the thermal energy, and Δv_R is the gain bandwidth (refer to Figure 3.26). The factor 2 in (6.8) accounts for two polarization modes within the spontaneous emission. The power of spontaneous emission is much smaller than that in EDFA, due to the distributed nature of the Raman amplification. One can see from (6.8) that the ASE noise is broadband and white noise–like.

Generation of DRB noise is also illustrated in Figure 6.2. This effect occurs due to Rayleigh scattering, which is normally present in all optical fibers, but to a much smaller extent. We should recall that the major part of the scattered light goes backward, while a smaller part takes a forward direction and adds to the signal. If there is a distributed amplification process, the photons due to the Rayleigh scattering are multiplied. In this case we can consider that the forward propagating portion of the Rayleigh scattering noise is added to the ASE noise, thus contributing to the increase of the total noise level. Generally speaking, even amplified, this portion is not a serious issue.

However, what becomes an issue relates to multiple reflections of the enhanced Rayleigh scattering. In this case, the optical fiber is acting as a weak distributed mirror. The backward scattering eventually turns to the forward one, thus creating crosstalk to the propagating optical signal. This crosstalk noise, which is referred to as the DRB noise, accumulates in long-haul transmission systems that have a large number of amplifiers. Such systems are, for example, submarine transmission systems, in which the DRB noise can reach a level that severely degrades the system performance. The power of the DRB noise can be calculated as [1]

$$P_{DRB} = P_S(0) G_R r^2 \int_0^L G^{-2}(0,z) \int_z^L G^2(0,\xi) d\xi dz \qquad (6.9)$$

where $P_S(0)$ is launched signal power, G_R is Raman gain, r is the Rayleigh backscattering coefficient, $G(0,z)$ is the net gain over distance ranging from 0 to z (the net gain equals the amplification minus attenuation), and L is the fiber span length that precedes to distributed Raman amplifier. In case of lumped Raman amplifiers, just the ASE noise will be present, while the DRB noise can be neglected. The DRB noise is not white-like, but it is rather Gaussian-like and related to the signal.

Finally, the Raman amplification process also introduces the signal impairment that comes from impact of the Kerr effect. That impact can be evaluated by calculating the integrated nonlinear phase along the length L, which is

$$\Phi_{DRA} = \int_0^L \gamma P(z) dz \qquad (6.10)$$

where γ is the nonlinear Kerr coefficient introduced by (3.105), and $P(z)$ refers to the evolution of the signal power. The power evolution in DRAs is more complex than a simple attenuation that occurs in transmission optical fibers, which are often referred to as passive optical fibers. In a general case, the distributed Raman amplifier is bidirectionally pumped, while the ratio between the forward and backward pump powers is a factor that determines the character of the optical signal evolution along the span length. If there was no pumping, the integrated phase given by (6.10) would take a linear form, which can be calculated as

$$\Phi_{pass} = P_S(0)\int_0^L \gamma \exp(-\alpha_s z)dz = \gamma P_S(0)\frac{1-\exp(-\alpha_s z)}{\alpha_s} \quad (6.11)$$

The index "pass" in (6.11) stands for passive, which means that phase is calculated when there is no pumping at all. The impact of the nonlinear index can be measured through the ratio

$$R_{NL} = \frac{\Phi_{DRA}}{\Phi_{pass}} \quad (6.12)$$

Both the ASE noise and the DRB noise will accompany the signal at the photodiode area. They will beat with the signal to produce the beat-noise electrical components during the photodetection process. The dominant components of the beat-noise will come from the signal-DRB beating and the signal-ASE beating. It was found in [1, 4] that the following approximations for noise variances can be used to evaluate the beat-noise components

$$\sigma^2_{S,ASE} = 2R^2 P_{ASE} P_{S,1} \Delta f \quad (6.13)$$

$$\sigma^2_{S,DRB} = 2R^2 P_{DRB} P_{S,1} \Delta f \quad (6.14)$$

where R is photodiode responsibility, $P_{S,1}$ is the signal power related to "1" bit, while P_{ASE} and P_{DRB} represent powers of the ASE and DRB noise components given by (6.8) and (6.9), respectively. The equivalent noise figure of a distributed Raman amplifier, such as the scheme shown in Figure 6.1, can be calculated as [1]

$$NF_{DRA} = \frac{SNR_{in}}{SNR_{out}} = \frac{1}{G_R \Gamma}\left(\frac{P_{ASE}}{h\nu_s} + \frac{5P_{DRB}}{9h\nu_s \sqrt{(\Delta f^2 + B_{op}^2/2)}} + 1\right) \quad (6.15)$$

where, ν_s is the optical frequency, Δf is the frequency bandwidth of the electrical receiver, B_{op} is the bandwidth of the optical filter, G_R is the Raman gain, and Γ is the loss at the optical fiber span $\Gamma = exp(-\alpha_s L)$.

The net result of the Raman amplifier deployment is the SNR enhancement, since an employment of Raman amplifiers will prevent the signal from coming closer to the noise level. The benefit of distributed Raman amplification can be estimated in a real-case scenario where a Raman amplifier is deployed in combination with EDFA, as shown in Figure 6.3. Since the Raman amplifier includes the entire fiber span, it is useful to introduce the noise figure of the fiber span section to make an "apple to apple" comparison between different cases.

The overall noise figure of the single fiber span in the case when only EDFA amplifiers are deployed can be expressed as

$$NF_{pass} = \frac{NF_{EDFA}}{\Gamma} + 1 - \Gamma_{loss} \quad (6.16)$$

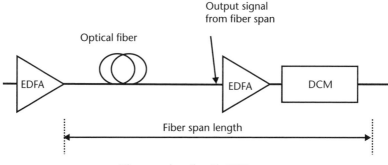

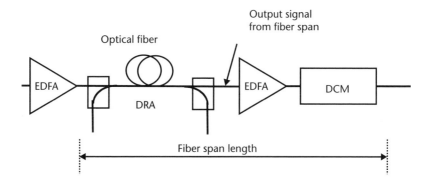

Figure 6.3 Deployment of Raman amplifiers.

where NF_{EDFA} is the noise figure of the EDFA amplifier, Γ is optical fiber span loss, and Γ_{loss} is the loss caused by some other elements, such as DCM inserted at the amplifier site. If DRA is activated in the previous fiber span, as shown in lower part of Figure 6.3, the following formula can be applied:

$$NF_{act} = NF_{DRA} + \frac{NF_{EDFA} - 1}{G_R \Gamma} + 1 - \Gamma_{loss} \qquad (6.17)$$

where NF_{act} refers to the case when EDFA is combined with the Raman pumping scheme (the index "act" stands for active, which means that the Raman pumping is in place), and NF_{DRA} is the equivalent noise figure of the DRA given by (6.15).

In a general case each span might operate at different launched powers due to different Kerr nonlinearities, but we can assume that the Kerr effect is the same for all spans. The improvement in the SNR after N spans can be calculated as

$$R_{DRA} = \frac{SNR_{pass}}{SNR_{act}} = R_{NL} \frac{N(NF_{act} - 1) + 1}{N(NF_{pass} - 1) + 1} \approx R_{NL} \frac{NF_{act}}{NF_{pass}} \qquad (6.18)$$

It was shown in [1] that the previous ratio can be optimized by optimizing the ratio of backward pumping to forward pumping, while it was found that the mix of 30% to 35% of forward pumping and 65% to 75% of backward pumping can bring the largest benefit. The improvement of the SNR is illustrated in Figure 6.4 for two different values of the Raman gain. One can see from Figure 6.4 that the signal does not come close to the noise level, as it would if just EDFA were deployed. In this case, the Raman amplifier acts as a low-noise preamplifier to EDFA and helps to reduce both the gain and the noise figure of EDFA. On the other hand, it is also beneficial for Raman amplifiers to operate in combination with EDFA since it allows them to stay below the DRS limit.

The total yield of Raman amplification is proportional to the improvement factor R_{DRA} from (6.18). This yield can also be transferred to the optical level and considered to be the optical power gain or a negative optical power penalty. For this purpose we can apply (5.34), which leads to the following relation:

$$\Delta P_{DRA} = R_{DRA} \frac{2\Delta f}{B_{op}} \tag{6.19}$$

where ΔP_{DRA} is the optical power gain that is brought to the system by employment of distributed Raman amplifiers, Δf is electrical filter bandwidth at the receiving side, and B_{op} is the bandwidth of an optical filter. The value of ΔP_{DRA} can be taken directly to the right side of (5.79) to relieve power margin requirements, which were in effect before the Raman amplifier was employed. In such a way, ΔP_{DRA} plays the role of a "positive" power margin.

The impact of the positive power margin is illustrated in Figure 6.5. As we see, with Raman amplification involved, the power per channel P'_{ch} can be much lower than the value P_{ch} otherwise needed to satisfy the original requirement with respect to the OSNR. In this scenario, the power per channel should be increased to value $P'_{ch,required}$ to compensate for the impact of impairments by using the allocated margin. On the other hand, the power per channel would increase to the value $P_{ch,required}$ if

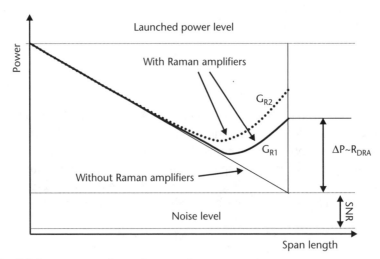

Figure 6.4 SNR improvement due to Raman gain.

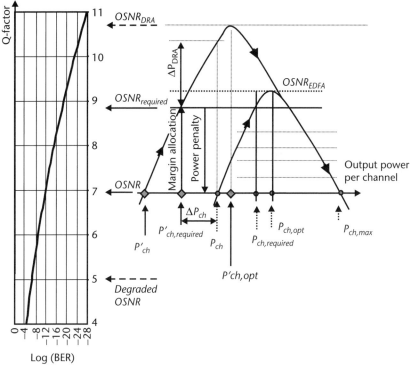

Figure 6.5 Margin allocation with DRA.

there were no Raman amplifiers. Therefore, there is a difference $\Delta P_{ch} = P_{ch,required} - P'_{ch,required}$ related to the channel power, which is very beneficial since it enables lower power operation and suppression of nonlinear effects.

We can also see from Figure 6.5 that the signal-to-noise ratio ($OSNR_{DRA}$), which corresponds to the optimum channel power $P'_{ch,opt}$, is considerably higher than the highest OSNR achievable in case when there were no Raman amplifiers. Therefore, the end result of the distributed Raman application scheme is that the Raman amplifier provides the power relieve that can be used in different ways. For example, it can be used to increase the number of spans or to add additional elements, such as optical add-drop multiplexers (OADMs) along the lightwave path.

As already mentioned, an efficient Raman amplification can also be applied over shorter lengths of the optical fiber if the cross-sectional area is smaller. In such a case, we are talking about lumped Raman amplifiers. This scheme can been used to compensate for the losses in a DCF by introducing backward pumping through it, to work in combination with distributed Raman amplifiers (so-called all-Raman combination) to enhance the power level, or to amplify the signals placed within the S band, as in [5]. The way the lumped Raman amplifiers are analyzed is similar to that applied to EDFA. For example, the gain value can be calculated by using (3.8) to (3.10), while the noise parameters can be handled by using (3.32) to (3.35) in combination with (6.8). It was shown in [5] that DCFs can be used as an effective fiber medium for lumped Raman amplification since they have a smaller effective area and a higher gain than transmission optical fibers.

Raman amplifiers can cover a wide area of optical wavelengths, while their application area is determined by the selection of pump wavelengths. This feature makes them suitable for WDM applications in different wavelength windows since proper combinations of pump wavelengths and their relative strengths can produce a broadband optical amplifier with flattened gain profile, as illustrated in Figure 6.6.

If there are several pumps acting together, the SRS effect might even occur among them, which should be taken into account during the pump design. The design parameters of broadband Raman amplifiers can be obtained by solving numerically a set of coupled propagation equations [refer to (7.32) to (7.34)]. As for the application area, the Raman amplification offers a wide range of possibilities, such as one proven through a three-band WDM transmission experiment with the total capacity of 10.9 Tbps, in which it provided a gain of about 10 dB over the 120-nm wavelength range [6].

6.1.2 Advanced Dispersion Compensation

It was mentioned in Chapter 5 that transmission systems could be either power budget (or signal loss) or bandwidth limited (i.e., dispersion limited). The fiber attenuation and other losses along the long-haul transmission line can be compensated by optical amplifiers to overcome the power-budget limit, which means that dispersion now becomes a major limiting factor in terms of the transmission distance that can be achieved. In general, dispersion will cause pulse spreading and power penalties due to the intersymbol interference effect.

Power penalties related to chromatic dispersion and PMD were studied in Sections 4.3 and 4.4. The power penalties due to dispersion effects can be alleviated to some extent by the optical power increase, or suppressed by applying some methods that would provide a relief against pulse dispersion effects.

Dispersion compensation is a commonly used method that goes back even to multimode optical fibers, where different graded-index profiles fibers were proposed to work together and to compensate for each other's dispersion. As for the chromatic dispersion compensation, it is applied just to single-mode fibers since it would not have any real merit if applied in systems with multimode optical fibers.

Therefore, when we talk about dispersion compensation, we have in mind just single-mode optical fibers. The dispersion compensation techniques related to both

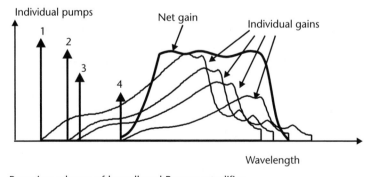

Figure 6.6 Pumping scheme of broadband Raman amplifier.

the chromatic dispersion and PMD will be discussed in this section. However, we will pay more attention to chromatic dispersion compensation since it is essential from the practical perspective; there are several methods that are commonly used today. On the other hand, the PMD compensation is more complex and has not found a wider application yet, although the situation might change with the introduction of 40-Gbps based transmission systems.

If there is a chromatic dispersion compensation scheme employed, the total power penalty will be related just to the impact of an uncompensated portion of dispersion, which is often referred to as the residual dispersion. In addition to compensation ability, the merit of any compensation scheme is evaluated with respect to the following parameters: the insertion loss, the width of the frequency band that can be compensated, the transparency with respect to the signal content (bit rate, modulation format), the ability to compensate nonlinear effects in parallel with dispersion compensation, and the physical size of the compensation equipment. There are several methods proposed so far that can be effectively used for chromatic dispersion compensation. They can be divided into the following groups:

- Predistortion, which is done at the transmitter side by controlling the frequency chirp of transmitted optical signal;
- Postdetection, which is done either by using adaptive feedback for nonlinear cancellation of dispersion effects, or by microwave equalization in a coherent optical receiver;
- In-line compensation, which is done along the lightwave path by using all-optical schemes that diminish the chromatic dispersion effect. There are several methods from this category that have been proposed so far, and they use one of the following: DCFs, fiber Bragg gratings, two-mode fibers, phase-conjunction modules, middle system spectral inversion modules, narrowband filters, and the Mach-Zehnder filters.

Some of methods listed above have not found wider application yet [7], while some others are commonly used. I will comment on each of them, paying more attention to the methods that are more commonly used.

Predistortion methods modify characteristics of the input optical pulses before they enter the optical transmission line. This pulse modification is related to both the amplitude and the phase, which eventually determine the pace of the pulse transformation during the propagation process. The most promising method that belongs to this category is the prechirp technique, which can be explained by using the concept of chirped Gaussian pulses [see (3.63) to (3.70)]. The electric field $E(z,t)$ of the input chirped Gaussian pulse at optical frequency ω_0 is given as

$$E(0,t) = A(0,t)\exp(-j\omega_0 t) = A(0)\exp\left[-\frac{(1+jC_0)t^2}{2\tau_0^2}\right]\exp(-j\omega_0 t) \qquad (6.20)$$

where $A(0) = A_0$ is the peak amplitude, and $2\tau_0$ represents the full width at the 1/e intensity point (FWEM). Please recall from Section 3.4.4 that the instantaneous frequency increases linearly from the leading to the trailing edge for a positive chirp parameter ($C_0 > 0$), and that is referred to as a positive chirp. On the other hand, the instantaneous frequency decreases linearly from the leading to the trailing edge for a negative chirp parameter ($C_0 < 0$), which is known as a negative chirp. We can also recall from Section 3.4.4 that when $C_0 \beta_2 < 0$, where β_2 is the GVD coefficient introduced by (3.61), the pulse goes through an initial compression as a dispersive optical fiber before it starts broadening again. The transmission distance that is associated with the maximum of pulse compression was estimated in [3] as

$$L = \left(\frac{\tau_0^2}{|\beta_2|}\right)\frac{C_0 + \sqrt{1+2C_0^2}}{1+C_0^2} = L_D \frac{C_0 + \sqrt{1+2C_0^2}}{1+C_0^2} \qquad (6.21)$$

where $L_D = \tau_0^2/|\beta_2|$ is the dispersion length, which was introduced in Section 3.4.4. The function $L(C_0)$ has a maximum value for $C_0 \sim 0.70$. However, this value is applicable just to Gaussian pulses, and it might be different for other pulse shapes. Therefore, a careful optimization is needed to achieve an optimum performance with pulses that do not have an ideal Gaussian shape. The optimizations of the chirp parameter can almost double the transmission distance, as compared with the distance associated with unchirped pulses.

The benefit of the pulse prechirping is pretty small for direct modulation schemes since semiconductor lasers impose relatively large negative chirp ($C_0 < 0$) through the amplitude-phase coupling parameter α_{chirp} [refer to (3.49)]. In addition, it can be observed only in the normal dispersion region (i.e., for $\beta_2 > 0$, or $D < 0$, where D is the chromatic dispersion coefficient). On the other hand, the external modulators offer larger flexibility. The condition $C_0 \beta_2 < 0$ can also be satisfied in the anomalous dispersion region (i.e, for $\beta_2 < 0$, $D > 0$) by programming the chirp parameter to be positive ($C_0 > 0$). The chrip programming can be done either by proper biasing and an external modulator, or by using a conversion of frequency modulation (FM) to amplitude modulation (AM).

The FM-AM conversion is based on using frequency modulation of the carrier emitted by the DFB laser. This is applied to the signal before it enters the external modulator to be modulated by amplitude. The FM modulation of the CW carrier, which is followed by AM modulation of the signal, will eventually produce a stream of chirped pulses. If we assume that frequency modulation is done by modulating the bias current by a small sinusoidal current, the optical frequency of the signal entering external modulator can be expressed as

$$\omega(t) = \omega_0 \left(1 + \delta \sin \omega_{FM} t\right) \qquad (6.22)$$

where ω_0 is the optical frequency of the CW signal, δ is the magnitude of the frequency change, and ω_{FM} represents the frequency of the sinusoidal component of the bias current. We can expect that a sinusoidal current of several milliamperes will produce frequency deviations of about 10 MHz since coefficient δ is of order of 2.5 MHz/mA.

If an external modulator produces pulses with the Gaussian shape, as given by (3.66), the electric field from (6.20) can be expressed as

$$E(0,t) = A(0)\exp\left[-\frac{t^2}{2\tau_0^2}\right]\exp\left[-j\omega_0(1+\delta\sin\omega_{FM}t)t\right]$$
$$\approx A(0)\exp\left[-\frac{(1+jC_0)t^2}{2\tau_0^2}\right]\exp(-j\omega_0 t) \tag{6.23}$$

The initial chirp parameter C_0 is determined by using approximation $\sin(\omega_{FM}t) \sim \omega_{FM}t$, which is acceptable for small frequency variations. This parameter can be expressed as

$$C_0 = 2\delta\omega_{FM}\omega_0\tau_0^2 \tag{6.24}$$

Therefore, both the magnitude and sign of the chirp parameter can be programmed by changing the frequency and amplitude of the small sinusoidal bias current.

In addition to changing chirp through frequency modulation of the CW signal, the chirp parameter can be changed from inside an external modulator by applying phase modulation of the optical carrier, which will accompany the intensity/amplitude modulation. The phase modulation is done through the change of the refractive index in modulator (refer to Section 2.2.2). In this specific case it is done by using a small sinusoidal voltage applied together with the ac bias voltage. By assuming that the phase of the optical signal changes as

$$\phi(t) = \delta\cos(\omega_{PM}t) \tag{6.25}$$

(6.20) becomes

$$E(0,t) = A(0)\exp\left(-\frac{t^2}{2\tau_0^2}\right)\exp\left[-j\omega_0 t + j\delta\cos(\omega_{PM}t)\right]$$
$$\approx A(0)\exp\left[-\frac{(1+jC_0)t^2}{2\tau_0^2}\right]\exp(-j\omega_0 t) \tag{6.26}$$

The initial chirp parameter C_0 can be determined by using approximation $\cos(\omega_{PM}t) \approx 1 - (\omega_{FM}t)^2/2$, which is applicable for small frequency variations ω_{PM}. It is expressed as

$$C_0 \approx \tau_0^2\omega_{PM}^2\delta \tag{6.27}$$

Therefore, by changing the parameters of the small bias voltage component, we can also change the chirp parameter. Both types of commonly used external modulators (i.e., MZ and EA modulators) can be used to produce a programmable chirp. The chirp parameter C_0 in external modulators is usually tuned to be in the range of 0.5 to 0.9. The prechirping has become quite a practical method for integrated versions of optical modulators.

In addition to pulse prechirping during the modulation stage, it can be also done afterwards by applying nonlinear prechirping methods. These methods involve some nonlinear medium that is used to prechirp the pulse. Such a medium might be, for example, an SOA working in saturation regime. In this case, the gain saturation in the SOA leads to variations of the carrier density, which in turn change the refractive index value and impose the frequency chirp. The process that occurs can also be recognized as the self-phase modulation induced by gain saturation. The frequency chirp can be found as [3]

$$\delta v(t) = -\frac{\alpha_{chirp}}{4\pi}\frac{P(t)}{E_{sat}}[G(t)-1] \quad (6.28)$$

where $G(t)$ is the gain of SOA, which operates in a saturated regime. It occurs that the chirp imposed in a saturation regime is almost linear over a large portion of the amplified pulse. Pulse compression can occur if the pulse propagates through an optical fiber with $\beta_2 < 0$, since the chirp parameter has a positive value ($C_0 > 0$).

A properly selected section of a special optical fiber, which is inserted in front of the optical fiber transmission line, can also be used to prechirp the pulse. The prechirping occurs due to the SPM effect that takes place in this section. The amplitude of the Gaussian optical pulse, which enters the special fiber section, can be expressed as

$$A(0,t) = \sqrt{P_0}\exp\left(-\frac{t^2}{2\tau_0^2}\right) \quad (6.29)$$

where $P(t) = P_0 exp(-t^2/\tau_0^2)$ is the input pulse power. The pulse amplitude at the output of the special fiber section can be expressed as

$$A(L_{fib},t) = \sqrt{P_0}\exp\left(-\frac{t^2}{2\tau_0^2}\right)\exp\left[-j\omega_0 t + j\gamma L_{fib}P(t)\right]$$
$$\approx \sqrt{P(0)}\exp\left[-\frac{(1+jC_0)t^2}{2\tau_0^2}\right]\exp\left[-j\omega_0 t - j\gamma L_{fib}P_0\right] \quad (6.30)$$

where L_{fib} is the length of the special fiber section, and γ is nonlinear fiber index introduced by (3.105). The chirp parameter from (6.30) is given as

$$C_0 = 2\gamma L_{fib}P_0 \quad (6.31)$$

The chirp parameter expressed by (6.31) will take positive values since the parameter γ is positive for silica-based optical fibers. Therefore, the prechirping will be beneficial if transmission is done in the anomalous chromatic dispersion region (i.e., for $\beta_2 < 0$, $D > 0$).

As a summary note, we can say that although the prechirping offers some benefits, it cannot be used as a standalone method for dispersion compensation in long transmission lines. Because of that, it is usually combined with another compensation scheme, such as one that employs DCFs. In such a combined scheme, prechirping offers some relief in terms of dispersion that needs to be compensated by DCF, thus lowering both the length of DCF and the DCF insertion losses. In addition, several prechirping techniques can be combined together to increase the overall benefit.

The optimization of the parameters in a combined prechirping scheme is rather complicated, and numerical simulation and computer-aided modeling might be necessary for this purpose.

Postdetection dispersion compensation is done in optical receivers, and it is applied on electrical signals. Suppression of the impact of chromatic dispersion is done through an electronic equalization filtering. In some cases, such as in coherent detection schemes described in Section 6.1.4, the dispersion can be compensated by linear filtering methods. A microwave filter that corrects the signal distortion caused by chromatic dispersion should have a transfer function [8],

$$H(\omega) = \exp\left[\frac{-j(\omega - \omega_{IF})^2 \beta_2 L}{2}\right] \tag{6.32}$$

where ω_{IF} is the intermediate frequency, β_2 is the GVD coefficient, and L is the fiber length over which chromatic dispersion has been accumulated. The postdetection technique was considered to be very promising for coherent detection schemes [8, 9]. The microstrip lines that are several tens of centimeters long were used to compensate for chromatic dispersion accumulated over several hundreds of kilometers.

The linear filtering method can be used for coherent detection schemes, since the information about signal phase is preserved. On the other hand, the direct detection scheme does not keep track of phase, since the photodiode responds to the optical signal intensity. In this case, some other approaches can be taken, such as changing decision criteria based on the presence of preceding bits. For example more "1" bits preceding the bit in question will mean that there is more intense intersymbol effect to be accounted for, and vice-versa. The other approach is based on making a decision only after examining the signal structure, which consists of several bits in a row, in order to estimate the amount of intersymbol interference that should be compensated. This method generally requires fast signal processing and logical circuits operating at the higher clock than the signal bit rate. In addition, the longer the examined bit sequence is, the faster the signal processing should be. We can say that this method is the most effective for bit rates up to 10 Gbps, and for transmission distances up to several hundred kilometers.

Another approach in electronic compensation is to use the transversal filters, which are well known from telecommunication theory [10]. In this case, the signal is split in several branches that are mutually delayed and multiplied by weight coefficients. These components are eventually joined again, and the decision is made on a summary signal.

All electronic methods for dispersion compensation that are mentioned above can improve the system characteristics if the signal bit rate is up to 10 Gbps. Such an improvement is larger for lower bit rates and shorter distances, and vice-versa. These methods may be combined with some other schemes to enhance the total dispersion compensation capabilities.

In-line, or all-optical, chromatic dispersion compensation methods can be applied at any point along the lightwave path. However, in practice, they are usually employed at in-line amplifier sites. Dispersion compensating modules are positioned as a middle-stage in in-line optical amplifiers. In some cases, however, they can be placed at the optical transmitter side to perform signal predistortion, or to be a part of the optical receiver to perform postcompensation of the residual dispersion. There are several compensation schemes that belong to this category:

- Compensation with DCFs, which are widely used in various transmission systems;
- Compensation with FBGs, which can be effectively used in a narrowband transmission systems;
- Compensation with two-mode fibers, which can be efficiently used for bit rates up to 10 Gbps and transmission distances up to several hundred kilometers;
- The other compensation methods, such as phase conjunction, middle system spectral inversion, compensation with narrowband filters and MZ interferometers, which can be used in some special situations.

The use of DCF is the most common method since chromatic dispersion can be fully compensated if nonlinear effects are negligible. If nonlinearities are small, compensation can be considered as a linear process that satisfies the condition

$$D_1 L_1 = -D_2 L_2 \qquad (6.33)$$

where D_1 and L_1 are the chromatic dispersion and the length of the transmission optical fiber, respectively, and D_2 and L_2 are related to dispersion and the length of DCF. Since both lengths (L_1 and L_2) are positive parameters, (6.33) can be satisfied only if dispersion coefficients D_1 and D_2 are different in sign. Therefore, the DCF should have a negative dispersion (i.e., $D_2 < 0$) if optical transmission is done in the anomalous dispersion region, while just the opposite applies for the normal dispersion region.

Transmission length L_1 and dispersion D_1 are parameters that are known in advance, which means that (6.33) can be satisfied by selecting parameters L_2 and D_2. The length L_2 should be as short as possible to minimize the insertion losses. This requires the dispersion D_2 to be as high as possible. The DCFs with large negative dispersion are used to compensate for the chromatic dispersion effect in a periodic manner along the transmission line. Dispersion compensating optical fibers are designed to operate at low V parameter, which is done through a smaller core diameter [refer to (2.6)]. In addition to smaller core diameter, the so-called depressed cladding refractive index profile is applied while producing DCF [see Figure 2.12(d)].

Such design of a DCF leads to a relatively large waveguide dispersion component, which becomes dominant and determines the total character of chromatic dispersion. The total value of the dispersion coefficient D_2 for commercially available DCFs lies in the region from –90 to –150 ps/nm·km. On the other hand, although providing a negative chromatic dispersion coefficient, a special design of DCF also introduces higher fiber attenuation due to weaker guidance of the propagating mode through the core area. In addition, the design of DCF is more sensitive to microbending losses (refer to Section 2.3). The total attenuation coefficient α_{DCF} of DCF is in the range of 0.4 to 0.7 dB/km, which is considerably higher than the attenuation in transmission fibers in the wavelength region around 1,550 nm. The ratio $FM = |D_2|/\alpha_{DCF}$, which is also called the *figure of merit*, is used to characterize different types of DCF. It is clear that the figure of merit should be as high as possible. DCFs that are produced today, usually have $FM > 200$ ps/nm-dB.

The DCF can be used for chromatic dispersion compensation in both single-channel and multichannel transmission systems. If used in a multichannel or WDM environment, the condition given by (6.33) should be satisfied for each individual channel, which means that the (6.33) now becomes

$$D_1(\lambda_i)L_1 = -D_2(\lambda_i)L_2 \qquad (6.34)$$

where λ_i ($i = 1, 2, ..., M$) refers to any optical channel out of the total number of M channels multiplexed together. Since $D_1(\lambda)$ is wavelength dependent, as illustrated in Figure 3.12, the accumulated dispersion $D_1(\lambda)L_1$ will be different for each individual channel. That puts more stringent requirements on DCF, which should also have a proper negative dispersion slope to satisfy (6.34). The dispersion slope of the DCF can be calculated by assuming that (6.34) is satisfied for a reference channel, which is often the central wavelength λ_c in a composite WDM signal. Dispersion for other channels can be expressed through the reference, which leads to the following set of equations:

$$D_1(\lambda_c)L_1 = -D_2(\lambda_c)L_2 \qquad (6.35)$$

$$[D_1(\lambda_c) + S_1(\lambda_i - \lambda_c)]L_1 = -[D_2(\lambda_c) + S_2(\lambda_i - \lambda_c)]L_2 \qquad (6.36)$$

$$S_2 = -S_1(L_1/L_2) = S_1(D_2/D_1) \qquad (6.37)$$

where λ_c is the wavelength related to the reference optical channel, and S_1 and S_2 are dispersion slopes of the fiber lengths L_1 and L_2, respectively.

The above set of equations governs chromatic dispersion compensation in multichannel optical transmission. The ratio S/D, often called the relative dispersion slope, should be the same for both the optical transmission and the DCFs. However, it is rather difficult in practice to satisfy (6.35) to (6.37) over a broader wavelength range, and that can lead to a nonperfect dispersion compensation for some channels, which is illustrated in Figures 6.7 and 6.8.

As we can see from Figure 6.7, the perfect compensation is achieved for the central wavelength λ_c, while there is a residual dispersion on both sides of the transmission bandwidth. The channels with shorter wavelengths are overcompensated, while the channels with longer wavelengths are undercompensated. It is possible to establish a reference with respect to a wavelength different than the central one. In such a case, the picture of residual dispersion will be different.

It is easier to find good matching for standard SMFs, which have $S/D \sim 0.003$ nm^{-1}, than for DSFs and NZDSFs, since both of them have much a higher S/D ratio (usually $S/D > 0.02$ nm^{-1} for both DSF and NZDSF). Because of that, the compensation scheme involving NZDSF might require postcompensation at the receiving side in order to eliminate the residual dispersion in some of channels. Proper dispersion compensation for all channels can be achieved much easier if the length L_2 is relatively long. If so, the DCFs are not just compensating elements but become a part of

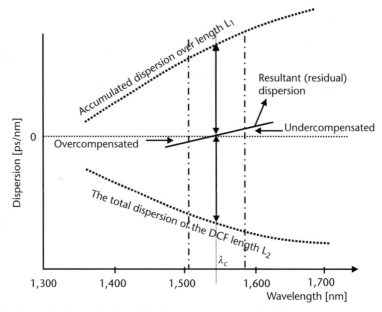

Figure 6.7 Example of nonideal matching between transmission fiber and DCF.

the transmission line. Such fibers, which are known as the reverse-dispersion fibers, are often used in combination with transmission fibers to enhance the overall transmission capability [6].

Transmission fibers and DCFs can be combined in a different manner, often referred to by terms such as precompensation, postcompensation, and overcompensation to describe the nature of the method used. The combination of different

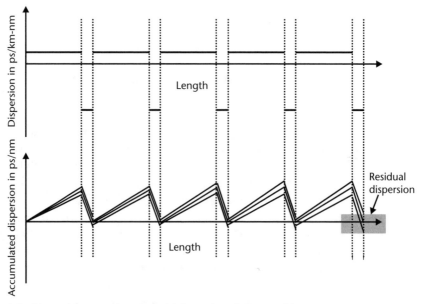

Figure 6.8 Dispersion compensation with imperfect slope matching.

sections is known as the *dispersion map*, which became one of the most relevant parameters from the engineering perspective. We will discuss the dispersion map schemes shortly.

Another type of DCF that operates near the cutoff normalized frequency (V_c = 2.4) and supports not just fundamental LP_{01} mode, but also the higher-order mode LP_{11}, can be also used for compensation purposes [refer to (2.6)]. If used for chromatic dispersion compensation, these fibers are combined with mode converters, as illustrated in Figure 6.9. An incoming optical signal is lead through the mode converter that transfers the fundamental mode LP_{01} to the next higher mode LP_{11}. The dual-mode fiber is specially designed so that the dispersion parameter D_2 has a large negative value for LP_{11}, and it can be as large as –550 to –770 ps/nm-km [11]. The total length of the dual-mode fiber is approximately 2% to 3% of the length of the transmission fiber. The higher-order mode is converted back to fundamental mode after passing through the dual-mode DCF. Mode conversion from LP_{01} to LP_{11} mode, and vice-versa, can be done by a long-period fiber grating, which efficiently couples these modes [12].

A fiber grating itself can be also effectively used for chromatic dispersion compensation. Such gratings, which belong to the class of so-called short-period gratings, are well known as fiber Bragg gratings [12, 13]. The application of FBGs for chromatic dispersion compensation as illustrated in Figure 6.10. They are mostly used in combination with optical circulators, which separate the reflected light from the forward going signal.

A fiber grating is a special piece of fiber with chirped Bragg grating imprinted inside. The grating can be imprinted in the core of photosensitive optical fibers by applying a strong UV laser source acting through a special phase mask, or by using special holographic methods [12]. The incident light power causes permanent changes and microvariations in the fiber refractive index, while the index maximums coincide with the diffraction fringes of the diffracted UV light. The refractive index varies along the fiber piece length as

$$n(z) = n_c + n_v \left(\frac{2\pi z G(z)}{\Lambda_0} \right) = n_c + n_v \cos\left(\frac{2\pi z}{\Lambda(z)} \right) \tag{6.38}$$

where n_c is the refractive index in the fiber core, n_v (~0.0001) is the magnitude of the index change due to grating imprint, $\Lambda(z) = \Lambda_0/G(z)$ is a variable grating period, Λ_0 is the maximum of the grating period, and $G(z)$ is a function that characterizes the grating chirp along the length of the fiber piece.

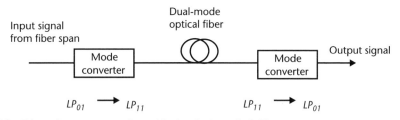

Figure 6.9 Dispersion compensation with the dual-mode DCF.

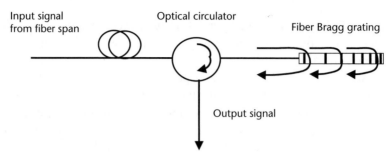

Figure 6.10 Application of fibber Bragg gratings for dispersion compensation.

The grating nature of the refractive index variations will cause coupling of the forward and backward propagating waves. The coupling occurs at so-called Bragg wavelengths—that is, for wavelengths that satisfy the following Bragg condition:

$$\lambda_B = 2n_c \Lambda(z) \tag{6.39}$$

Since the Bragg wavelength varies along the grating length, different wavelengths will be coupled to backward wavers at different places along the grating. The wavelength coupling will occur at the places where the Bragg condition is satisfied locally. Depending on the chirp, which can increase or decrease along the fiber piece, either shorter or longer wavelengths will be coupled first. For example, longer wavelengths will be coupled first in Figure 6.10. Chromatic dispersion in the anomalous dispersion region can be compensated with Bragg gratings having a negative (decreasing) chirp along the length, while the dispersion from normal dispersion region can be compensated by gratings imposing increasing chirp. Please notice that classification to gratings imposing positive and negative chirp is quite irrelevant since each grating can impose any chip if input and output ports are reversed.

Therefore, the FBG filters different wavelengths at different points along its length and introduces wavelength-dependent delays. The dispersion parameter of the FBG D_{grat} can be obtained from round-trip time delay $\Delta\tau$ between two wavelengths, which are coupled at the very beginning and at the very end of the grating. Since the round-trip time delay is

$$\Delta\tau = D_{grat} \Delta\lambda L_{grat} = \frac{2n_c L_{grat}}{c} \tag{6.40}$$

the dispersion parameter becomes

$$D_{grat} = \frac{2n_c}{c\Delta\lambda} \tag{6.41}$$

where c is the light speed in vacuum, n_c is the refractive index in the fiber core, L_{grat} is the length of the grating, and $\Delta\lambda = \lambda_1 - \lambda_2$ is difference in wavelengths reflected at the opposite ends of the grating.

The FBG is, in fact, an optical bandpass filter that reflects the wavelengths from the specified band, which is defined by the grating period, fiber length, and chirping rate. The compensation capability is proportional to the grating length. As an example, the fiber length of 10 cm is needed to compensate for dispersion of 100 ps/nm, while the length of 1m is needed to compensate for dispersion of 1,000 ps/nm. At the same time, the dispersion compensation ability is inversely proportional to the bandwidth, which means that a larger dispersion can be compensated over a smaller bandwidth, and vice-versa. For example, 1,000 ps/nm can be compensated over the bandwidth of 1 nm, while 100 ps/nm can be compensated over the wavelength bandwidth of about 10 nm.

FBGs impose some ripples with respect to reflectivity and time delay along the wavelength bandpass. The ripples are associated with the discrete nature of grating imprinting and present one of the biggest problems related to the grating application. This effect can be minimized by the so-called apodization technique, in which the magnitude of the index change, which is given by n_v in (6.33), is made nonuniform across the grating length. The apodization is usually made in such a way that the refractive-index variation is maximal at the middle of the fiber piece, while it decreases towards its ends.

Wider application of FBGs is restricted by their limited bandwidth, since it is rather difficult to maintain chirping stability over longer lengths. On the other hand, they can be used in a number of applications since they have relatively small insertion loses and lower cost. In addition, compensation by FBGs can be combined with other dispersion compensation methods to enhance the overall compensation ability.

The *phase conjugation* is an all-optical nonlinear dispersion compensation method in which the input optical signal is converted to its mirror image. It is done in such a way that the amplitude stays unchanged, while the output signal spectrum is a complex conjugation, or a phase reversal, of the input spectrum. Accordingly, the Fourier transforms of the input and output signals are related as

$$A_{in}(\omega_0 - \delta\omega) = A_{out}(\omega_0 + \delta\omega) \tag{6.42}$$

where A_{in} and A_{out} are amplitudes of the input and output pulse, respectively, ω_0 is the central optical frequency, and $\delta\omega$ presents frequency deviation from the central frequency. As we can see, the upper spectral components of the input spectrum have been converted to lower spectral components of the output spectrum.

The phase conjugation can be used for chromatic dispersion compensation in a way illustrated in Figure 6.11. In the case where the transmission link is characterized by a uniform chromatic dispersion, the phase conjugation should be performed at the distance $z = L/2$, where L is the total length of the transmission line. This method is well known as the mid-span spectral inversion. On the other hand, if transmission line is a heterogeneous one, the phase conjugator should be placed at the point where

$$D_1 L_1 \approx D_2 L_2 \tag{6.43}$$

where D_1 and L_1 are referred to as the incoming section length and corresponding chromatic dispersion, while D_2 and L_2 represent the outgoing section and its associated chromatic dispersion, respectively. Equations (6.42) and (6.43) imply that chromatic dispersion accumulated along the fist part of the transmission line can be

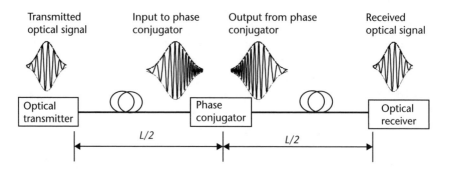

Figure 6.11 Phase conjugator as a dispersion compensator.

exactly compensated in the second part. This is just partially true since it can be done for the second-order group velocity dispersion but not for the third-order chromatic dispersion, which is represented by coefficient β_3 in (3.65).

The phase conjugation can be utilized not just for the chromatic dispersion compensation, but also for compensation of the SPM effect. In such a case, the phase conjugation is performed at the middle of a homogenous transmission line, while the heterogeneous transmission line requires that the following requirement be satisfied:

$$n_2(L)\Pi(L)L_1\big|_{L=L_1} \approx n_2(L)\Pi(L)L_2\big|_{L=L_2} \tag{6.44}$$

where n_2 is the nonlinear Kerr index, $\Pi = P/A_{eff}$ is the optical power density, P is the optical power, and A_{eff} is the effective cross-sectional area [refer to (3.103)].

There are several physical effects that have been considered for an effective application of phase conjugation, such as the FWM process that occurs either in optical fibers, special nonlinear waveguides (for example, LiNbO3 waveguide), or SOAs. Several lab experiments have been successfully demonstrated to compensate chromatic dispersion over several hundreds of kilometers [14, 15].

On the other hand, numerical simulations have shown that the phase conjugation method is feasible over several thousand kilometers if very precise systems engineering is used. The phase conjugators can eventually be combined with optical amplifiers to provide both the amplification and chromatic dispersion compensation at the same place. These amplifiers, known as parametric amplifiers, are still in an early experimental stage, but they show a significant application potential [7].

Optical filters can also be used for chromatic dispersion compensation. The application is based on the fact that chromatic dispersion changes the phase of the signal in proportion to the GDV coefficient β_2. Therefore, the chromatic dispersion can be compensated for if there is an optical filter that reverses the phase change. For this purpose, the optical filter should have a transfer function that is just inverse to the transfer function of the optical fiber in question. However, it is very difficult to achieve this in practice since there is no such optical filter that can fully satisfy this requirement.

It is possible, instead, to use special optical filters for partial compensation of chromatic dispersion effect. The transfer function of such a filter can be expressed as

$$H_{fil}(\omega) = |H_{fil}(\omega)| \exp[j\Phi(\omega)] \approx |H_{fil}(\omega)| \exp\left[j\left(\Phi_0 + \Phi_1\omega + \frac{1}{2}\Phi_2\omega^2\right)\right] \quad (6.45)$$

where $\Phi_i = d_i\Phi/d\omega^i$ ($i = 1, 2...$) are the derivatives evaluated at the center frequency ω_0. We should recall that a similar expansion was used for the signal phase [see (3.61) to (3.64)]. Coefficients with indexes $i = 0$ and $i = 1$ in (6.45) are responsible for an inherent phase shift and for time delay, respectively, and they are not relevant with respect to the phase change. On the other side, the coefficient with an index $i = 2$ is the most important one since it is associated with the phase change. This coefficient should be matched with the GVD coefficient by relation $\Phi_2 = -\beta_2 L$. The amplitude of the optical filter should be $|H(\omega)| = 1$, which is necessary in order to prevent attenuation of any spectral components within the incoming optical signal.

Optical filters for chromatic dispersion compensation can be realized as a cascaded structure of either Fabry-Perot, or Mach-Zehnder interferometers. Cascaded structures that contain multiple optical filters can be produced by using planar lightwave circuits (PLC) [16]. Optical filters, which are designed for chromatic dispersion compensation, can bring some additional benefits since they filter the optical noise and limit its power. In general, optical filters can compensate for more than 1,000 ps/nm, but over relatively narrow bandwidth.

Postcompensation schemes are mainly used to eliminate the residual dispersion that remains at the receiving side. The residual dispersion becomes an issue in WDM systems since it might exceed a critical level for some of channels. Therefore, the postcompensation should be applied to just a specified number of channels, usually on a per-channel basis. However, this approach can be rather cumbersome and expensive if there are a large number of transmission channels.

In addition, the employment of dispersion compensating modules with fixed negative dispersion is a static approach that might not be suitable for an environment in which the exact amount of inserted chromatic dispersion is not known. It happens due to variations in the group velocity dispersion coefficient that might be caused by temperature, or may appear due to dynamic reconfiguration of lightwave paths in an optical networking environment. The fluctuations in chromatic dispersion are more critical for high bit rates, such as 40 Gbps. Therefore, it is highly desirable to have an adaptive compensation scheme that can be adjusted automatically. Such a scheme can either employ tunable dispersion compensating elements, or utilize an electronic equalization of dispersion effects [17, 18]. Tunable dispersion compensation, which can be remotely controlled and adjusted per channel base, is something that would enhance the transmission characteristics and provide benefits from both the cost perspective and the systems engineering perspective.

Tunable dispersion compensation can be easily understood if we recall that any dispersion compensator can be considered as a filter, which reverses the phase deviation occurring due to the impact of chromatic dispersion. Therefore, tunability is related to a fine adjustment of the filter characteristics. Several methods of tunable dispersion compensation have been proposed so far and demonstrated in the lab. Most of them use the FBG as a tunable filter. The controllable chirp, which relates to the compensation capability, is changed through the distributed heating. By this approach, the achieved dispersion range can be larger than 1,500 ps/nm over subnanometer wavelength band. Generally speaking, any tunable optical filter that follows requirements given by (6.45) could be considered for dynamic dispersion compensation.

As a summary, a comparative review of several commercially available modules for chromatic dispersion compensation is given in Table 6.1. These methods are compared with respect to the figure of merit, dispersion compensation ability and fine-tuning capability, the frequency band, and sensitivity to polarization.

PMD compensation becomes very important for high bit rates in an environment where the transmission lightwave path can include some sections with relatively high PMD value. The PMD effect was discussed in Section 3.5, and the transmission penalty related to PMD impact were evaluated in Section 4.4. The main difference between PMD and chromatic dispersion is in the stochastic nature of the PMD effect, which makes it more difficult to deal with. The PMD compensation schemes are generally considered independently from chromatic dispersion schemes.

Several methods have been proposed and demonstrated so far to compensate for the PMD effect [19, 20]. They can be classified as being related to the receiver side, or to the optical transmitter side. In some cases these two schemes can even be combined together. In addition, the PMD compensation can be related to either optical or electrical signals. The basic idea behind the PMD compensation is related either to correction and equalization of the delay between two polarization modes or to changing the polarization state of the incoming optical signal to achieve more favorable detection conditions.

The equalization of the PMD effect by optical means is illustrated in Figure 6.12. It can be done by using a feedback loop to change the polarization state of the incoming optical signal by some polarization alternating devices. The feedback can be established within the optical receiver side, which provides relatively fast operation, but with rather limited dynamic range. On the other hand, the feedback can be established all the way to the transmitter side to find and launch signals in the principal polarization state. Such a method provides larger dynamic range; however, the operation can be relatively slow since there is transmission delay due to feedback that is established over longer distances.

PMD equalization can be also done by splitting the incoming optical signal in two polarization modes in accordance with the principal polarization states, as illustrated in lower part of Figure 6.12. Two polarization modes are separated by the polarization splitter and combined together through the polarization combiner. In the meantime, the faster polarization mode is delayed with respect to the slower one, which means that the first-order PMD is alleviated.

Table 6.1 Review of Dispersion Compensating Techniques

Method	Dispersion Compensating Fiber (DCF)	Fiber Bragg Grating	Dual-Mode Conversion
Figure of merit, ps/dB	50–300	100–200	50–170
Mean PMD	0.06–0.1 ps/km$^{0.5}$	0.5–1.5 ps (for the bandwidth range below)	0.05–0.08 ps/km$^{0.5}$
Bandwidth	More than 30 nm	0.5–6 nm	~30 nm
Tunability	No	Yes	Yes

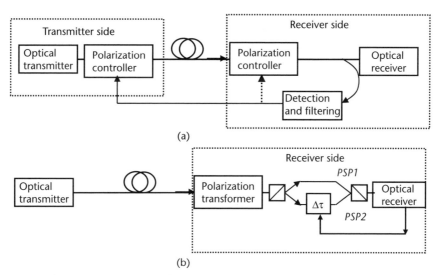

Figure 6.12 The PMD compensation by optical methods: (a) controlling polarizion states, and (b) correcting differential group delay.

The electrical PMD compensation relies on fast electronic signal processing. It is done through the transversal filter scheme, which is well known from telecommunication theory [10]. The filter splits the detected electrical signal into multiple parts (branches), as illustrated in Figure 6.13. Each branch is then multiplied by a weight coefficient (A_1, A_2, A_3 in Figure 6.13) and delayed by a specified time amount through the delay lines. All branches are then combined into a unified output signal.

In general, the PMD compensation should be applied on a per channel basis, which increases the total cost of the system. The PMD compensators, which are now commercially available, are usually placed close to the optical receivers and handle the incoming optical signal by equalizing the delay between the two principal polarization states. In fact, the approach taken so far is to alleviate the PMD effect rather than to compensate for it. Two major factors that determine the design of PMD compensators are the power penalty that can be tolerated and the cost of the compensation scheme. The PMD compensation schemes will become

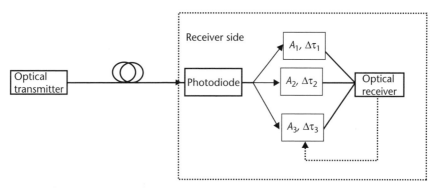

Figure 6.13 The PMD compensation by electrical methods.

indispensable for future high-speed applications since the average length of lightwave paths will increase gradually and the amount of accumulated dispersion will be higher.

6.1.3 Advanced Modulation Schemes

Throughout this book we implicitly assumed that an on-off keying (OOK) scheme has been applied while modulating the optical signal. The OOK scheme is the simplest yet effective method to utilize the properties of an optical signal. In addition, it is the only one widely used in practice. The other modulation methods, which are mentioned in Section 7.8, can be also used, but they have not yet found wider application. The properties of different modulation formats that belong to the OOK group will be discussed in this section, while the other methods of interest from the application perspective, which do not belong to the OOK group, will be mentioned in Section 6.1.4.

The simplest way to perform the OOK is by using the binary NRZ coding, in which the content occupies the entire time slot, as shown in Figure 6.14. Therefore, the 1 bit is associated with high optical power level through the entire duration of the bit, while the 0 bit is associated with lower power level. The 0 bits are associated with a nonzero optical level since there is some optical power radiated in these intervals (refer to Section 5.3.1).

The NRZ format is the simplest from the realization point of view and occupies the bandwidth proportional to $1/T$, where T is the time duration of the bit slot. On the other hand, the NRZ format is bit-pattern dependent, which is important when

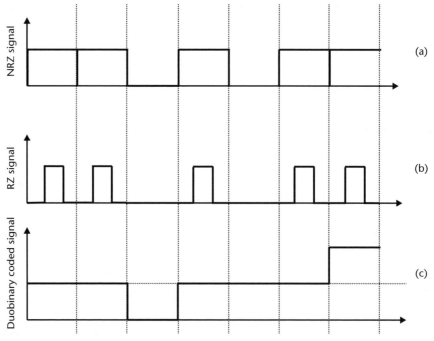

Figure 6.14 OOK Modulation formats: (a) NRZ, (b) RZ, and (c) duobinary.

considering the implications of impairments such as chromatic dispersion, cross-phase modulation, and stimulated Raman scattering.

The impact of chromatic dispersion can be suppressed by using modulation schemes that produce the bandwidth that is smaller than the bandwidth of the NRZ-coded signals. Such a scheme is duobinary coding [10], where the bandwidth is reduced by 50% as compared to bandwidth of the NRZ format. The duobinary modulation scheme is a three-level format, at half the bit rate of the NRZ code. There might be different schemes of how to produce the three-level codes, and one of them is shown in Figure 6.14. Since the bandwidth of the duobinary code is smaller than the one of NRZ code, the total impact of chromatic dispersion is suppressed. However, although the duobinary coding can offer some benefit from the chromatic dispersion point of view, it suffers from increased impairments due to nonlinearities, since the power level of three-level codes is generally higher than the one associated with two-level OOK modulation formats. In addition, the filtering of signal spectrum is needed at the transmitting side, which makes the transmitter design more complex. Duobinary coding can be attractive for some high-speed applications since it can be combined with some other advanced methods that enhance transmission system characteristics (i.e., with Raman amplification, forward-error correction, and so forth).

Another type of the OOK modulation scheme is known as RZ coding, which is also shown in Figure 6.14. The main characteristic of RZ coding is that the duration of the 1 bit is shorter than the duration of the bit slot. Since there are more transitions between higher and lower power levels, the bandwidth of the RZ code is twice as broad as the spectrum of the NRZ format. The main advantage of RZ coding comes from the fact that it is more robust to the impact of nonlinear effects, since the total energy per pulse is lower that the energy of the NRZ pulses.

The three conventional coding schemes shown in Figure 6.14 are related just to manipulation of the optical power level. However, each of them can be accompanied by an additional manipulation of other signal parameters (phase, frequency, and polarization) on a per-bit basis. This means that some signals can follow the same bit pattern and have the same information content but still be different from the signal spectrum perspective. The modified modulation schemes are more sophisticated since they include a combination of the OOK scheme and some other modulation methods. Such schemes are more often applied to the RZ coding to take a full advantage of its robustness against nonlinear effects, while improving its resilience to the chromatic dispersion effect.

Two examples of the modified RZ coding are shown in Figure 6.15. In the chirp RZ format (CRZ) [21], the pulses of the each bit are prechirped before they are introduced to the optical link. In some cases the chirping might not be uniform for all bits, but performed on an alternative basis (alternative CRZ coding). The other modifications of RZ codes include techniques well known in microwave communications [22]. Such methods are, for example, carrier suppressed RZ coding, in which the optical carrier frequency is suppressed to decrease the power level, or single sideband (SSB) and vestigial sideband (VSB) methods, which are performed to limit the total bandwidth and suppress the chromatic dispersion impact [23]. The RZ coding schemes can also be combined with the differential phase shift keying (DPSK) modulation scheme to further improve the system performance [24].

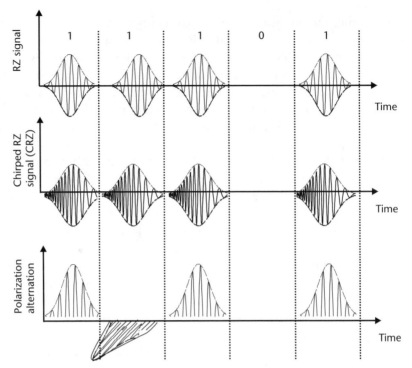

Figure 6.15 Examples of modified RZ coding.

The OOK scheme can also be combined with manipulation of polarization states. It can be done either on a per-bit basis [as in Figure 6.15(b)] or a per-channel basis if related to WDM systems [6]. If it is done on a per-bit basis, the overall modulation scheme is a combination of OOK and polarization shift keying (PoSK). This scheme is more robust to the impact of chromatic dispersion, but its realization is relatively complex. On the other hand, polarization multiplexing can increase the system spectrum efficiency if it is applied on a per-channel basis, but the impact of PMD effect may diminish the overall benefit.

Each of these methods can be used in special occasions to help with chromatic dispersion or with nonlinear effects. There is also a method that uses the chromatic dispersion to compensate for SPM and vice-versa, which should result in a transmission distance increase. This method is well known as soliton transmission. The soliton regime is established in an optical fiber if effects caused by chromatic dispersion and SPM offset each other under certain circumstances. It is not difficult to understand intuitively that these two effects can balance each other if they produce just opposite effects, so that the optical pulse reaches an equilibrium state and retains its shape.

As a reminder, it is possible to offset pulse spreading due to chromatic dispersion by the pulse compression during the early stage of its propagation though an optical fiber, which occurs if the product of the GVD coefficient β_2 and the chirp parameter C_0 is negative. The condition $\beta_2 C_0 < 0$ can be achieved by programming the chirp parameter to be just opposite in sign with respect to the GVD coefficient β_2. Accordingly, the pulse compression can be achieved either in normal or in anomalous dispersion region. The initial chirp programming can be done by following (6.24) or (6.27).

The SPM effect can also be effectively used for the pulse prechirping. In such a case, it produces a positive pulse chirp parameter C_0, which can be expressed by (6.31). In addition, the same character of the pulse chirping can be achieved at any stage of the pulse transmission throughout the optical fiber. It means that an effective suppression of chromatic dispersion by the SPM effect can also be done along the transmission line, but only in the anomalous dispersion region (i.e., for $D > 0$, or $\beta_2 < 0$). The condition $\beta_2 < 0$ is just an initial requirement for the solution regime to occur. The most important requirement is to keep the right balance between chromatic dispersion and SPM imposed on a specific pulse shape. The right balance means that the chirp induced by SPM is just enough to counterattack the chromatic dispersion effect. Any deviation from an optimum chirp value will result in an imbalance and will destroy the equilibrium state.

It is not so easy to keep the proper balance in practical situations, since the SPM-induced chirp is power dependent and any signal attenuation during the pulse propagation will affect the chirp value. On the other hand, an optical amplifier can restore the pulse power and regain the required chirp value. A distributed amplification, such as the one done by Raman amplifiers, is more favorable to preserve the soliton regime then the discrete amplification scheme. However, the restored pulses will experience some deviations around their original positions. These deviations, which are observed as a timing jitter, are caused by the ASE noise generated during the amplification process [25]. Since EDFAs have larger ASE noise than DRAs, the total jitter accumulated along the line employing a chain of EDFAs will be larger that the jitter resulting from the chain of Raman amplifiers.

The properties of solitons have been studied by using the nonlinear Schrodinger equation [3, 26]. For this purpose we can recall (7.29):

$$\frac{\partial A(z,t)}{\partial z} + \frac{j\beta_2}{2}\frac{\partial^2 A(z,t)}{\partial t^2} - \frac{\beta_3}{6}\frac{\partial^3 A(z,t)}{\partial t^3} = j\gamma |A^2(z,t)|A(z,t) - \frac{1}{2}\alpha A(z,t) \qquad (6.46)$$

where $A(z, t)$ is the slowly varying pulse amplitude as a function of distance z and time t, α is the fiber attenuation parameter, and $\gamma = 2\pi n_2/(\lambda A_{eff})$ is the nonlinear coefficient introduced by (3.105). Recall that β_2 is the second-order GVD coefficient, β_3 is the third-order GVD coefficient, n_2 is the nonlinear Kerr coefficient, and A_{eff} is the effective cross-sectional area of the fiber core. One should also recall that any solution of (6.46) is expressed delayed in time by amount $\beta_1 z$, where β_1 is proportional to the propagation delay [refer to (3.61) and (3.65)].

Equation (6.46) is a partial differential equation that produces analytical solutions [26], which serves as a great advantage as compared with other similar equations. There could be two types of the solutions. The fist solution type is pulse-like, and exists for $\beta_2 < 0$ (anomalous dispersion region). This solution is referred to as the bright soliton, since the soliton pulse brings the energy, or brightness. The second solution type is obtained for $\beta_2 > 0$ (normal dispersion), and it is dip-like in a constant intensity backgound. This solution is referred to as the dark soliton. The bright solitons are the only one that are interesting from the transmission point of view. As for the dark solitons, there were several experiments to use them for transmission purposes, but the scale of eventual practical application is not yet clear.

The solution of (6.46) that refers to the bright solitons can be explained as follows. The input pulse with amplitude

$$A(0,t) \sim N \operatorname{sech}\left(\frac{t}{\tau_0}\right) \tag{6.47}$$

will remain unchanged in shape while propagating through the fiber if $N = 1$. The function "sech" is known as the hyperbolic secant [27]. The parameter τ_0 is related to the pulse width. The FWHM of the sech-function shape is connected with parameter τ_0 as $T_{FWHM} \sim 1.763\, \tau_0$.

The pulse will change from its original shape if $N > 1$ but will eventually recover it in a periodical manner. An optical pulse is called the fundamental soliton if it is engineered to satisfy the condition

$$N = (\gamma P_0 L_D)^{1/2} = \left(\frac{2\pi n_2}{\lambda A_{eff}} \frac{\tau_0^2}{|\beta_2|} P_0\right)^{1/2} \tag{6.48}$$

By definition, the fundamental soliton is chirp-free. Therefore, the following parameters define the condition for the generation of the fundamental solution: the nonlinear Kerr coefficient n_2, the effective cross-sectional area of the fiber core A_{eff}, signal wavelength λ, the pulse with τ_0, the GVD coefficient β_2, and the pulse peak power P_0. The fundamental soliton is the only one that remains unchanged in shape and it is chirp-free, which means that the design goal is to generate and maintain the fundamental solitons. On the other hand, the high-order solitons will recover their original shape after going though period of uncertainty, and it happens at the distance

$$z_0 = \frac{\pi}{2} \frac{\tau_0^2}{|\beta_2|} \tag{6.49}$$

It was shown that optical solitons are relatively stable with respect to various perturbations of parameters in (6.48). The solitons can be created even if the initial pulse deviates from the "sech" shape, which means that the Gaussian shape could eventually evolve to the soliton form. As for deviations of the peak power, it was also proven that the number N could be in the range from 0.5 to 1.5 to still produce the solitons afterwards [3]. The transformation of the initial pulse to the soliton shape is accompanied by formation of a dispersion wave, which arises from the energy spread out of the formed soliton shape due to the difference between the "sech" shape and an original pulse waveform. The dispersive waves will continue propagation together with solitons and will have a negative impact to the system characteristics. Therefore, the dispersive waves should be minimized by proper matching of the initial pulse shape to an ideal soliton form expressed by (6.47).

Soliton pulses have a tendency to destruct each other if they overlap during the propagation process. That is the reason why pulse overlapping should be prevented by applying a proper pulse spacing in addition to the RZ coding. The separation between neighboring soliton pulses in the data stream is dependent on the soliton pulse width and the signal bit rate. The higher bit rate should be engineered with the faster repetition of soliton pulses, and with the selection of narrower pulses. The requirement for soliton spacing is usually expressed through the ratio T/τ_0, where $T = 1/B$ is the bit-slot duration, B is the bit rate, and τ_0 is the soliton pulse width, as defined in (6.47) and (6.48). This ratio should be around 10 to 12 to be on the safe side in terms of preventing any soliton interaction, as illustrated in Figure 6.16. It was shown, though, that the soliton interaction can be prevented even if the spacing ratio is 7 to 8 if the neighboring soliton pulses differ in amplitude by about 10% [3].

The power of the launched pulse, which is able to form a fundamental soliton shape, can be expressed as

$$P(t) = |A(0,t)| = P_0 \sec h^2 \left(\frac{t}{\tau_0} \right) \qquad (6.50)$$

where the peak power P_0 is found from (6.48) as

$$P_0 = \frac{\lambda A_{eff}}{2\pi n_2} \frac{|\beta_2|}{\tau_0^2} \qquad (6.51)$$

The total energy carried by a soliton pulse can be calculated by the integration of (6.50), and it is

$$E_0 = \int_{-\infty}^{\infty} P(t) dt = 2 P_0 \tau_0 \qquad (6.52)$$

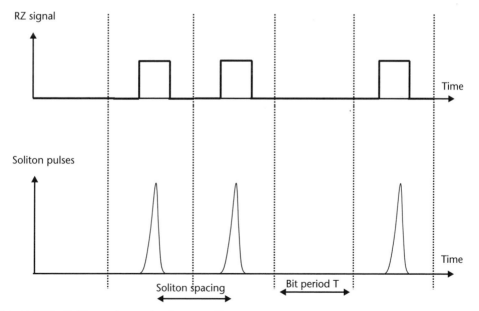

Figure 6.16 Soliton pulses and pulse separation.

Soliton pulse generation is more complex that the generation of other RZ pulses. First, the pulses should be narrow enough for a specified bit rate to keep the pulse spacing and to prevent any eventual soliton interaction. For example, 10-ps-wide pulses are needed to provide the soliton regime for 10-Gbps bit rate, while 2- to 3-ps pulse width is required for the 40-Gbps bit rate. Second, the generated pulses should have the "sech" shape and be chirp-free. It was mentioned before that this requirement is not critical for the soliton pulse formation since solitons can be formed even if the pulse does not follow the "sech" shape, and if there is a nonzero chirp parameter. However, the problem might arise afterwards since the dispersive waves will eventually be generated, and part of the soliton energy will be lost to a dispersive wave during the soliton formation. As an example, if the initial chirp parameter is $C_0 = 0.5$, about 17% of the total energy will go to dispersive waves, which is quite a considerable amount. Therefore, the initial pulse should have the chirp as low as possible and be as close as possible to an ideal "sech" shape.

Solitons can be generated by sources that can impose a programmable chirp. In such a case, the chirp parameter is adjusted to be as close as possible to the zero value. Another possibility is to use lasers that generate some smaller frequency chirp and use nonlinear pulse shaping to compensate for the finite value of the chirp parameter. Such a shaping can be achieved by using a section of a special optical fiber.

The soliton regime should be preserved over longer distances in order to be an efficient technique for communication purposes. The biggest obstacle to such a goal is the fiber attenuation, since it will decrease the optical power otherwise needed to keep the equilibrium between chromatic dispersion and SPM. Accordingly, the soliton pulses should be amplified along the transmission line, and that is known as the loss management of the soliton pulses. The loss management can be done by both lumped and distributed amplification.

The lumped amplification is performed by conventional in-line optical amplifiers. The amplified soliton pulse regains the energy during the amplification and adjusts its width and shape dynamically while propagating through the fiber section placed after the amplifier in question. The process of the shape adjustment is accompanied by the creation of the dispersive wave, which continues propagation together with soliton pulses. Both the soliton pulses and the created dispersive wave will be amplified by subsequent amplifier stages. The dispersive wave can be minimized if the soliton pulse does not experience a dramatic change in the power. This means that it should be amplified more often with a smaller gain per amplifier stage to prevent bigger perturbation during its transmission. It was shown that the amplifier spacing should be kept below the dispersive length $L_D = \tau_0^2/|\beta_2|$, where β_2 is the GVD parameter, in order to minimize the power of the dispersive waves.

The power launched at the amplification stage should be higher than the peak power P_0 needed in an ideal case when no attenuation is present. That ratio between these two powers, which is called the power enhancement factor due to amplification, is defined as [3]

$$F_{lm} = \frac{P_{S,lm}}{P_0} = \frac{E_{S,lm}}{E_0} = \frac{G \ln G}{G-1} \qquad (6.53)$$

where $P_{S,lm}$ is the launched power at the amplifier output, $E_{S,lm}$ is the energy of the launched amplified soliton, G is the amplifier gain, and E_0 is the energy of the soliton pulse at the transmitter output, which is given by (6.52). Equation (6.53) shows that parameter F_{lm} should be in the range from 2 to 4, which may not be quite practical for commercial systems.

Distributed amplification is much more favorable for soliton transmission than lumped amplification, since it inherently prevents large perturbation of both the optical power and soliton width. It seems that bidirectional pumping of Raman amplifiers is very convenient for this purpose since it provides gain at the points where the signal is weak and the solution regime might be otherwise threatened. In such case, any amplifier spacing L_A lower than $4\pi L_D$ will do reasonably well and will prevent the soliton regime destruction.

The soliton regime can be maintained much more easily if a dispersion management accompanies the loss management along the transmission line. Such a conclusion is based on (6.51), which shows that the pulse width will be unchanged if a power decrease is accompanied by the GVD decrease. Since the power decreases exponentially with the length in proportion to the attenuation coefficient, the soliton pulse shape will be maintained if the GVD parameter follows the same way; that is, if it is

$$|\beta_2(z)| = |\beta_2(0)|\exp(-\alpha z) \tag{6.54}$$

Although appealing, (6.54) is not easily achievable in practice, since optical fibers with a decreasing dispersion are not readily available due to manufacturing difficulty. In addition, they might be relatively expensive even if there is an improvement in the manufacturing process. As an alternative, a periodic dispersion map that contains optical fiber sections with the opposite GVD parameters may be beneficial in supporting the soliton regime (refer to Figure 6.17). The dispersion map keeps the average value of the GVD parameter low, but there is enough dispersion in each section to prevent the appearance of a strong FWM effect.

The soliton regime is preserved during the pulse propagation through the dispersion-managed line even if some of the fiber sections operate in the normal dispersion region. Therefore, the solitons can still be effectively supported if the SPM effect is balanced by the average value of chromatic dispersion, which is calculated over the entire transmission length. Accordingly, the solitons proved themselves to be robust enough to survive not just perturbation in the peak power, but also the perturbations in the pulse shape. The soliton pulse width will oscillate periodically along the line, but the pulse destruction is avoided. An optimized design of dispersion-managed solitons was, in fact, the step that bought them to an application level. It is important to notice, however, that the higher optical peak power is required to preserve dispersion-managed solitons, as compared with the case that applies to standard soliton systems associated with an uniform GVD coefficient. The increase in energy needed to a support soliton regime is expressed by the enhancement factor, given as

$$F_{dm} = \frac{E_{S,dm}}{E_0} \tag{6.55}$$

where $E_{s,dm}$ and E_0 are the energies related to the dispersion-managed and standard case, respectively. The enhancement factor F_{dm} is higher than the power enhancement factor F_{lm} from (6.53).

The conditions for managed solitons can be examined by solving numerically the nonlinear Schrodinger equation (NSE) [refer to (7.30)]. The pulse input parameters (peak power, chirp, pulse width) can be optimized in accordance with specified dispersion map parameters (map period, and the length of different sections), and vice-versa. The NSE is often solved by assuming that the input pulse has a Gaussian, rather than the "sech" shape. It was found that a well-designed dispersion-managed map would support the soliton regime under the following condition [3]:

$$\tau_0 = \sqrt{\frac{1+C_0^2}{|C_0|}} \sqrt{\frac{|\beta_{2,I}\beta_{2,II}L_{d,I}L_{d,II}|}{\beta_{2,I}L_{d,I} - \beta_{2,II}L_{d,II}}} = C_{equ}T_{map} \qquad (6.56)$$

where τ_0 and C_0 represent the initial values of the pulse width and chirp parameter, respectively, while $\beta_{2,I}$ and $\beta_{2,II}$ are referred to as the GVD parameters of optical fiber section with normal and anomalous dispersion, respectively. These sections have lengths $L_{d,I}$ and $L_{d,II}$, respectively, as shown in Figure 6.17. Parameter T_{map}, which is related to the second square root on the right side of (6.56), is called the map parameter. This parameter should be shorter for higher bit rates, which could limit the practicality of dispersion management for higher bit rates.

Any amplification of soliton pulses will be followed by the generation of ASE, which will be mixed with dispersive waves during the soliton regeneration. The overall degradation of transmission performance is caused by the generated noise and by fluctuations in the pulse energy, since both of these effects will decrease the SNR. In addition, the performance degradation will come through the timing jitter, which is induced by frequency fluctuations. The total energy fluctuations during the soliton amplification process were estimated in [7]. It was found that the OSNR could be expressed as

$$OSNR_{soliton} = \sqrt{\frac{E_0}{2NS_{sp}}} = \sqrt{\frac{E_0}{2Nn_{sp}h\nu(G-1)}} \qquad (6.57)$$

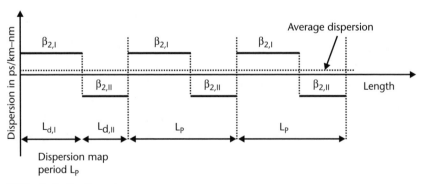

Figure 6.17 Periodic dispersion map.

where E_0 is the energy of the pulse calculated as an integral of the optical power over time, N is the number of optical amplifiers along the line, S_{sp} is the spectral density of the ASE noise, n_{sp} is the spontaneous emission factor, h is the Planck's constant, and ν is the optical frequency.

Equation (6.57) applies to standard soliton systems with in-line amplification along the transmission line. This situation also implies that there is not any real dispersion management in place. On the other hand, the dispersion management will improve OSNR since the pulse energy is higher than the energy of standard solitons. Such OSNR improvement is proportional to $F_{dm}^{1/2}$, where F_{dm} is the energy enhancement factor given by (6.55).

The timing jitter is caused by the change in soliton frequency, since the frequency affects the speed at which soliton propagates through the optical fiber. The timing jitter is the most serious problem that limits the overall capabilities of soliton-based transmission systems. This jitter is referred to as the Gordon-Haus jitter. The standard deviation σ_t^2 of the timing jitter introduced in standard soliton systems is [7]

$$\left.\frac{\sigma_t^2}{\tau_0^2}\right|_{lm} \approx \frac{S_{sp} L_T^3}{9 E_{S,lm} L_D^2 L_A} \tag{6.58}$$

where τ_0 is the initial soliton pulse width, L_T is the total transmission length that is equal to $L_T = NL_A$, where L_A is the amplifier spacing, N is the number of in-line amplifiers, $E_{S,lm}$ is the soliton energy as given by (6.53), and L_D is the dispersion length equal $L_D = \tau_0^2/|\beta_2|$. On the other hand, the standard deviation of the timing jitter introduced in dispersion-managed soliton systems can be expressed as

$$\left.\frac{\sigma_t^2}{T_m^2}\right|_{dm} \approx \frac{S_{sp} L_T^3}{3 E_0 L_{D,dm}^2 L_A} \tag{6.59}$$

where E_0 is the soliton energy at the output of transmitter, T_m is the minimum pulse width occurring during transmission, and $L_{D,dm} = T_m^2/|\beta_2|$. Recall that the soliton pulse width oscillates in a dispersion-managed case and that its minimum occurs at the middle of the anomalous GVD section. By comparing (6.58) and (6.59), and by assuming that $T_m \sim \tau_0$, one can see that the timing jitter is reduced while using dispersion-managed solitons by factor $(F_{dm}/3)^{1/2}$.

It is important to mention that the soliton self-frequency shift is also observed in soliton-based transmission systems. This effect occurs due to interpulse Raman scattering effect, which is enabled by the broadband spectrum of the soliton pulse. The high-frequency components of the pulse shift energy to lower frequency components and cause a continuous downshift in the soliton carrier frequency. The soliton self-frequency shift is more intense for higher bit rates since narrower soliton pulses produce much broader pulse spectrum. This can be better understood if we recall that solitons are the transform-limited pulses that satisfy (3.70).

The timing jitter that occurs in soliton transmission should be carefully controlled to minimize its negative impact. This can be done by placing optical filters after each amplifier since this will limit the ASE noise power, while simultaneously increasing the OSNR. The growth of the timing jitter can be reduced by sliding optical filters, which shift the central frequency along the line [28]. Such frequency sliding follows the self-frequency shift and ensures that the maximum portion of the noise power has been removed.

The soliton regime can also be supported in WDM transmission systems. However, it is necessary to account for the effect of pulse collisions among solitons belonging to different WDM channels. During such a collision, the cross-phase modulation effect induces the time-dependent phase shift, which leads to a change in the soliton frequency. Such a frequency change will either speed up or slow down the pulses. At the end of the collision process, each pulse eventually recovers in both frequency and speed, but the timing jitter that has been induced will remain in place. The timing jitter would not be a factor if the data stream just consisted of 1 bits, since the change will affect all pulses equally. However, since there is a random content of 1 bits, the shift in position will occur on a per-bit basis.

A practical realization of the WDM transmission soliton-based system, which contains a large number of high-speed channels, would not be possible with standard solitons. On the other hand, the managed solitons help to reduce the effect of pulse overlapping in WDM systems, which makes them attractive for practical applications. It is also important to mention that dispersion management through the dispersion maps can be combined with the CRZ pulse coding to be used as an alternative to soliton transmission. Such an approach does not fully utilize the compensation of chromatic dispersion by SPM but does offer a less complex design and systems engineering as compared with the soliton-based systems.

6.1.4 Advanced Detection Schemes

All topics discussed in previous chapters were related to optical systems that employ OOK modulation schemes. Such approaches are combined with direct detection of the optical power by the receiver photodiode. However, there are alternative schemes, which are well known from radio communication systems, that might bring some additional benefits to the optical transmission systems. Since these methods utilize the coherent nature of the optical carrier, the associated optical systems have been recognized as coherent lightwave systems. These systems have been studied intensively in the literature, before EDFA employment became a massive event [8, 9].

A coherent lightwave system is characterized by the coherent detection scheme, which takes place within an optical receiver, as illustrated in Figure 6.18. An input optical signal is mixed with a much stronger signal from the local laser before photodetection takes place. The generated photocurrent contains a signal component that is proportional to the incoming optical signal. However, it also contains a component generated through the beating process that takes place between the incoming optical signal and light coming from the local optical oscillator. This beating component contributes to the total signal level, thus enhancing the receiver sensitivity.

The optical power at the transmitting side is modulated by using a digital modulation scheme, such as the amplitude shift keying (ASK), frequency shift keying (FSK), or phase shift keying (PSK). PSK is a favorable technique since it provides the highest receiver sensitivity, while the FSK format is considered the most feasible one since it is less complex and provides the second highest receiver sensitivity [9]. The

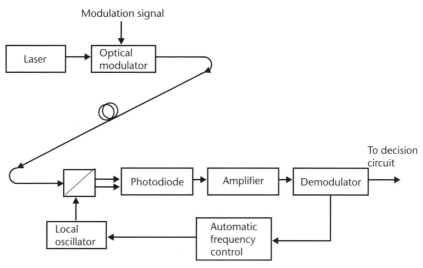

Figure 6.18 Coherent lightwave system.

ASK scheme does not have real merit as compared with the FSK and PSK formats, but it is the simplest among these three schemes and can be considered for some applications. It is also important to mention that there are several modifications of FSK and PSK schemes, such as minimum shift keying (MSK) or differential PSK (DPSK), respectively, that could be considered for practical applications [9].

The incoming optical signal at the receiver side is mixed with a signal coming from the local optical oscillator (laser) before the signal-mix is converted to photocurrent. The electric fields of the incoming optical signal and local optical oscillator can be expressed as

$$E_S(t) = A_S(t)\exp\{-j[\omega_S(t)t + \phi_S(t)]\} \tag{6.60}$$

$$E_{LO}(t) = A_{LO}\exp\{-j[\omega_{LO}t + \phi_{LO}]\} \tag{6.61}$$

where A_s and A_{LO} refer to amplitudes, ϕ_s and ϕ_{LO} to phases, and ω_s and ω_{LO} to frequencies of the incoming optical signal and local oscillator, respectively. As we can see, the parameters of the incoming optical signal may vary in time and depend on the modulation scheme that has been applied. On the other hand, the parameters of the local optical oscillator are considered to be constant. It is important to mention, however, that parameters of the local oscillator can vary in a stochastic manner, which causes an additional noise. The assumption is that both waves have the same polarization, since an effective beating between them can occur only if the polarizations are matched.

The photodiode will respond to an optical intensity equal to the square of the summary electric field amplitude, while the generated photocurrent is [9]

$$I(t) = R\{P_S + P_{LO} + 2\sqrt{P_S P_{LO}}\cos[\omega_{IM}(t)t + \Delta\phi(t)]\} \tag{6.62}$$

where P_s and P_{LO} are the powers of the optical signal and local oscillator, respectively, and R is the photodiode responsivity. Parameter

$$\omega_{IM}(t) = \omega_S(t) - \omega_{LO} \tag{6.63}$$

is called the intermediate frequency. The parameter $\Delta\phi = \phi_s - \phi_{LO}$ is related to the difference in phases of the incoming optical signal and local oscillator, respectively.

The major enhancement to the signal photocurrent in (6.62) has come from the component $I \sim 2RP_sP_{LO}\cos(\omega_{IM}t)$. The sensitivity of the coherent receiver is considerably enhanced as compared to the receiver sensitivity of direct detection receivers since there is a relatively strong contribution from the local oscillator side. Therefore, the major signal component is a radio frequency signal of intermediate frequency (IF). This signal should be demodulated to bring the information to base frequency band, which is done by using methods well known in commercial radio techniques [22].

The details of the detection scheme depend on the modulation format of the incoming optical signal and can be either heterodyne or homodyne in nature. In the heterodyne detection scheme, the carrier frequency ω_s and the frequency ω_{LO} of the local oscillator signal differ for the value $\delta\omega = 2\pi f_{IF} = |\omega_s - \omega_{LO}|$, where the intermediate frequency f_{IF} is usually up to several gigahertz. On the other hand, in the homodyne detection scheme, the optical frequencies and the phases of the signal and the local oscillator are completely matched, while the photodiode produces a baseband electrical signal.

The frequency bandwidth of the photodiode employed in the heterodyne detection scheme should be two times wider than the bandwidth of the photodiode that belongs to the homodyne detection scheme. There might be more than one photodiode within the same detection scheme, such as in the case of the balanced receiver, to generate multiple electrical signals for afterward processing.

The sensitivity of a coherent receiver is enhanced as much as 7 to 18 dB in comparison with conventional systems based on intensity modulation and direct detection (IM/DD) [8, 9]. This enhancement is caused by a strong contribution from the local optical oscillator, which suppresses the thermal noise impact. In addition to improved receiver sensitivity, the coherent detection offers some other benefits, such as more precise channel filtering at the electrical level, high-density multichannel transmission, and chromatic dispersion compensation at the electrical level. However, the realization of the coherent detection scheme involves the necessity of having a stable relationship between the phases and frequencies of the incoming optical signal and the local laser signal, which brings an additional complexity to the system design. Furthermore, the polarization of the incoming optical signal should be matched with the polarization of the local optical oscillator.

The SNR, which applies to coherent detection schemes, is obtained as the ratio of an average signal power divided by the sum of shot and thermal noise powers; that is,

$$SNR = \frac{\langle I^2(t) \rangle}{\sigma_{noise}^2} \approx \frac{2RP_sP_{LO}}{2q(RP_{LO} + I_d)\Delta f + \frac{4k\Theta F_{ne}\Delta f}{R_L}} \tag{6.64}$$

where q is the electron charge, R is the photodiode responsivity, Δf is the receiver bandwidth, k is the Boltzmann's constant, Θ is temperature, F_{ne} is the noise figure of the front-end amplifier, R_L is the load resistance, and I_d is the dark current. Equation (6.64) is obtained under assumptions that the power of the local oscillator is much larger than the incoming optical signal. Therefore, the signal and shot noise levels are determined by the local oscillator power.

The shot noise, which is represented by the first term in the denominator, and the thermal noise given by the second term in the denominator have the same nature as in the direct detection case [see (5.16) and (5.17)]. Equation (6.64) is related to the heterodyne detection case, while it should be multiplied by a factor of two if it is applied to homodyne detection [9]. The shot noise is the dominant noise factor since the local oscillator power is relatively high, and it is much higher than the thermal noise and the dark current. As a result, both the thermal noise and the dark current contributions can be neglected, so that the equation for SNR takes the form

$$SNR \approx \frac{RP_S}{q\Delta f} \quad (6.65)$$

The main difficulty in the realization of coherent detection schemes is related to the necessity of having a stable relationship between the phases and frequencies of the incoming optical signal and the local oscillator. Accordingly, an efficient control scheme must be employed to achieve a stable relationship, as shown in Figure 6.18. In addition, coherent detection schemes are polarization sensitive, and either polarization control should be applied, or diversity reception of two polarizations is necessary to prevent the signal-to-noise degradation. Finally, only the lasers with relatively narrow linewidth can be used as signal transmitters and local oscillators.

It is expected that some form of coherent detection might be attractive in the near future when better and smaller components will be available. The eventual integration of several components in a multifunctional element at the receiver side will improve the cost-effectiveness of coherent detection schemes. Meanwhile, some quasi-coherent detection schemes can be used to enhance the system transmission characteristics.

6.1.5 Forward Error Correction

FEC has become an integral part in the design of high-speed optical transmission systems and in the systems engineering process. There are a number of FEC codes that can be used to improve overall performance of the transmission systems, but an optimum choice might depend on the bit rate used, the design parameters, and the cost-effectiveness [29–32]. The major benefits brought by FEC are:

- Improvement in BER, while keeping a lower SNR. Therefore, the FEC helps to reduce the requirements for the optical power level needed to keep the specified BER.
- Reduction of impairments caused by nonlinearities since the total optical power is decreased. Therefore, if the signal power level is reduced, the total impact of nonlinearities will be also reduced.

The philosophy behind the FEC application is in the insertion of some number of redundant bits that will be correlated to the information content of the digital data signal. This process might be accompanied with some other signal processing, such as code concatenation, code interleaving, the iteration procedure, and the receiver threshold adjustment. Although the coding procedure is applied to the information data stream, the stream itself is chopped into blocks, and the coding process is applied to a block of bits.

The code is typically characterized as the (n,k)-type, where n is the number of output symbols per block after the coding took place and k is the number of information data bits per block. Therefore, the number of k-bites per block after chopping is enlarged by $n - k$ "check bites" that are used for error correction purposes. This resembles well-known parity codes, in which additional bites help to recognize deviations from a reference case. The output code blocks, which are sometimes called code words, are generated by mathematical algorithmic methods. The insertion of the correction bites will increase the transmission bit rate by a factor equal to $(n - k)/k$.

In general, the FEC code efficiency can be characterized by a number of correctable errors. It is obvious that the codes with larger block size will be more powerful in terms of number of correctable errors. However, the large overhead formed by additional bits will be a burden from the transmission point of view since it will increase the bit rate and make dispersion influence more critical. There are several types of the FEC codes:

- Systematic codes, where the information bits appear unchanged in the output code blocks. These codes are relatively simple and can be turned off if necessary.
- Linear codes, where the coding process is structured by using matrix multiplication. These codes are widely used today.
- Cyclic codes, where the code words are generated by polynomials.

The two most commonly used code types today, which have the above attributes, are the Reed-Solomon (RS) and the Bose-Chaudhuri-Hocquenghem (BCH) codes. The Reed-Solomon code is nonbinary, which means that it can deal with symbols, not just with bits. The total number n of the output symbols per block is expressed through an arbitrary integer m as $n = 2^m - 1$. The number of correctable symbols per block when using the Reed-Solomon code is $(n-k)/2$. The BCH codes are binary in nature. The total number of output bits per BCH block can be expressed through an arbitrary integer p as $n = 2^p - 1$, while the number of errors that can be corrected per block is $(n - k)/p$.

The most widely used FEC code so far has been the RS (255,239) code, which can correct up to eight symbol errors per block by inserting an overhead equal to 6.7% of the original bit rate. In some situations, code shortening is used to align block lengths to a specific purpose. In such a case, the FEC (n, k)-code becomes $(n-s, k-s)$ code, which means that the block size is effectively reduced by s bits. The most important parameter to characterize FEC codes is the net effective coding gain, ΔG, which is expressed in decibels. This gain is related to the difference in Q-factors associated with corrected and uncorrected cases, and to the number of check bits.

The net effective gain is usually quoted at BER = 10^{-15}, which corresponds to $Q = 8$ (or $Q = 18$ dB) (refer to Figure 5.3). Therefore, it is

$$\Delta G = 10 \log \left(\frac{k Q^2_{out}}{n Q^2_{in}} \right) \qquad (6.66)$$

where k and n represent the number of bits per block before and after coding, respectively, while Q_{in} and Q_{out} are the Q-factor values at the receiving side before and after correction, respectively. The achievable net effective coding gain at $BER = 10^{-15}$ for the RS (255,239) code is 4 to 5 dB [31].

The decoding process of the FEC codes starts with identifying the beginning of the code block, which is usually done by using a fixed pattern that repeats itself at known intervals. The most powerful FEC codes that are used today are coupled to the signal as an extra overhead to the SONET/SDH frames, as illustrated in Figure 6.19. This type of control bit attachment is also known as out-of-band FEC, since redundancy bytes are being inserted out of the SONET/SDH frame. In OCh frame, which was defined in [33], the FEC bits are placed within the superframe structure with a size of 4 × 4,080 bytes, in which 4 × 16 bytes are for overhead, 4 × 3,808 bytes are for payload, and 4 × 256 bytes are for FEC.

There are still some applications where the FEC control bits use unoccupied space within SONET/SDH overhead, as illustrated in Figure 6.19. This type of attachment is also known as inband FEC. For this purpose, the code shortening (n-s, k-s) type code is used to align the block lengths to a limited overhead space. The net effective code gain for inband FEC is smaller than the gain associated with more efficient out-of-band coding.

The elementary block-coding schemes are sometimes used as a base for the more powerful concatenated FEC codes. The concatenated coding scheme, which is illustrated in Figure 6.20, is a combination of an inner coder/decoder and an outer coder/decoder. The inner decoder corrects most of the errors, with only a few still

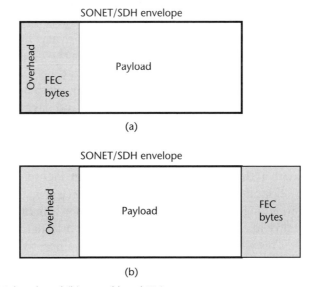

Figure 6.19 (a) Inband, and (b) out-of-band FEC.

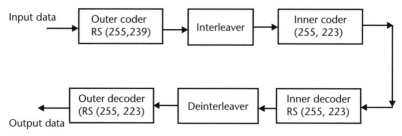

Figure 6.20 Concatenated RS FEC.

uncorrected. The deinterleaver takes the remaining errors and spreads them into different code blocks. The outer decoder corrects the fraction of errors that came from the deinterleaver. From the realization perspective, this coding scheme has an advantage of multiple parallel tributaries and larger memories while using the column-write and row-read method. The concatenation of the Reed-Solomon codes shown in Figure 6.20 adds the 22% overhead in bit rate, but brings more than 8-dB net-effective coding gain. This scheme can be enhanced by an additional iteration in which the output data needs to be resent to the interleaver in order to perform the inner decoding and deinterleaving once more.

The FEC process can be combined with so-called soft decision in the optical receiver. The soft decision process means that the decision circuit returns a floating (or analog) value at the sampling moments, rather than two distinguished logical levels, which are associated to "1" and "0" bits (recall Figure 5.2). The soft decision circuits used at high bit rates usually support from three to seven decision levels, which inherently improves the error correcting capability. The total yield from the soft decision receiver can be as much as 3 dB in comparison with a hard decision receiver, but not all FEC codes are equally good for use in combination with the soft receiver. The group of codes that seems to be best suited for this purpose is known as the turbo convolutional codes [30].

The turbo convolutional codes (TCC) use two coders at the transmission side, as shown in Figure 6.21. At the receiving side, the output data goes through decoder-one, deinterleaver, decoder-two, interleaver, and again through decoder-one. The number of iterations through the closed loop can vary, but three to four iterations are usually performed in practical realizations. Turbo convolutional codes insert an overhead of about 25% of bit rate since the k/n ratio is 0.33. Therefore, they might not be quite suitable for high-speed optical systems with bit rates above 10 Gbps. In addition, the corrected BER is higher than 10^{-15} and has strong error floors around 10^{-7} BER level, which also limits their application area.

The modification of the TCC is a turbo-product code, in which the encoder forms a multidimensional block of systematically encoded data. In this case, the decoder is either the same as the one shown in Figure 6.21 or a simplified one obtained by skipping both interleaver and deinterleaver. Turbo-product codes can be applied to all high bit rates since the error floor is manageable.

The total yield of the FEC application is proportional to the net coding gain ΔG from (6.66). This yield can also be considered as a negative power penalty (positive margin) from the systems engineering perspective. However, since the coding gain is related to Q-factor, it should be converted to an equivalent associated with the

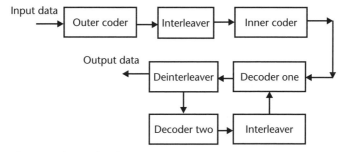

Figure 6.21 Turbo convolutional codes.

change in the OSNR. It can be done by applying (5.34), if we assume that the signal is optically amplified along the lightwave path. Therefore, it is

$$\Delta P_{FEC} = \Delta G \frac{2\Delta f}{B_{op}} \qquad (6.67)$$

where ΔP_{FEC} is the optical power gain (or positive power margin) brought to the system by employment of the FEC, Δf is electrical filter bandwidth, and B_{op} is the bandwidth of an optical filter. The value of ΔP_{FEC} can be taken directly to the right side of (5.79) and (5.80) to decrease the value of the overall power margin otherwise needed.

The impact of a negative power penalty is illustrated in Figure 6.22. As we can see, with the FEC employed, $OSNR_{required}$ decreases below the original $OSNR$ value. Because of that, the power per channel P'_{ch} is much lower than the value P_{ch} otherwise needed to satisfy the original $OSNR$ ratio. The power per channel with FEC employed should be increased to the value $P'_{ch,required}$ if the power margin is allocated to compensate for the impact of impairments. On the other hand, the power per channel would increase to the value $P_{ch,required}$ if there were no FEC. Therefore, there is the difference $\Delta P_{ch} = P_{ch,required} - P'_{ch,required}$ in channel power levels, and that is very beneficial from the system point of view since it helps to reduce the impact of the nonlinear effects. We can also see from Figure 6.22 that the $OSNR_{required}$, which corresponds to the optimum channel power $P'_{ch,opt}$, is considerably lower than the ratio that would be required if there were no FEC deployed.

6.1.6 Wavelength Conversion and Optical 3R

The role of a wavelength converter has become more important as we enter into an era of optical networking. Wavelength converters are used to translate one wavelength to the other for three different reasons:

1. To convert the input wavelength to the output one for adaptation purposes, either to improve the overall transmission capability or to make it more convenient for optical networking. A typical example of such a conversion is the translation from the 1,310-nm wavelength window to the 1,550-nm wavelength window in order to increase the transmission distance and to make it compatible with the optical wavelength grid.

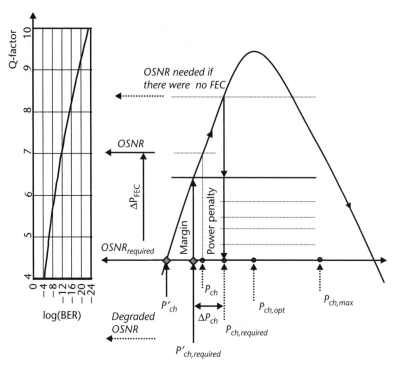

Figure 6.22 Margin allocation with FEC.

2. To convert a wavelength within one administrative domain to a different wavelength within another administrative domain in order to facilitate network management, as shown in upper part of Figure 6.23. This example can be applied to networks that belong to different carriers since the wavelength assignment is not coordinated.
3. To convert a wavelength within one network to improve utilization of the optical channels. Wavelength congestion may prevent the establishment of a lightwave path between the source and destination since the transmission link may already be occupied by the same wavelength, as shown in the lower part of Figure 6.23. In this figure, the wavelength conversion is needed for Path 8 at Node 4, since wavelength λ_1 has been already in place from Node 4 to Node 1.

The wavelength conversion is sometimes correlated with the need for full regeneration of the optical signal. A full regeneration—or reamplification, reshaping, and retiming (3R)—is needed when signal quality is compromised and there is no other way to clean the signal and to improve the BER. Full regeneration can be done either electrically or by some optical means. Some of the methods applied for wavelength conversion can also be used for the optical signal regeneration. Some advanced methods used for wavelength conversion and signal regeneration at the optical level are presented below.

The simplest way for both the wavelength conversion and signal regeneration is by applying the O-E-O conversion. In such a scheme, an incoming optical signal is

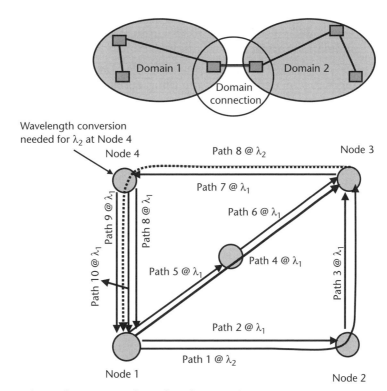

Figure 6.23 The application area of wavelength conversion.

converted to an electrical signal by the photodiode and then reconverted back by using another optical source at a wavelength that is different than the one associated with the incoming optical signal. Meanwhile, some electronic signal processing, usually recognized as 1R/2R/3R scheme, is commonly done between two conversions. The 1R processing includes just reamplification of the signal; 2R processing adds signal reshaping through the gating function to reamplification of the signal; and full 3R processing also includes a retiming function by a clock reference.

A full signal transparency is preserved only if the 1R scheme is employed, with an additional noise added during the amplification process. The signal can be cleaned from the noise by using the 2R scheme, which is transparent for all digital signals. However, the timing jitter is introduced through this process, which limits the number of 2R stages that can be cascaded. A full 3R function is not a signal-transparent process since retiming is a bit-rate specific, but is serves to clean the signal entirely while restoring its original shape.

The O-E-O wavelength conversion and signal regeneration is a more traditional or classical approach, which serves the purpose very well from the operational point of view. However, both the cost and complexity of the method are relatively high, and that is the reason why some advanced methods of optical wavelength conversion are needed. Optical wavelength converters should satisfy the following key functional requirements in order to be considered for practical applications:

- Provide high-speed operation and optical signal transparency (in terms of the modulation format and bit rate);
- Provide signal reshaping capability and the possibility of cascading a few stages together to perform multihop connections, which are often needed in optical networks;
- Provide smaller frequency chirp and high extinction ratio of the converted optical signal, and be polarization insensitive.

There are several methods that could be used for wavelength conversion purposes, which are based either on the employment of the optical gating or on the new frequency generation through nonlinear effects [34–37].

The converters based on the optical gating include semiconductor optical amplifiers to stimulate either cross-gain modulation or cross-phase modulation, as a mean to change the value of an input wavelength. The operational principle of a wavelength converter, which is based on cross-gain modulation (XGM), is illustrated in Figure 6.24. The semiconductor optical amplifier in Figure 6.24 plays the role of an optically controlled gate, whose characteristics are changed with the intensity of the input optical signal. Any increase in the input signal will cause a depletion of the carriers in the active region, which effectively means that the amplifier gain drops. That change occurs very quickly and follows dynamics of the input signal on a bit-per-bit basis. If there is a probe signal with lower optical power, it will experience a low gain during 1 bits and a high gain during 0 bits.

Since the probe wavelength λ_p is different than the wavelength of the input signal, the information content will be effectively transferred from the input signal to the probe. It is also possible to use a tunable probe, which will produce a tunable output signal. One should notice that the output signal is out of phase, or inverted, with respect to the input signal. This is because the gain of a semiconductor amplifier structure decreases with the level of the input optical signal.

The cross-gain modulation wavelength converter can also operate in two modes: the counterpropagation mode and the copropagation mode. The input signal and the probe come from different sides of the semiconductor amplifier structure in the counterpropagation mode, while the converted signal exits from the input side. Such an operational scheme does not require any optical filter. In addition, it is

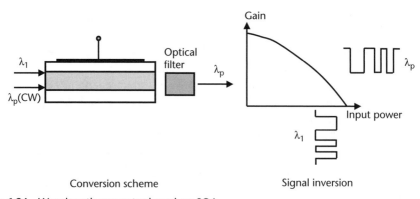

Figure 6.24 Wavelength converter based on SOA.

possible to generate an output signal at the same wavelength as the input one, which means the signal regeneration is achievable with a single stage. However, this scheme is not suitable for high-speed operations.

The copropagation scheme, which is illustrated in Figure 2.24, needs an optical filter at the output to eliminate the remaining portion of the input signal. It is not possible to have the output signal at the same wavelength as the input one, which means that another stage is needed to perform the optical regeneration. However, wavelength converters based on a copropagation design can operate at higher modulation speeds, which is currently more than 40 Gbps [35].

Each wavelength conversion is accompanied by an optical power penalty, which is dependent on the wavelength converter design and the levels of the input optical signal and probe. It seems that the input signal ranging from −6 to −4 dBm is the most convenient for the SOA-based wavelength conversion since it introduces a minimal power penalty. The power penalty ranges from 0.5 to 1.5 dB if the probe level is around −10 dBm [35]. The power penalty also depends on the nature of the conversion process, in such a way that a downward conversion in wavelength introduces a lower power penalty than the upward conversion. As a summary, we can say that wavelength converters based on the cross-gain modulation effect offer some good features such as high-speed operation, polarization insensitivity, and simple implementation. However, they suffer from frequency chirping of the output signal, a limited extinction ratio (lower than 10 dB), and a limited level of the input signal that can be efficiently converted.

The second group of converters based on the optical gating includes semiconductor optical amplifiers that stimulate the cross-phase modulation as a mean to change the value of input wavelength. These amplifiers are placed in the arms of a Mach-Zehnder interferometer, as illustrated in Figure 6.25. Any variation in carrier density due to variation of the input signal will change the refractive index in the active region of the semiconductor amplifiers, while the changes in the refractive index will change the phase of the probe signal.

The phase modulation induced by this process can be converted to amplitude modulation by the MZ interferometer through the process of constructive or destructive interference, which was explained in Section 2.2. The power of a converted optical signal can be expressed as [38]

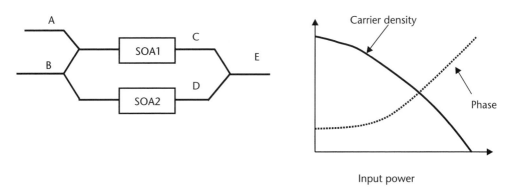

Figure 6.25 MZ SOA wavelength converters.

$$P_{out} = P_{probe} \frac{\left[G_1 + G_2 + 2\sqrt{G_1 G_2} \cos(\Delta\Phi)\right]}{8} \tag{6.68}$$

where G_1 and G_2 are the gain coefficients of the amplifiers in arms 1 and 2, respectively, and $\Delta\Phi$ is the phase difference between signals at the output of arms 1 and 2. The phase difference is directly proportional to the level of input signal P_{in}. The maximum output level is achieved when there is no phase shift at all, while the minimum level is obtained when the phase shift reaches value of $-\pi$ radians.

The main advantage of the XPM design lies in fact that it needs less signal power to achieve the same effect as compared with converters based on XGM. This means that a lower signal power can be used in combination with a higher probe level to produce the better extinction ratio. There are several design schemes proposed so far to optimize characteristics of the wavelength converters based on the XPM effect. Some of them are all-active versions of semiconductor optical amplifiers and MZ resonators [37].

As a summary, we can say that the XPM-based wavelength converters offer the following benefits: fast operation (above 40 Gbps), high extinction ratio (few decibels higher than the extinction ratio of the input signal), operation at medium input optical powers (ranging from −11 to 0 dBm), high SNR so that they can be cascaded, and polarization insensitivity. On the other hand, their implementation is relatively complex since they need a precise control of the interferometer arms. The total power penalty that occurs due to wavelength conversion is lower than 1 dB.

Wavelength converters based on XPM can also serve as optical regenerators. The 2R functionality is basically the same as the one presented in Figure 6.25, except that two stages are needed for conversion to the same optical wavelength as the input one. The operation of the XPM-based optical regenerator is illustrated in Figure 6.26. The gating signals in this scheme are introduced through separate input control ports, while the energy contained in the gating pulses determines the beginning and the end of the process [38].

In Figure 6.26, two SOAs are differently biased to achieve an initial phase difference of π radians. The control pulses are applied to inputs B and C, while the incoming signal is brought to input A. Destructive interference will happen if there are no control signals at gates B and C since there is a phase difference of π radians between output ports D and E. Therefore, the outcome does not depend on the signal level at input A. The time difference Δt between two control pulses applied at gates B and C will determine the gating window. Since the scheme from Figure 6.26 produces an

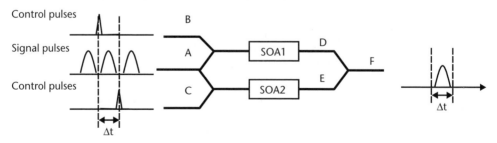

Figure 6.26 Optical 3R scheme based on SOA and MZ interferometer, After [38].

output wavelength different than the input one, two stages are needed for full 3R functionality at the specified wavelength.

The second major group of wavelength converters is based on the FWM process, which is intentionally stimulated in some semiconductor structures or in optical fibers. Recall from Section 3.3.8 that the FWM process involves three optical wavelengths λ_1, λ_2, and λ_3 to produce the fourth one (λ_4). The FWM process can be quite detrimental from the transmission point of view, especially if there are conditions that stimulate the mixing process. Such conditions are created in an optical fiber if transmission of high-power optical channels is done in the wavelength region with a minimum value of chromatic dispersion.

Any intentionally induced FWM process can be used for wavelength conversion [39]. Such a situation can be easily created in a highly nonlinear medium, such as DSF. In addition, the FWM process can be induced if a semiconductor optical amplifier is used as an active nonlinear medium, as shown in Figure 6.27.

The FWM process is centered around a strong probe signal, which produces a mirror-like replica of the input wavelengths that interact with the probe. The three wavelengths from Figure 6.27 are connected by the following relation:

$$1/\lambda_{pout} = 1/\lambda_p + \left(1/\lambda_p - 1/\lambda_{in}\right) \tag{6.69}$$

where λ_{in}, λ_{out}, and λ_p are wavelengths related to the input, output, and the probe signals, respectively. The wavelength converters based on the FWM process are fully transparent and capable of operating at extremely high speeds, which is the major competitive advantage. These devices are also polarization insensitive. However, they need high power levels for high efficiency, while implementation is relatively complex since they require tunable pumps and optical filtering of the output signal.

6.2 Transmission System Engineering Trade-offs

As we have seen in previous sections, optical systems engineering includes a large number of different parameters. In general, the engineering goal can be achieved through several different ways, while the system characteristics can be optimized by the fine-tuning of some parameters. The engineering process involves not just calculation and fine-tuning, but also the selection and trade-offs with respect to both the system parameters and the components deployed in the system.

In this section we will explain possible trade-offs related to the systems engineering process. The trade-offs can be related to selection of key optical elements (fiber, light source, photodiode) or to the selection of methods that have an impact

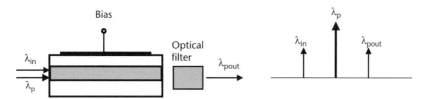

Figure 6.27 Wavelength conversion based on the FWM process.

on the spectral efficiency, chromatic dispersion management, optical power level, and optical path length.

6.2.1 Optical Fiber Type Selection

All optical fibers can be either multimode or single-mode. The application of single-mode optical fibers became widespread since they offer considerably wider bandwidth and larger application flexibility. The multimode optical fibers provide smaller application flexibility, but they are generally cheaper and can be used for all applications where the transmission bandwidth is not an issue. Such applications are, for example, the LAN connections with bit rates up to 1 Gbps over distances up to 2 km. Multimode optical fibers can be also used for very short connections between terminals with bit rates up to 10 Gbps over distances up to several hundred meters. The systems employing multimode optical fibers are either power-budget or bandwidth limited, and the systems engineering is a straightforward and relatively simple. In some cases multimode optical fibers are already in place, and systems engineering will involve the steps that were explained in Section 5.4. In the case where new optical fibers should be employed, the engineering might include an assessment whether the multimode fibers are good candidates or not. The main criteria in such an assessment will be the available bandwidth for specified length and the cost per available bandwidth and distance.

As for single-mode optical fibers, the situation is much more complex since there are several fiber types. If fibers are already deployed, the engineering process will be mainly related to analysis of the transmission characteristics (loss, wavelength windows, dispersion, dispersion slope, effective area, effective length, and polarization mode dispersion). The engineering goal would be to find out the following: what the system is capable of, what a critical limitation is, and what options are available to achieve requested transmission capacity. This may include a comparison and trade-off between two different scenarios. The first scenario is related to a higher number of optical channels loaded with lower bit rates, while the second scenario involves fewer channels but with higher bit rates.

The engineering process is different if it is related to new fiber deployment. Any such deployment should support the system requirements and maximize the usage of the transmission link. The comprehensive analysis should plan for the future increase in the transmission capacity even if the single optical channel transmission is envisioned, which is very important from the fiber type selection perspective. In general, any single-mode fiber type can be used for single channel transmission, while the decision is usually made based on the cost of the transmission line. Accordingly, the selection of DSF might be a good choice since it can provide long distance and high capacity for a single channel, with a minimum cost for dispersion compensation.

The situation is different if projected transmission capacity cannot be satisfied with a single channel transmission. The DSF fiber can still be a good choice for some number of channels. However, in terms of the overall transmission capacity, the DSF is inferior as compared to NZDSF since these fibers have been designed to satisfy requirements related to multichannel transmission systems.

When deciding what fiber type is the best choice, one should consider the overall requirements in terms of transmission length and the total capacity, the impact of chromatic dispersion in a specified wavelength band, and the impact of nonlinear effects. The only distinction is in the case of power-budget-limited high-capacity systems where standard SMF is the best choice since it is the most cost-effective solution.

As for the bandwidth-limited transmission, the major question is not just the fiber type, but also the overall transmission line design. Each fiber needs some additions that will help with the dispersion compensation. Therefore, the pairing of transmission fiber with a dispersion compensating module is a critical point, especially in cases with a large number of optical channels. That is because dispersion compensation should be equally effective for all channels.

The pairing of transmission fiber and dispersion compensating module is much easier if standard single-mode fibers are used. In addition, the combination of SMF and DCF keeps chromatic dispersion above the critical level at all local points along the transmission line, which helps to suppress nonlinear effects. As for NZDSF, their main advantage lies in the fact that the amount of dispersion that should be compensated is much smaller than that associated with the SMF case. However, pairing of NZDSF with dispersion compensating module might be an issue for large number of transmission channels. In addition, nonlinear effects might have a much stronger impact than in the SMF case. There are several different types of NZDSF, which are designed to either minimize the nonlinear effects through the larger effective area, or to minimize dispersion over larger wavelength range (recall Figure 3.11 and Table 3.2). We can expect that there will be a number of new fiber types in the future that will be even more advantageous in terms of the application area.

6.2.2 Spectral Efficiency

The transmission capacity is closely related to the spectral efficiency, which measures the spectrum utilization. The spectral efficiency is expressed in bits per hertz, and it is dependent on the bit rate, channel spacing, and coding scheme that is used. If, for example, 10-Gbps channels are spaced apart for 50 GHz, the spectral efficiency is just 0.2 bit/Hz. However, if 40-Gbps channels use the same channel spacing, the spectral efficiency increases to 0.8 bit/Hz. The spectral efficiency can be higher than 1 bit/Hz if more sophisticated modulation schemes are applied.

Spectral efficiency can be increased by either increasing the bit rate for the specified channel spacing or by decreasing the channel spacing for the specified bit rate. The increase in bit rate will have a negative implication in terms of the overall dispersion impact and will generally lead to shorter distances. Generally speaking, the chromatic dispersion can be alleviated with proper dispersion compensation, but polarization mode dispersion might become a major issue with respect to a higher bit rate (i.e., for 10 Gbps and over). In addition, the bit rate increase would bring the channels closer to each other even without any decrease in channel spacing, and that would increase the chance of having larger intrachannel crosstalk and four-wave mixing.

The negative implications of an eventual decrease in channel spacing can be suppressed by applying polarization interleaving of neighboring channels [6, 23].

The interaction between neighboring channels will be minimized in this case, while the selection of a specified channel will be more efficient. It was reported that polarization interleaving could increase the Q-factor for about 3 dB.

The spectral efficiency can also be increased by special coding and filtering methods, such as duobinary coding, in which filtering is applied to limit the spectral bandwidth. In addition, it can be done in the SSB and VSB modulation schemes, in which parts of the signal spectrum are filtered out to narrow the spectrum and to provide denser channel spacing [23].

Multilevel coding offers another possibility for spectral efficiency increase since the information content is increased for a specified number of symbols if each of them can take several distinctive levels. The relationship between the number of symbols (or bauds) M_{baud} and the number of bits M_{bit} can be expressed as $M_{bit} = M_{band} \log_2(N)$, where N is the number of signal levels. For example, if the number of levels is 2, then the number of bauds and bits are the same, while the signal line rate is equal to the bit rate. However, is the number of levels is 3, the equivalent bit rate will be eight times higher than the actual baud rate. Therefore, by keeping the same symbol duration and the same channel spacing, the information content and the spectral efficiency can be substantially increased.

Each of the methods mentioned above has its drawbacks. If we increase the bit rate at the specified channel spacing, the nonlinear effects and chromatic dispersion will increase. As a countermeasure, a well-designed dispersion map and polarization interleaving can be considered. If the number of channels is increased, the nonlinear effects will increase. As a countermeasure, the optical power per channels can be decreased to suppress nonlinear effects, but it should be combined with methods that would compensate for the negative impact of the power decrease to the SNR. Such methods are Raman amplification or forward error correction (or both of them). If the number of signal levels is increased, the SNR becomes a critical factor. As a countermeasure, one should consider the use of more sophisticated modulation/detection schemes that improve receiver sensitivity and employment of advanced forward error correction methods.

The spectral efficiency is related to both the system design and the systems engineering. The engineering will be simpler and less expensive if there are a smaller number of channels, wider channel spacing, and lower bit rates per optical channel. This is related to simplified dispersion management, relaxed requirements in terms of nonlinear effects, relaxed requirements for wavelength stability, and easier upgrade in the future with new channels added.

On the other hand, a limited amplifier bandwidth may impose a requirement that channels should be tightly spaced in order to be amplified by a single amplifier. The smaller channel spacing will reduce the SRS effect but will increase the FWM effect. The smaller channel spacing will also have an implication to the wavelength stability requirements, since it will need tighter and more expensive frequency control. In addition, the smaller channel spacing will have an impact on the power per channel since the total power at the amplifier output will be divided among larger number of channels. Therefore, all this should be carefully measured when analyzing options how to increase the spectral efficiency of the optical transmission system.

6.2.3 Chromatic Dispersion Management

A proper chromatic dispersion management is one of the most important engineering goals since it minimizes the impact of chromatic dispersion. In addition, it helps to manage the impact of nonlinear effects. Chromatic dispersion cannot generally be compensated just at the receiving side since nonlinear effects disrupt the linear nature of the pulse spreading that causes the intersymbol interference. It was proven in practice that periodic dispersion compensation is the best way to deal with chromatic dispersion for long-haul dispersion-limited transmission systems. The periodic dispersion compensation means that each transmission section is followed by some dispersion compensation module. The module can be either a section of the optical fiber having the opposite dispersion to the dispersion of the transmission section, or some other dispersion-compensating element discussed earlier in this chapter.

If a section of another optical fiber is used for compensation, such fiber can be considered as a part of the transmission line. The overall transmission line is designed in a periodic manner as a composite structure of fiber sections with positive and negative chromatic dispersion. The average dispersion of the two-section fibers length can be calculated as

$$D_{ave} = \frac{D_I L_{d,I} + D_{II} L_{d,II}}{L_{d,I} + L_{d,II}} \quad (6.70)$$

where $L_{d,I}$ and D_I are the fiber length and chromatic dispersion with respect to the first section, respectively, and and $L_{d,II}$ and D_{II} are the fiber length and chromatic dispersion related to the second section, respectively. The summary length $L_P = L_{d,I} + L_{d,II}$ is called the dispersion map period (refer to Figure 6.17). The map can start with either positive or negative dispersion. Dispersion precompensation can be performed, for example, if transmission is done in an anomalous dispersion region, but transmission begins with a negative dispersion section. The length ratio $L_{d,I} / L_{d,II}$ could be as high as 5 to 10 if transmission fiber is combined with DCF. However, this ratio is much smaller if two different transmission fiber types are combined together. In some situations such a ratio is close to one since the lengths are almost identical.

The total dispersion compensation is achieved if the average dispersion from (6.70) is zero. Such a setting is suitable for linear transmission systems, where the impact of SPM can be neglected. However, if the SPM is not negligible, it is advisable to have only about 90% to 95% of chromatic dispersion compensated after each map period. Such a dispersion map design is the best suited for transmission systems with CRZ format employed. On the other hand, the average dispersion from (6.70) should be kept below the level of about 0.08 ps/km-nm if it is designed for dispersion-managed solitons. In addition to the dispersion map, the dispersion management in CRZ and soliton-based transmission systems should include so-called dispersion strength in order to achieve the optimum performance [7]. The dispersion strength, which is a measure of the chromatic dispersion variations between two sections within each map period, is expressed as

$$S = \frac{\beta_{2I} L_{d,I} + \beta_{2II} L_{d,II}}{T_{Min}^2} \tag{6.71}$$

where T_{Min} is FWHM pulse width associated with the half of the anomalous dispersion section (section $L_{d,I}$). The value of the dispersion strength can be optimized with respect to a specific transmission scenario, which is related either to CRZ coding or solitons. Such an optimization includes numerical solving of the propagation equations to find the parameter T_{Min}. It was found, for example, that the map strength should be below 3.9 in systems with dispersion-managed solitons [3]. As for the length of the dispersion period, it is usually aligned with the amplifier spacing, which is typically 80 to 120 km for terrestrial systems and 40 to 60 km for submarine systems. If the dispersion compensation element is not a fiber section, but a discrete scheme, it is usually placed as an intermediate module in the in-line amplifier stage.

As a summary, the trade-offs related to dispersion compensation are associated with the placement of the dispersion compensation module and with the design of the dispersion compensation map. The dispersion compensation is considered as a straightforward standalone process if the transmission line does not exceed several hundred kilometers. However, if the transmission distance is longer (usually in excess of 1,000 km), the dispersion compensation is a sophisticated process that needs a very precise engineering of the dispersion map with respect to planned transmission scenario (i.e., if CRZ coding, solitons, NRZ coding, and so forth is going to be applied). In addition, some adaptive dispersion compensation elements might be required at the specific points along the lightwave path.

6.2.4 Optical Power Level

The transmitted power per optical channel determines the signal level, as explained in Section 3.1. Therefore, it should be as high as possible to achieve better SNR and BER. However, there are some factors that limit the optical power increase. First, the available optical power from lasers and modulators is limited by the type and physical design of the device. Second, any power increase above an optimum level may be quite detrimental due to an increase of the impact of nonlinear effects.

The optical power level can be enhanced by employing optical amplifiers. The amount of the optical power available per channel is determined by the number of channels and by the amplifier saturation power. Therefore, the first trade-off in such a scheme might be related to the number of channels with respect to the optical power per channel. The engineering goal is to provide each channel with enough power to keep SNR and BER at the specified level for a given transmission distance. However, if it not achievable, either the number of channels or the transmission distance should be decreased.

The increase in optical power will increase SNR up to the point where demerit becomes bigger than merit, since the penalties due to nonlinear effects become dominant and overcome the benefit due to the SNR increase. Such a point is related to an optimum power level that should be launched in order to optimize the systems characteristics. It is not easy, however, to determine that optimum power value since there is a number of parameters involved in each specific case. The computer-aided engineering and numerical simulation tools can help to identify the optimum value for each specific case.

The optimum power value is illustrated in Figure 6.28. As we see from Figure 6.28, OSNR, which is otherwise proportional to the transmission distance, increases until power P_{opt} is reached, and then starts decreasing due to the impact of nonlinear effects. The required OSNR can be achieved by lower optical power per channel if Raman amplification is applied. At the same time, the maximum OSNR that can be achieved with Raman amplification becomes considerably higher, while the optimum optical power per channel shifts towards lower levels, as shown in Figure 6.28. The employment of the FEC scheme relaxes an initial requirement for OSNR, which means that a lower optical power per channel is required. In this case, the benefit from the FEC will gradually decrease with the power increase. Both the Raman amplifier and FEC are really aimed to help with the SNR, while they cannot cope with the impact of nonlinear effects. However, they do decrease the impact of nonlinearities by decreasing the required power per channel.

The optimization of the optical power per channel is one of the key steps in the systems engineering process. The empirical evaluation presented in Chapter 5 can help to establish reference points, and to understand the overall picture. It could also serve as guidance on how to utilize numerical methods and computer-aided engineering if necessary.

6.2.5 Optical Path Length

The optical path length is determined either by the most critical impairment or by the join impact of several impairments. Optical fiber attenuation is the most critical factor in unamplified power-budget-limited systems, and the engineering process does not include any trade-offs with respect to the transmission length. On the other hand, any case with optical amplification, or a case related to bandwidth-limited

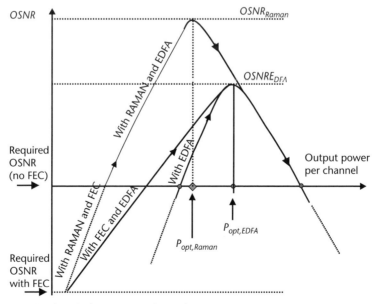

Figure 6.28 Optimal optical power per channel.

systems, opens the door for possible trade-offs with respect to the transmission length. The trade-offs are related to the optical power and chromatic dispersion/nonlinear effects management.

The total transmission length that could be achieved under specified conditions can be calculated from (5.79), (6.19), and (6.67) as

$$l = \frac{-OSNR + P_{ch} - F_{no} - 10\log(N) - 10\log(h\nu B_{op}) - \Delta P + \Delta P_{DRA} + \Delta P_{FEC}}{\alpha} \quad (6.72)$$

where $OSNR$ is the required value of the OSNR, P_{ch} is optical power per channel, F_{no} is the noise figure of optical amplifiers, N is the number of optical fiber spans, h is the Planck's constant, ν is optical frequency, ΔP is optical power margin that needs to be allocated to cover the power penalties due to different impairments, ΔP_{DRA} is the optical power gain (positive power margin) brought by Raman amplifiers, and ΔP_{FEC} is the optical power gain (positive power margin) brought by application of forward error correction scheme.

As we can see from (6.72), there are a number of parameters that determine the value of the maximum transmission length. In the general case, the trade-off can be done between the transmission capacity and the total link length. This means that the longer distance is associated with the smaller transmission capacity, and vice-versa. It is also possible to trade the transmission distance versus the amplifier spacing. The smaller amplifier spacing is used for longer transmission lines (such as submarine transmission), and vice-versa. It is important to mention that the system cost factor may be an ultimate criterion when considering the optimum transmission scenario.

The transmission length is often determined with respect to the channel with the lowest OSNR. This means that the channel with the lowest OSNR will have a commanding role in the WDM system, unless there are different performance quality criteria for different channels. It is quite possible that such a differential approach will be taken in a future optical networking environment since there will be a mix-and-match of different bit rates with different quality requirements. The dynamic performance monitoring and evaluation of optical parameters along the lightwave path will play a crucial role in future optical transmission systems.

If the required performance quality cannot be achieved, the optical signal should be regenerated. It can be done optically by applying 3R technology, in which case the lightwave path is not broken, or by the O-E-O regeneration, in which case the lightwave path is effectively broken and the engineering process should start from the beginning. The timing jitter will be the only relevant factor that will influence the total transmission length in the case where the full 3R regeneration is employed. The total accumulated jitter for each specified bit rate should be within the specified borders.

6.3 Summary

There are a number of technologies and methods that could improve the overall system performance. Some of them, such as Raman amplification or forward error correction, became in indispensable part of the high-speed optical transmission system design, while others, such as advanced modulation methods or advanced detection schemes, can be used in conjunction with others to bring some additional engineering advantages. The reader is advised to pay a special attention to formulas introduced by following equations: (6.18), (6.19), (6.35) to (6.37), (6.59), (6.66), and (6.72). The reader is also advised to consider Section 6.2 as the most useful from a practical perspective.

References

[1] Essiambre, R. J. et al., "Design of Bidirectionally Pumped Fiber Amplifiers Generating Double Rayleigh Scattering," *IEEE Photon. Techn. Lett.*, Vol. 14, 2002, pp. 914–916.

[2] Smoth, R. G., "Optical Power Handling of Low Loss Optical Fibers as Determined by Stimulated Raman Scattering and Brillouin Scattering," *Applied Optics*, Vol. 11, 1972, pp. 2489–2494.

[3] Agrawal, G. P., *Fiber Optic Communication Systems*, 3rd ed., New York: Wiley, 2002.

[4] Kin, C. H., et al., "Reflection Induced Penalty in Raman Amplified Systems," *IEEE Photon. Techn. Lett.*, Vol. 14, 2002, pp. 573–575.

[5] Puc, A., et al., "Long Haul WDM NRZ Transmission at 10.7 Gbps in S-Band Using Cascade of Lumped Raman Amplifiers," in *Proc. of Optical Fiber Conference OFC*, Anaheim, CA, 2001, PD-39.

[6] Fukuchi, K., et al., "10.92 Tb/s(273x40 Gbps) Triple Band Ultra Dense WDM Optical-repeated Transmission Experiment," *Proc. of Optical Fiber Conference OFC*, Anaheim, CA, 2001, PD 26.

[7] Radic, S., et al, "Selective Suppression of Idler Spectral Broadening in Two-Pump Parametric Architectures," *IEEE Photon. Techno. Lett.*, Vol. 15, 2003, pp. 673–678.

[8] Kazovski, L., S. Benedetto, and A. Willner, *Optical Fiber Communication Systems*, Norwood, MA: Artech House, 1996.

[9] Cvijetic, M., *Coherent and Nonlinear Lightwave Communications*, Norwood, MA: Artech House, 1996.

[10] Proakis J. G., *Digital Communications*, 3rd ed., New York: McGraw-Hill, 1995.

[11] Poole, C. D., et al., "Optical Fiber Based Dispersion Compensation Using Higher Order Modes Near Cutoff," *IEEE/OSA Journal Lightwave Techn.*, Vol. LT-12, 1994, pp. 1746–1751.

[12] Kashyap, R., *Fiber Bragg Gratings*, San Diego, CA: Academic Press, 1999.

[13] Hill, K. O., et al., "Chirped In-Fiber Bragg Grating Dispersion Compensators: Linearization of the Dispersion Characteristics and Demonstration of Dispersion Compensation in a 100 km, 10 Gbps Optical Fiber Link," *Electronics Letters*, Vol. 30, 1994, pp. 1755–1757.

[14] Watanabe, S., et al., "Compensation of Chromatic Dispersion in a Single-Mode Fiber by Optical Phase Conjugation," *IEEE Photon. Techn. Lett.*, Vol. 5, 1993, pp. 92–94.

[15] Watanabe, S., and M. Shirasaki, " Exact Compensation for Both Chromatic Dispersion and Kerr Effect in a Transmission Fiber Using Optical Phase Conjuction," *IEEE/OSA Journal Lightwave Techn.*, Vol. LT-14, 1996, pp. 243–248.

[16] Kashima, N., *Passive Optical Components for Optical Fiber Transmission*, Norwood, MA: Artech House, 1995.

[17] Vohra, S. T., et al., "Dynamic Dispersion Compensation Using Bandwidth Tunable Fiber Bragg Gratings," *Proc. of European Conference on Optical Communications ECOC*, Munich, 2000, Vol. 1, pp. 113–114.

[18] Yonenaga, K., et al., "Automatic Dispersion Equalization Using Bit-Rate Monitoring in a 40 Gbps Transmission System," *Proc. of European Conference on Optical Communications ECOC*, Munich, 2000, Vol. 1, pp. 119–120.

[19] Bulow, H., et al., "PMD Mitigation at 10 Gbps Using Linear and Nonlinear Integrated Electronic Equalizer Circuits," *Electron Letters*, Vol. 36, 2000, pp. 163–164.

[20] Karlsson, M., H. Sunnerud, and P. A. Andrekson, "A Comparison of Different PMD-Compensation Techniques," in *Proc. of European Conference on Optical Communications ECOC*, Munich, 2000, Vol. 1, pp. 119–120.

[21] Liu, F., et al., "A Novel Chirped Return to Zero Transmitter and Transmission Experiments," in *Proc. of European Conference on Optical Communications ECOC*, Munich, 2000, Vol. 3, pp. 113–114.

[22] Shanmugam, K. S., *Digital and Analog Communication Systems*, New York: Wiley, 1979.

[23] Bigo, S., et al., "Transmission of 125 WDM Channels at 42.7 Gbps (% Tbit/s capacity) over 12x100 km of TeraLight™ Ultra Fibre," in *Proc. of European Conference on Optical Communications ECOC*, Amsterdam, 2001, paper Th. M.4.9.

[24] Cai, J. X., et al., "A DWDM Demonstration of 3.73 Tb/s over 11000 km Using 373 RZ-DPSK Channels at 10 Gbps," in *Proc. of Optical Fiber Conference OFC*, Atlanta, GA, 2003, PD-22.

[25] Gordon, J. P., and H. A. Haus, "Random Walk of Coherently Amplified Solitons in Optical Fiber Transmission," *Optical Letters*, Vol. 11, 1986, pp. 665–667.

[26] Miwa, T., *Mathematics of Solitons*, New York: Cambridge University Press, 1999.

[27] Abramovitz, M., and I. A. Stegun, *Handbook of Mathematical Functions*, New York: Dover, 1970.

[28] Merlaud, F., and T. Georges, "Influence of Sliding Frequency Filtering on Dispersion Managed Solitons," *Proc. of European Conference on Optical Communications ECOC*, Nice, France, 1999, Vol. 1, pp. 228–229.

[29] Vasic, B., and I. Djordjevic, "Low-Density Parity Check Codes for Long-Haul Optical Communication Systems," *IEEE Photon. Techn. Lett.*, Vol. 14, 2002, pp. 1208–1210.

[30] Berrou, C., and A. Glavieux, "Near Optimum Error-Correcting Coding and Decoding: Turbo Codes," *IEEE Trans. Communic.*, Vol. 44, 1996, pp. 1261–1271.

[31] Sab, O. A., "FEC Technique in Submarine Systems," *Proc. of Optical Fiber Comm. Conference OFC*, Anaheim, CA, 2001, paper TuF1.

[32] Grower, W. D., "Forward Error Correction in Dispersion-Limited Lightwave Systems," *IEEE/OSA J. Lightwave Techn.*, Vol. LT-6, 1988, pp. 643–654.

[33] ITU-T Rec. G.709/Y1331, "Interfaces for the Optical Transport Network (OTN)," ITU-T (02/01), 2001.

[34] Yoo, S. J. B., "Wavelength Conversion Technologies for WDM Network Applications," *IEEE/OSA J. Lightwave Techn.*, Vol. LT-14, 1996, pp. 955–966.

[35] Deming, L., et al., "Wavelength Conversion Based on Cross-Gain Modulation of ASE Spectrum of SOA," *IEEE Photon. Techn. Lett.*, Vol. 12, 2000, pp. 1222–1224.

[36] Spiekman, L. H., "All Optical Mach-Zehnder Wavelength Converter with Monolithically Integrated DFB Probe Source," *IEEE Photon. Techn. Lett.*, Vol. 9, 1997, pp. 1349–1351.

[37] Digonnet, M. J., (ed.), *Optical Devices for Fiber Communications*, Bellingham, WA: SPIE Press, 1998.

[38] Ueno, Y., et al., "Penalty Free Error Free All-Optical Data Pulse Regeneration at 84 Gbps by Using Symmetric Mach-Zehnder Type Semiconductor Regenerator," *IEEE Photon. Techn. Lett.*, Vol. 13, 2001, pp. 469–471.

[39] Girardin, F., et al., "Low-Noise and Very High Efficiency Four-Wave Mixing in 1.5 mm Long Semiconductor Optical Amplifiers," *IEEE Photon. Techn. Lett.*, Vol. 9, 1997, pp. 746–748.

CHAPTER 7
Optical Transmission Systems Engineering Toolbox

This chapter contains material related to physical phenomena and mathematical treatment of the topics discussed in previous chapters. It should help the reader to better understand the subjects associated with optical transmission systems engineering. The material was prepared by using well-established reference literature [1–15]. The reader is advised to consult the reference material if more detailed explanation is needed. There is not a strong logical connection between different sections presented in this chapter. Each of them is rather a self-contained and self-explanatory topic.

7.1 Physical Quantities, Units, and Constants Used in This Book

Physical quantities mentioned throughout this book, and the units used for their measurements, are presented in Table 7.1. Please notice that all quantities are expressed in the international system of units (SI system).

Table 7.1 Physical Quantities and Units

Physical Quantities		Units		
Quantity	Symbol	Unit	Unit Symbol	Dimensions
Length	L, l	Meter	m	m
Mass	m	Kilogram	kg	kg
Time	t	Second	s	s
Temperature	Θ	Kelvin	K	K
Electric current	I	Ampere	A	A
Frequency	f, ν (if optical)	Hertz	Hz	1/s
Wavelength	λ	Micron	μm	10^{-6} m
Force	F	Newton	N	Kg•m/s^2
Energy	E	Joule	J	Nm
Power	P	Watt	W	J/s
Electric charge	q	Coulomb	C	As
Electric voltage	V	Volt	V	J/C
Resistance	R	Ohm	W	V/A
Capacitance	C	Farad	F	C/V
Magnetic flux	Φ	Webber	Wb	Vs
Magnetic inductance	B	Tesla	T	Wb/m^2
Electric inductance	D	Henry	H	Wb/A
Electric field	E	—	—	V/m
Magnetic field	H	—	—	A/m

253

The physical constants related to optical transmission systems engineering are summarized in Table 7.2. They are also expressed in the international system of units. Please note that typical value might mean the exact value whenever it implies. The typical value can be used in general engineering considerations if other more specific values are not available.

7.2 Electromagnetic Field and the Wave Equation

Electromagnetic field is specified by its electric and magnetic field vectors, usually denoted by $\mathbf{E}(\mathbf{r},t)$, and $\mathbf{H}(\mathbf{r},t)$, respectively, where $\mathbf{r}$ represents the space position vector, and t is the time coordinate. The flux densities of the electric and magnetic fields, usually denoted by $\mathbf{D}(\mathbf{r},t)$, and $\mathbf{B}(\mathbf{r},t)$, respectively, are given as [1–3]

$$\mathbf{D} = \varepsilon_0 \mathbf{E} + \mathbf{P} \tag{7.1}$$

$$\mathbf{B} = \mu_0 \mathbf{H} + \mathbf{M} \tag{7.2}$$

where ε_0 and μ_0 are permeability and permittivity in the vacuum, respectively, and vectors $\mathbf{P}$ and $\mathbf{M}$ represent induced electric and magnetic polarizations, respectively. The vectors $\mathbf{M}$ and $\mathbf{P}$ are material specific. The evolution of the electric and magnetic fields in space and time defines the electromagnetic wave.

The light propagation in a specific medium can be evaluated by using the electromagnetic wave theory. In our case it can be applied to an optical fiber that serves as a guiding medium for light waves. Since the fiber does not posses any magnetic properties, the vector $\mathbf{M}$ becomes zero, while the vectors $\mathbf{P}$ and $\mathbf{E}$ are mutually connected by relation

$$\mathbf{P}(\mathbf{r},t) = \varepsilon_0 \int_{-\infty}^{\infty} \chi(\mathbf{r}, t - t') \mathbf{E}(\mathbf{r}, t') dt' \tag{7.3}$$

Table 7.2 Physical Constants and Their Units

Constant	Symbol	Typical Value
Electron charge	q	1.61×10^{-19} C
Boltzmann's constant	k	1.38×10^{-23} J/K
Plank's constant	h	6.63×10^{-34} J/Hz
Raman gain coefficient, at $\lambda = 1.55\ \mu m$	g_R	7×10^{-13} m/W
Brillouin gain coefficient, at $\lambda = 1.55\ \mu m$	g_B	5×10^{-11} m/W
Nonlinear refractive index coefficient	n_2	2.2–$3.4 \times 10^{-8}\ \mu m^2/W$
Nonlinear propagation coefficient	γ	0.9–2.75 (Wkm)$^{-1}$, at $\lambda = 1.55\ \mu m$
Fiber attenuation	α	0.2 dB/km, at $\lambda = 1.55\ \mu m$
Photodiode responsivity	R	0.8 A/W
Chromatic dispersion in SMF	D	(16–17) ps/nmkm, at $\lambda = 1.55\ \mu m$
GVD parameter	β_2	-20 ps^2/km at $\lambda = 1.55\ \mu m$
Permeability of the vacuum	μ_0	$4\pi \times 10^{-7}$ H/m
Permittivity of the vacuum	ε_0	8.854×10^{-12} F/m
Light speed in vacuum	$c = (\mu_0 \varepsilon_0)^{-1/2}$	2.99793×10^{8} m/s

7.2 Electromagnetic Field and the Wave Equation

where parameter χ is called the linear susceptibility. In a general case, χ is the second-rank tensor but becomes a scalar for an isotropic medium. Since optical fiber can be considered as an isotropic medium, electric polarization vector $\mathbf{P}(\mathbf{r}, t)$ from (7.3) will have the same direction as the electric field vector $\mathbf{E}(\mathbf{r}, t)$. Accordingly, they become the scalar functions that can be denoted as $P(\mathbf{r}, t)$ and $E(\mathbf{r}, t)$, respectively. Equation (7.3) now becomes

$$P_{IS}(\mathbf{r},t) = \varepsilon_0 \int_{-\infty}^{\infty} \chi^{(1)}(t-t')E(\mathbf{r},t')dt' \tag{7.4}$$

where $\chi^{(1)}(t)$ is now a scalar function, instead of being the second-rank tensor. Please note that the scalar polarization function was dented by $P_{IS}(\mathbf{r}, t)$, which refers to the linear isotropic case. Equation (7.4) is valid for smaller values of the electric field, while the following equation should be used if the electric field becomes relatively high [1]

$$P(\mathbf{r},t) = P_{IS}(\mathbf{r},t) + \varepsilon_0 \chi^{(3)} E^3(\mathbf{r},t) \tag{7.5}$$

Parameter $\chi^{(3)}$ is known as the third-order nonlinear susceptibility. It is worth mentioning that, in general, there are also nonlinear susceptibilities with the ith ($i = 2, 4, 5, 6...$) order, but they are either zero in the optical fiber material or can be neglected.

The electromagnetic wave is characterized by a change in electric and magnetic fields in space and time, and it is governed by the Maxwell's vectors equations [1]:

$$\nabla \times \mathbf{E} = -\partial \mathbf{B} / \partial t \tag{7.6}$$

$$\nabla \times \mathbf{H} = \partial \mathbf{D} / \partial t \tag{7.7}$$

$$\nabla \cdot \mathbf{D} = 0 \tag{7.8}$$

$$\nabla \cdot \mathbf{B} = 0 \tag{7.9}$$

where

$$\nabla = \partial / \partial x + \partial / \partial y + \partial / \partial z \tag{7.10}$$

denotes the Laplacian operator applied to Cartesian coordinates x, y, and z.

The set of equations (7.1) to (7.2) and (7.6) to (7.9) eventually leads to the wave equation defined as

$$\nabla \times \nabla \times \mathbf{E} = -\mu_0 \varepsilon_0 \frac{\partial^2 \mathbf{E}}{\partial t^2} - \mu_0 \frac{\partial^2 \mathbf{P}}{\partial t^2} \tag{7.11}$$

Equation (7.11) can be transferred from the time to the frequency domain by using the Fourier transform that connects the variables in time and frequency domains. The Fourier transforms applied to the vectors $\mathbf{E}(\mathbf{r}, t)$ and $\mathbf{P}(\mathbf{r}, t)$ are

$$E(r,t) = \frac{1}{2\pi} \int_{-\infty}^{\infty} \tilde{E}(r,\omega) \exp(-j\omega t) d\omega \qquad (7.12)$$

$$P(r,t) = \frac{1}{2\pi} \int_{-\infty}^{\infty} \tilde{P}(r,\omega) \exp(-j\omega t) d\omega \qquad (7.13)$$

Please note that superscript ($\sim$) above a specific variable denotes the frequency domain. By applying the Fourier transform to (7.11), it becomes

$$\nabla \times \nabla \times \tilde{E} = \mu_0 \varepsilon_0 \omega^2 \tilde{E} + \mu_0 \omega^2 \tilde{P} = \mu_0 \varepsilon_0 \omega^2 \tilde{E} + \mu_0 \varepsilon_0 \omega^2 \tilde{\chi} \tilde{E} \qquad (7.14)$$

It should also be noticed that (7.3) has been used in the above relation to express the Fourier transform of the electric field vector by the electric polarization vector. Equation (7.14) can be rewritten as

$$\nabla \times \nabla \times \tilde{E} = \frac{\varepsilon(r,\omega)\omega^2 \tilde{E}}{c^2} \qquad (7.15)$$

where $\varepsilon(\mathbf{r}, \omega)$ represents a complex function related to dielectric permittivity in the transmission medium. It is important to notice that the light speed in vacuum is defined as $c = (\varepsilon_0 \mu_0)^{-1/2}$ [1].

Dielectric permittivity $\varepsilon(\mathbf{r}, \omega)$ is connected to the linear susceptibility by the functional relation

$$\varepsilon(r,\omega) = 1 + \tilde{\chi}(r,\omega) \qquad (7.16)$$

On the other hand, it is well known that function $\varepsilon(\mathbf{r},\omega)$ can be defined through its real part, which represents the refractive index n in the medium, and the imaginary part associated with the attenuation coefficient $\alpha(\mathbf{r},\omega)$ in the medium; that is, [4, 5]

$$\varepsilon(r,\omega) = [n(r,\omega) + j\alpha(r,\omega)c/2\omega]^2 \qquad (7.17)$$

The real (Re) and imaginary (Im) parts of the dielectric constant can be found from (7.16) and (7.17) as

$$n(r,\omega) = \left[1 + \operatorname{Re}\tilde{\chi}\right]^{1/2} \qquad (7.18)$$

$$\alpha(r,\omega) = \frac{\omega}{cn(r,\omega)} \operatorname{Im}\tilde{\chi} \qquad (7.19)$$

Therefore, the refractive index and attenuation coefficient are not constant, but depend on the space position and frequency. The frequency dependence of the refractive index causes the chromatic dispersion effect, as explained in Chapter 3.

The wave equation, (7.15), can be simplified if applied to optical fibers. First, the coefficient α can be neglected by assuming that the attenuation in fibers is relatively small. Second, the refractive index can be considered as a parameter that is independent on the spatial position. This approximation is just partially true, but it is justified by the fact that the index changes occur over the lengths much longer than the wavelength. After the simplification process, (7.15) takes the form

$$\nabla \times \nabla \times \tilde{\mathbf{E}} = -\nabla^2 \tilde{\mathbf{E}} = \frac{n^2(\omega)\omega^2 \tilde{\mathbf{E}}}{c^2} \qquad (7.20)$$

(Please note that the vector identity represented by the left-side part was used in the above equation.) Equation (7.20) can be rewritten as

$$\nabla^2 \tilde{\mathbf{E}} + n^2(\omega) k_0^2 \tilde{\mathbf{E}} = 0 \qquad (7.21)$$

The parameter k_0 is the wave number in vacuum, defined as

$$k_0 = \omega/c = 2\pi/\lambda \qquad (7.22)$$

where λ is the wavelength, and ω is the radial frequency of the lightwave.

The wave number can be associated with any specific medium, through the propagation vector $\mathbf{k}(\mathbf{r}, \omega)$ defined as

$$\mathbf{k}(\mathbf{r}, \omega) = k_0 n(\mathbf{r}, \omega) \qquad (7.23)$$

The component of the propagation vector $\mathbf{k}(\mathbf{r}, \omega)$ along the z axis is called the propagation constant. The propagation constant, usually denoted by Greek letter β, is one of the most important parameters when considering the wave propagation through an optical fiber.

7.3 The Propagation Equation for Single-Mode Optical Fiber

Optical fiber can be considered as a guided medium propagating along the z axis. In the general case, the propagating light pulse is a composite optical signal that contains a number of monochromatic spectral components. Each spectral component behaves differently in a dispersion medium, such as the optical fiber, which eventually leads to the light pulse distortion.

Each axial component of the monochromatic electromagnetic wave can be represented by its complex electric field function as

$$E(z,t) = E_a(z,t) \exp[j\beta(\omega)z] \exp(-j\omega_0 t) \qquad (7.24)$$

where $E_a(z,t)$ is the amplitude of the wave as a function of time t and the distance z. It is also referred to as the pulse envelope. The parameter $\omega_0 = 2\pi\nu_0$ is the radial carrier optical frequency, ν_0 is the linear carrier optical frequency, and $\beta(\omega)$ is the propagation constant of the optical wave (refer to Section 3.4.4).

The light pulse distortion, which is observed through the pulse broadening along the optical fiber, can be evaluated by knowing the frequency dependence of the propagation constant $\beta = \beta(\omega)$ at a specified distance z along the fiber. Each spectral component within the launched optical pulse will experience a phase shift proportional to $\beta(\omega)z$. The amplitude spectrum observed at point z along the fiber length is given in the frequency domain as

$$\tilde{E}_a(z,\omega) = \tilde{E}_a(0,\omega)\exp[j\beta(\omega)z] \qquad (7.25)$$

Please recall that superscript ($\sim$) denotes the frequency domain of the associated parameter.

The behavior of the pulse envelope during the propagation process can be evaluated though the inverse Fourier transform of (7.25). The exact calculation of the inverse Fourier transform is very complex and cannot be generally carried out without additional simplifications. The simplification can be done by the following several steps, as presented in [5]. First, the propagation constant $\beta(\omega)$ can be expressed by a Taylor series

$$\beta(\omega) \approx \beta(\omega_0) + \frac{d\beta}{d\omega}\bigg|_{\omega=\omega_0}(\omega-\omega_0) + \\ + \frac{1}{2}\frac{d^2\beta}{d\omega^2}\bigg|_{\omega=\omega_0}(\omega-\omega_0)^2 + \frac{1}{6}\frac{d^3\beta}{d\omega^3}\bigg|_{\omega=\omega_0}(\omega-\omega_0)^3 + \ldots \qquad (7.26)$$

The first derivative at the right side of (7.26), often denoted as $\beta_1 = d\beta/d\omega$, is just the inverse value of the pulse group velocity v_g, which was introduced in Section 3.4.4. The term $\beta_2 = d^2\beta/d\omega^2$ is commonly known as the GVD parameter, which determines the extent of the pulse broadening during the propagation. It is easy to relate this parameter to the chromatic dispersion parameter D [recall (3.56)].

Secondly, the concept of slowly varying amplitude $A(z,t)$ can be introduced to express the pulse field function from (7.24) as

$$E(z,t) = E_a(z,t)\exp[j\beta(\omega)z]\exp(-j\omega_0 t) = A(z,t)\exp(j\beta_0 z - j\omega_0 t) \qquad (7.27)$$

Equations (7.26) and (7.27) can be now used to calculate the inverse Fourier transform of (7.25) and find the slowly varying amplitude $A(z,t)$, as explained in Section 3.4.4 [see (3.64) and (3.65)]. The propagation equation that governs the pulse propagation through single-mode optical fibers can be found by calculating the partial derivative of the slowly varying amplitude with respect to axial coordinate z, and has the form

$$\frac{\partial A(z,t)}{\partial z} = -\beta_1 \frac{\partial A(z,t)}{\partial t} - \frac{j\beta_2}{2}\frac{\partial^2 A(z,t)}{\partial t^2} + \frac{\beta_3}{6}\frac{\partial^3 A(z,t)}{\partial t^3} \qquad (7.28)$$

Equation (7.28) is a basic propagation equation that refers to the pulse propagation in single-mode optical fibers. This equation can be generalized by including the impacts of optical signal attenuation and the nonlinear Kerr index, so it becomes [5]

$$\frac{\partial A(z,t)}{\partial z} + \beta_1 \frac{\partial A(z,t)}{\partial t} + \frac{j\beta_2}{2}\frac{\partial^2 A(z,t)}{\partial t^2} - \frac{\beta_3}{6}\frac{\partial^3 A(z,t)}{\partial t^3}$$
$$= j\gamma |A(z,t)|^2 A(z,t) - \frac{1}{2}\alpha A(z,t) \tag{7.29}$$

where α is the fiber attenuation coefficient, and $\gamma = 2\pi n_2/(\lambda A_{eff})$ is the nonlinear coefficient introduced by (3.105). Please recall that n_2 is the nonlinear Kerr coefficient, and A_{eff} is the effective cross-sectional area of the optical fiber core.

Since the second term on the left side of (7.29) defines pulse delay, it can be omitted. In fact, (7.29) can be transformed to a new equation by introducing a new time coordinate $t_1 = t - \beta_1 z$ and a new axial coordinate $z_1 = z$. Since the notation does not have any specific meaning, we can still keep the notations z and t. However, we should always recall that the output pulse shape will not be represented in a real time, but will be delayed by amount $\beta_1 z$. The consolidated form of the propagation equation becomes

$$\frac{\partial A(z,t)}{\partial z} + \frac{j\beta_2}{2}\frac{\partial^2 A(z,t)}{\partial t^2} - \frac{\beta_3}{6}\frac{\partial^3 A(z,t)}{\partial t^3} = j\gamma |A(z,t)|^2 A(z,t) - \frac{1}{2}\alpha A(z,t) \tag{7.30}$$

Let us state again that any solution of (7.30) will not be represented in a real time, but will carry a delay equal to $\beta_1 z$, which means that the solution in a real time can be written as $A(z, t-\beta_1 z)$.

Equation (7.30), known as the nonlinear Schrodinger equation, is the fundamental equation in the evaluation of various effects that can occur during the pulse propagation (such as dispersion and self-phase modulation). In addition, the coupled equations can be established to treat more complex effects that are related to the energy exchange between individual optical pulses. In such a case, the total electric filed is created by the sum of slowly varying amplitudes. It is common to consider interaction among three optical pulses that jointly create the electric field proportional to amplitude

$$A(z,t) = A_1(z,t) + A_2(z,t) + A_3(z,t) \tag{7.31}$$

where $A_1(z, t)$, $A_2(z, t)$, and $A_3(z, t)$ are associated with three separate signal pulses. The propagation equation (7.30) is now converted to three coupled equations, as shown in [5]

$$j\frac{\partial A_1(z,t)}{\partial z} - \frac{\beta_2}{2}\frac{\partial^2 A_1(z,t)}{\partial t^2} = -\gamma \left[\left(|A_1|^2 + 2|A_2|^2 + 2|A_3|^2\right)A_1 + A_2^2 A_3^*\right] \tag{7.32}$$

$$j\frac{\partial A_2(z,t)}{\partial z} - \frac{\beta_2}{2}\frac{\partial^2 A_2(z,t)}{\partial t^2} = -\gamma \left[\left(|A_2|^2 + 2|A_1|^2 + 2|A_3|^2\right)A_2 + 2A_1 A_2^* A_3\right] \tag{7.33}$$

$$j\frac{\partial A_3(z,t)}{\partial z} - \frac{\beta_2}{2}\frac{\partial^2 A_3(z,t)}{\partial t^2} = -\gamma \left[\left(|A_3|^2 + 2|A_1|^2 + 2|A_2|^2\right)A_3 + A_2^2 A_1^*\right] \tag{7.34}$$

The first terms on the right sides of the above equations account for the SPM effect, the second and third terms are related to XPM, and the forth terms refer to the

FWM effect. Please notice that the impact of third-order chromatic dispersion, which is associated with coefficient β_3 in (7.28), has been neglected in (7.32) to (7.34).

Generally, the Schrodinger equation is solved numerically, although there are some special cases where an analytical solution can be readily found. Such an example is associated with soliton pulses that are discussed in Section 6.1.3. As for the numerical solution of the Schrodinger equation, it is often carried by the software package used for modeling and simulation of transmission system characteristics.

7.4 Frequency and Wavelength of the Optical Signal

There is the following relation between frequency v and wavelength λ of an optical signal:

$$\lambda = c / v \qquad (7.35)$$

where c is the light speed in vacuum. The relation between the frequency band and the wavelength band (i.e., the spectral width) of an optical signal can be obtained by expanding (7.35) in a Taylor series around the central value $\lambda_0 = c/v_0$ of the wavelength band and by keeping just two terms of the expansion; that is,

$$\lambda - \lambda_0 = \Delta\lambda \approx \frac{\lambda^2 (v - v_0)}{c} = \frac{\lambda^2 \Delta v}{c} \qquad (7.36)$$

This equation can be used to establish a relation between the wavelength band $\Delta\lambda$ and the frequency band Δv, for any specific carrier wavelength λ. For example, frequency band $\Delta v = 100$ GHz corresponds to the wavelength band $\Delta\lambda = 0.8$ nm in vicinity of wavelength $\lambda = 1.55$ μm.

7.5 Stimulated Emission of Light

The process of the emission and absorption of light can be explained by using the basic quantum mechanics model presented in Figure 7.1. The electrons are associated with their energy levels, which means that each electron can be attributed with a specific energy level [6, 7]. The population of electrons is associated with the energy band that contains multiple energy levels.

The lower energy levels are better populated in normal conditions than higher energy levels. Therefore, in normal conditions it is valid that $N_1 > N_2$ for $E_1 < E_2$, where N_1 and N_2 are the number of electrons at the energy levels E_1 and E_2, respectively (see Figure 7.1). The electrons can move in energy from level to level, by either absorbing the photons (when moving from lower to higher levels) or by radiating photons (while moving from higher to lower levels).

The normal conditions are related to so-called thermal equilibrium, where the ratio between the numbers N_1 and N_2 can be expressed by the Boltzmann's formula [1]

$$\frac{N_2}{N_1} = \exp\left(-\frac{E_2 - E_1}{k\Theta}\right) = \exp\left(-\frac{hv}{k\Theta}\right) \qquad (7.37)$$

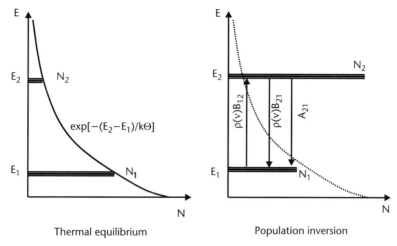

Figure 7.1 Energy levels and population inversion.

where k is the Boltzmann's constant, Θ is the absolute temperature, h is the Planck's constant, and ν is the optical frequency that is proportional to the difference between the energy levels.

Electrons can move from lower to upper energy levels through absorption of the incoming photons, while the movement from upper to lower energy levels can occur through either stimulated or spontaneous emission, or both of them simultaneously. During the stimulated emission, downward transition occurs under the influence of the outside photons that penetrate to the region. Newly radiated photons are at the same frequency, phase, and polarization as the incoming ones. The spontaneous emission occurs without any outside influence, through a stochastic downward energy transition.

The rates of light absorption, stimulated emission, and spontaneous emission, all with respect to the two-level system from Figure 7.1, can be written as

$$\frac{dN_{1,abs}}{dt} = B_{12}\rho(\nu)N_1 \tag{7.38}$$

$$\frac{dN_{2,stim}}{dt} = B_{21}\rho(\nu)N_2 \tag{7.39}$$

$$\frac{dN_{2,sp}}{dt} = A_{21}N_2 \tag{7.40}$$

where $\rho(\nu)$ is the spectral density of the electromagnetic energy, while coefficients B_{12}, B_{21}, and A_{21} characterize the absorption, stimulated emission, and spontaneous emission, respectively. The subscripts "abs," "stim," and "sp" stand for "absorption," "stimulated," and "spontaneous," respectively. In thermal equilibrium, the upward and downward transitions ere equalized, while the following equation is valid:

$$A_{21}N_2 + B_{21}\rho(\nu)N_2 = B_{12}\rho(\nu)N_1 \tag{7.41}$$

The spectral density $\rho(v)$ can be obtained from (7.37) and (7.41) as

$$\rho(v) = \frac{A_{21}/B_{21}}{(B_{12}/B_{21})\exp(hv/k\Theta) - 1} \quad (7.42)$$

The coefficients A_{21}, B_{21}, and B_{12} can be evaluated by comparing (7.42) with the following equation, otherwise known as the Planck's equation [1]:

$$\rho(v) = \frac{8\pi h v^3/c^3}{\exp(hv/k\Theta) - 1} \quad (7.43)$$

It is, therefore,

$$A_{21} = \frac{8\pi h v^3 B_{21}}{c^3} \quad (7.44)$$

$$B_{21} = B_{12} \quad (7.45)$$

Equations (7.44) and (7.45) were first established by Einstein a while ago, and that is the reason why these coefficients are known as Einstein's coefficients.

Spontaneous emission dominates over stimulated emission in the normal situation known as thermal equilibrium. Stimulated emission can become the dominant process only if it overcomes the absorption, which can be achieved if it is $N_2 > N_1$. This condition, known as the population inversion, is not a normal state that can be achieved by the thermal equilibrium. The population inversion, which is a prerequisite for stimulated emission of the light, can be achieved through an external process. This means that an external energy, usually in optical or electrical form, is introduced to excite the electrons and to lift them from the lower to the upper energy levels.

The scheme presented in Figure 7.1 is the simplest two-level scheme. However, more complex three- and four-level energy schemes are commonly used in practice. In these schemes, the electrons are lifted from the lowest level, often called the ground state, to one of the upper energy levels by skipping intermediate energy levels. The electrons do not stay at the upper energy level, but rather move to lower intermediate levels through a nonradiative energy decay. Therefore, the intermediate energy level, known as a metastable one, serves as a base for population inversion, and the number N_2 is commonly associated with the metastable level.

7.6 Semiconductors as Basic Materials for Lasers and Photodiodes

Semiconductors are materials that can easily accommodate the population inversion, but the total picture is more complex than the simplified two-level atomic system presented in Section 7.5. The atoms in semiconductors are close enough to each other to interact and to shape the distribution of energy levels in a way that is semiconductor compound specific. As for semiconductor properties, just the two highest energy bands, called valence band and conduction band, are of real importance. These bands are separated by the energy gap or forbidden band where no energy levels exist.

If electrons are excited by some means (for example, optically or thermally), they move in energy from the valence to the conduction band. For each electron lifted to the conduction band, there is one empty spot in the valence band. This vacancy is called a hole. Both the electrons from the conduction band and holes can move under the influence of an external electric field and contribute to the current flow through the semiconductor crystal. If semiconductor crystal contains no impurities, it is called intrinsic material. Such materials are, for example, silicon and germanium, which belong to the IVth group of elements and have four electrons in their outer shell. With these electrons an atom makes the covalent bonds with neighboring atoms. In this environment, some electrons can be excited to the conduction band due to thermal vibrations of the atoms in the crystal. Therefore, the electron-hole pairs can be generated by pure thermal energy. There might be an opposite process as well, when an electron releases its energy and drops into a free hole in the valence band. This process is known as the electron-hole recombination. The generation and recombination rates are equal in thermal equilibrium.

The conduction capability of intrinsic semiconductor materials can be increased through the doping process by adding some impurities from either group V or group III of the elements. Group V is characterized by having five electrons in the other shell, while there are just three electrons in outer shell for the elements from group III. If group V elements are used, four electrons are engaged in covalent bonding, while the fifth electron is loosely bound and available for conduction. The impurities from group V elements are called donors since they give an electron. They are also known as n-type semiconductors, since the current is carried by electrons (n stands for negative charge.)

On the other hand, by adding atoms from the group III of elements, three electrons will create the covalent bonds, while a hole will be created in the environment where other atoms have four electrons engaged in covalent bonds. The conduction property of the hole is the same as the property of the electron at the conduction level. This is because an outside electron will eventually occupy the hole. That electron will leave the hole at its original place, which will be occupied by some other electron leaving the hole at its original place, and so on. Therefore, the process of occupation and reoccupation is moving along the crystal. The impurities from group III elements are called acceptors, since they accept an electron. They are also known as p-type semiconductors, since the current is carried by holes (p stands for positive charge characterized by the missing electron).

By adding donor or acceptor impurities, en extrinsic semiconductor is formed. Each extrinsic semiconductor has majority carriers (electrons in n-type and holes in p-type) and minority carriers (holes in n-type and electrons in p-type). While both n-type and p-type semiconductors can serve as conductors, the true semiconductor-based device is formed by joining both types in a single continuous crystal structure. The junction between two regions, known as the p-n junction, is responsible for all useful electrical characteristics of semiconductors. The p-n junction is characterized by several important phenomena. First, the holes in p-type will diffuse towards the p-n junction to neutralize the majority of electrons present at the other side of the border. This will leave a small region in p-type that is close to the p-n junction with smaller number of holes than in the rest of the p-type region. Therefore, the electrical neutrality is effectively disrupted, since bonded electrons are still in place.

Consequently, a small negatively charged region close to the p-n border is created in p-type semiconductor. Just the opposite will happen at the n-side of the p-n junction, where the electron shortage will occur and a positively charged region will be created.

The carrier's transition just described will effectively establish a potential barrier at the p-n junction. The potential barrier can be characterized by the electric field vector $\mathbf{E}_{bar}$ with the direction from n-type to p-type, as shown in Figure 7.2. Any positive charge in n-type, which happens to be close to the p-n junction, is attracted to move towards the p-side under the impact of the electric field $\mathbf{E}_{bar}$. The electric field $\mathbf{E}_{bar}$ will restrict the further diffusion of the majority carriers from their native regions (electrons from n-type and holes from p-type). If there is no external voltage applied, this situation will stay, and there is no organized current flow through the junction. The width of the potential barrier, and the strength of the electric field $\mathbf{E}_{bar}$, is determined by the concentration of dopants in n-type and p-type semiconductors.

The situation will change if an external bias voltage is applied to the p-n junction. If the reverse bias voltage is applied, the external electric field $\mathbf{E}_{bias}$ has the same direction as an internal electric field $\mathbf{E}_{bar}$ and enhances the restriction already imposed. Therefore, the width of the depletion region is increased when p-n junction is under reverse bias, and there is not any current flow. (The exception is a small leaky current due to the presence of minority carriers in p-type and n-type regions.) If some electron-hole pairs appear in the depletion region due to thermal activity, they are immediately separated by the electric field $\mathbf{E}_{bias}$. The end result will always be such that electrons go towards n-type and holes towards p-type. This reverse bias of the p-n junction corresponds to the photodiode mode. The static situation will change if there is the illumination of the depletion region with the incoming light signal, since the generation of the carriers (both the electrons and holes) will be

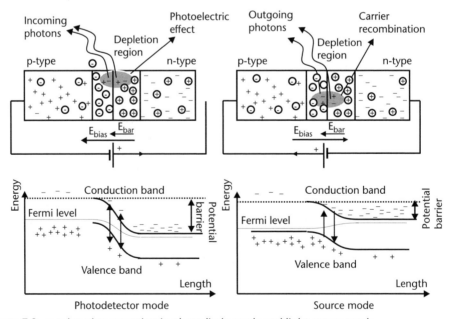

Figure 7.2 p-n junction operating in photodiode mode and light source mode.

initiated. The pairs of electron-holes will be created due to photoelectric effect, as shown in Figure 7.2, but will immediately be separated by a strong electric field $\mathbf{E}_{bias}$.

The opposite situation occurs if a forward bias voltage is applied to the p-n junction since the external electric field $\mathbf{E}_{bias}$ will be opposite to the internal field $\mathbf{E}_{bar}$. This situation will result in the reduction in the potential barrier, which means that the electrons from n-type and holes from p-type can flow and cross the p-n junction much easier. Consequently, the free electrons and holes can be simultaneously present in the depletion region. They can also recombine together and generate the photons due to energy release, as shown in Figure 7.2. The forward biasing corresponds to semiconductor light sources (LED and lasers). The rate and the character of recombination will eventually determine the character of the output light signal.

Therefore, from the energy perspective, the bias voltage can either enhance the difference in energy bands between p-type and n-type semiconductor compounds, which happens in photodiode mode, or flatten the energy band difference, which occurs in the light source mode. The energy levels for these two operational modes are shown in the lower part of Figure 7.2. Notice that there is an energy level called the Fermi level, which is also shown in Figure 7.2. The meaning of this level will be explained shortly.

The conditions related to the light generation and laser emission in semiconductors can be better understood by explaining the conditions for population inversion. The probability that an electron occupies a specified energy level E when the system is in thermal equilibrium is given by the Fermi-Dirac distribution as

$$p(E) = \frac{1}{\exp\left(\dfrac{E - E_F}{k\Theta}\right) + 1} \tag{7.46}$$

where Θ is the absolute temperature, k is the Boltzmann's constant, and E_F is the Fermi energy level.

The Fermi energy level is a parameter that indicates the distribution of electrons and holes in the semiconductor. This level is located at the center of the band gap in an intrinsic semiconductor that is in thermal equilibrium, which means that there is a smaller probability that an electron will occupy the conduction band. The probability will be higher if the temperature Θ is increased.

The position of the Fermi level varies in different semiconductor types. The Fermi level in n-type semiconductors is raised to a higher position, which increases the probability that electrons will occupy the conduction band. On the other hand, in p-type semiconductors, the Fermi level is lowered to be below the center of the band gap, as shown in Figure 7.2.

The probability that electrons will occupy the energies E_1 and E_2, associated with valance and conduction bands, respectively, can be expressed by using (7.46) as

$$p_{val}(E_1) = \frac{1}{\exp\left(\dfrac{E_1 - E_{F,val}}{k\Theta}\right) + 1} \tag{7.47}$$

$$p_{cond}(E_2) = \frac{1}{\exp\left(\dfrac{E_2 - E_{F,cond}}{k\Theta}\right) + 1} \tag{7.48}$$

The Fermi levels $E_{F,cond}$ and $E_{F,val}$ are related to the conduction and valence bands, respectively. The Fermi level can be moved up, which is relevant to the n-type semiconductors, or moved down in p-type semiconductors. In both cases, it is done by increasing the dopant concentration. In some situations, the Fermi level can be positioned within the conduction band for n-type or within the valence band for p-type semiconductors, if they are heavily doped or if the population inversion is achieved.

The population inversion can be achieved by an electrical current that populates the conduction band with a rate that is higher that the rate the band is emptied. The population inversion and position of Fermi levels in valence and conductance bands is illustrated in Figure 7.3. It is represented through the so-called E-k diagram, where E is the energy and **k** is the wavespace vector [1]. The situation shown in Figure 7.3 represents the so-called direct bandgap semiconductors, where the minimum energy point in the conduction band coincides with the maximum energy point in the valence band. Such semiconductors are, for example, GaAs, InP, InGaAsP, and AlGaAs, all of which are used for semiconductor laser manufacturing.

There is another type of semiconductor in which the minimum energy point in the conduction band does not coincide with the maximum energy point in the valence band. These semiconductors are known as the indirect bandgap semiconductors. Both silicon and germanium belong to this group. The electron-hole recombination in this case cannot be done without an adjustment in momentum (i.e., in the wavespace vector **k**). This is done through the crystal lattice vibration and the creation of phonons (or the thermal energy). The efficiency of the light generation will be smaller since part of the energy is lost, and that is the reason why these semiconductors are not good materials for light sources.

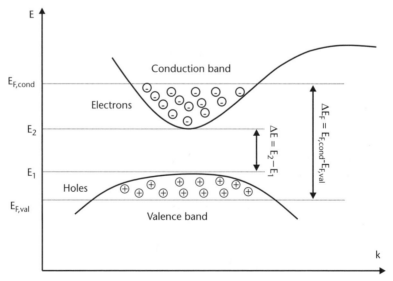

Figure 7.3 Energy bands and population inversion in semiconductors.

Once the population inversion in semiconductor material is achieved, the stimulated emission can take place. The electrons will fall back in energy to the valence band and recombine with holes. Each recombination should produce a photon of the light signal, due to the energy release. The stimulated emission is achieved if the difference between the Fermi levels $E_{F,con} - E_{F,val}$ is larger than the energy bandgap $E_2 - E_1$. The difference between these two Fermi levels becomes larger with an intensive pumping by the forward bias current that flows through the p-n junction.

The condition $(E_{F,con} - E_{F,val}) > (E_2 - E_1)$, which is required for stimulated emission, is observed if the direct bias current is higher than the threshold current, as shown in Figure 2.6. Any specific doping of p-type and n-type layers will have an impact on the position of the Fermi levels. Several special doping schemes can be applied to influence the stimulated radiation process (please recall that multistructured semiconductor lasers mentioned in Section 2.2 are manufactured for this purpose).

The difference $\Delta E = E_2 - E_1$ between the energy levels in conduction and valence bands will determine the wavelength and the frequency of the optical radiation. It is expressed through the well-known equation [1]

$$h\nu = \Delta E = E_2 - E_1 \tag{7.49}$$

where h is the Plank's constant, and ν is the frequency of optical radiation. Therefore, the output frequency is determined by the position of the semiconductor energy bands, which are specific for each semiconductor structure. It is also important to notice that the laser regime should be maintained through the proper confinement of the energy in the active region. This is often done through heterojunctions that form a complex waveguide structure in the active region (recall Figures 2.3 and 2.5).

7.7 Laser Rate Equations

The light radiation from semiconductor lasers can be characterized through the rate equations describing the change in number of photons and electrons with the time. The rate equations can be expressed as [5–7]

$$\frac{dP}{dt} = GP + Gn_{sp} - \frac{P}{\tau_p} \tag{7.50}$$

$$\frac{dN}{dt} = -GP - \frac{N}{\tau_e} + \frac{I}{q} \tag{7.51}$$

where P is the radiated optical power measured by the number of photons involved in the process, N is the number of carriers (i.e., electrons) involved in the process, G is the net gain related to the stimulated emission, $n_{sp} = N_2/(N_2 - N_1)$ is the spontaneous emission factor expressed through the electron populations at the upper and lower energy states, respectively, I is the bias current, τ_p and τ_e are the lifetimes of electrons and photons, and q is the electron charge.

The electron lifetime τ_e represents the time that an electron can live at the upper metastable level before being recombined, while the photon lifetime τ_p is related to the time the photon can spend in the laser resonant cavity. Equation (7.51) shows that the number of generated photons is directly proportional to the gain and inversely proportional to the photon lifetime. The gain and photon lifetime are dependent on the material structure of the resonant cavity, and can be expressed as

$$G = \Gamma v_g g \tag{7.52}$$

$$\tau_p = v_g \alpha_{cavity} \tag{7.53}$$

where v_g is the group velocity of the light, Γ is the cavity confinement factor, g is the material gain, and α_{cavity} is the loss in the resonant cavity.

The number of electrons is enhanced by direct bias current, while it is being depreciated by the recombination process. The deprecation is expressed by the fist term on the right side of (7.51). The rate of depreciation is faster if the lifetime is smaller. Equations (7.50) and (7.51) can be solved either for the CW regime in order to evaluate the P-I curve, or for a dynamic regime related to the optical signal modulation. A solution of the rate equations is commonly done numerically, often as a part of the modeling and simulation software package.

There is one more rate equation that can be established to express the phase ϕ of the output radiation. It is [5]

$$\frac{d\phi}{dt} = \frac{\alpha_{chirp}}{2}\left(G - \frac{1}{\tau_p}\right) \tag{7.54}$$

where α_{chirp} is the chirp factor introduced by (3.49). Equation (7.54) reflects the fact that any change in carrier population N causes a change in the refractive index within the laser resonant cavity. The change in the refractive index means that some amount of phase modulation always accompanies intensity modulation of an optical signal.

Equations (7.50) to (7.51) and (7.54) can be generalized by including terms that represent the noise impact, so they become

$$\frac{dP}{dt} = GP + Gn_{sp} - \frac{P}{\tau_p} + F_P(t) \tag{7.55}$$

$$\frac{dN}{dt} = -GP - \frac{N}{\tau_e} + \frac{I}{q} + F_N(t) \tag{7.56}$$

$$\frac{d\phi}{dt} = \frac{\alpha_{chirp}}{2}\left(G - \frac{1}{\tau_p}\right) + F_\phi(t) \tag{7.57}$$

where $F_P(t)$, $F_N(t)$, and $F_\phi(t)$ are Langevin forces related to fluctuations of the intensity, number of carriers, and the phase of output optical radiation, respectively. It is often assumed that the Langevin forces are Gaussian random processes.

7.8 Modulation of an Optical Signal

A monochromatic electromagnetic wave, which is used as a signal carrier, can be represented through its electric field as [8–10]

$$E(t) = \mathbf{p} A \cos(\omega t + \varphi) \tag{7.58}$$

where A is the wave amplitude, ω is the frequency, and ϕ is the phase of the carrier, while $\mathbf{p}$ presents the polarization orientation. Each of these parameters (amplitude, frequency, phase, and the polarization state) can be utilized to carry information. This is done by making them time dependent and related to the information content. Accordingly, four modulation types can be recognized. They are amplitude modulation (AM), frequency modulation (FM), phase modulation (PM), and polarization modulation (PoM). If the information is in digital form, the modulation is referred to as shift-keying. Therefore, there are the amplitude shift keying, the frequency shift keying, the phase shift keying, and the polarization shift keying.

The power of an optical signal is proportional to the square of its electric field amplitude. The intensity modulation of the optical signal can be performed if the power level is changed in accordance with the modulation signal. The modulation is referred as on-off keying (OOK) if the information is in digital form. The OOK modulation scheme is the one commonly used in practice. This is because it is the simplest yet effective method that utilizes the properties of an optical carrier.

7.9 Optical Receiver Transfer Function and Signal Equalization

The detection of an optical signal by the photodiode in the optical receiver is followed by the amplification process and signal filtering. These functions enhance the signal level and limit the noise power. In addition, waveform equalization is usually applied to recover the pulse shape and to suppress the intersymbol interference. Mathematical treatment of these processes is often done in the frequency domain. The photocurrent signal $I(t)$ can be transferred to the frequency domain by Fourier transform as [8–10]

$$\tilde{I}(\omega) = \int_{-\infty}^{\infty} I(t) \exp(j\omega t) dt \tag{7.59}$$

Recall that superscript ($\sim$) denotes the frequency domain of a specific variable. The inverse Fourier transform converts signals from frequency to time domain as

$$I(t) = \frac{1}{2\pi} \int_{-\infty}^{\infty} \tilde{I}(\omega) \exp(-j\omega t) d\omega \tag{7.60}$$

The current signal is converted to voltage signal at the front-end amplifier (see Figure 5.1). The voltage signal is further amplified by the main amplifier and processed by the filter before it goes to the clock recovery and decision circuits. The output voltage $V_{out}(t)$ that comes to the decision circuit can be expressed in the frequency domain as [11, 12]

$$\tilde{V}_{out}(\omega) = \frac{\tilde{I}(\omega)}{Y(\omega)} H_{F-end}(\omega) H_{amp}(\omega) H_{filt}(\omega) = \tilde{I}(\omega) H_{rec}(\omega) \qquad (7.61)$$

where $Y(\omega)$ is the input admittance, which is determined by load resistor and input of the front-end amplifier. Transfer functions $H_{F-end}(\omega)$, $H(\omega)_{amp}$, and $H(\omega)_{filt}$ are referred to as the front-end, main amplifier, and filter/equalizer, respectively. Transfer function $H_{rec}(\omega)$ is related to the total transfer properties of the optical receiver.

The intersymbol interference from the neighboring pulses is minimized, or possibly removed, if the output voltage signal takes the shape of the raised cosine function,

$$\tilde{V}_{out}(\omega) = \frac{1}{2}\left[1 + \cos\left(\frac{\omega}{2\beta B}\right)\right] \quad \text{for } \omega/2\pi = f < B \qquad (7.62)$$

The output pulse shape given by (7.62) becomes zero for frequencies larger than the bit rate B. The parameter β is selected to be within the range from 0 to 1.

An impulse response in the time domain can be found by taking the inverse Fourier transform of (7.61), which leads to

$$v_{out}(t) = \frac{\sin(\pi Bt)}{\pi Bt} \frac{\cos(\pi \beta Bt)}{1 - (2B\beta t)^2} \qquad (7.63)$$

The output voltage will have the shape represented by (7.62) and (7.63) only if the filter transfer function is chosen to satisfy the equation

$$H_{filt}(\omega) = \frac{\tilde{V}_{out}(\omega) Y(\omega)}{\tilde{I}(\omega) H_{F-end}(\omega) H_{amp}(\omega)} \qquad (7.64)$$

The function given by (7.64) is shown in Figure 7.4 for $\beta = 0$. It is important to notice that the only bit in question will define the decision at the decision points $t = n/B (n = 1, 2, 3, \ldots)$, since the neighboring pulses take the zero value at these specific moments. Any other transfer function will not be so favorable from the decision perspective, since intersymbol interference might have a significant impact on the decision at any given moment. Please notice that the received current pulse shape has been affected by the processes occurring during modulation, propagation, and photodetection of the signal. The functions given by (7.62) and (7.63) are related to an ideal case. In practice, the transfer functions are calculated numerically by using common mathematical methods [13–15]. A significant signal processing might be involved in the practical realizations in order to recover the signal shape at the receiving side.

The quality of the received optical signal is often estimated by using the eye diagram, which is also shown in Figure 7.4. The eye diagram, which is obtained by summation of several bit sequences of the received signal on top of each other, resembles the eye opening. An ideal eye diagram is quite open and clean, as shown in Figure 7.4. However, it might be severely degraded due to the impact of various

7.10 Summary

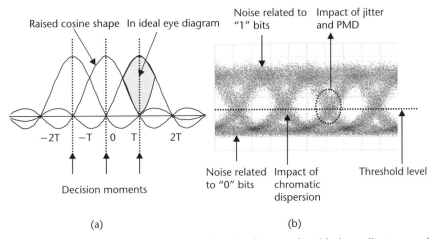

Figure 7.4 Received pulse shape: (a) the raised-cosine shape and an ideal eye diagram, and (b) the impact of different impairments to the eye diagram.

impairments. For example, chromatic dispersion and nonlinear effects will cause a partial closing of the eye. Timing jitter and polarization mode dispersion will cause jittery crossing and lack of sharpness, while the noise will have an impact on the cleanness of the diagram, as illustrated in Figure 7.4.

7.10 Summary

The information contained in this chapter should serve as a source of better understanding the physical background of effects involved in optical transmission. The reader may refer to Tables 7.1 and 7.2 for a snapshot of physical quantities and physical constants used in previous chapters. The reader is also advised to consult the references listed below whenever there is un urge to have a more comprehensive picture about any specific general topic mentioned in this chapter.

References

[1] Born, M., and E. Wolf, *Principles of Optics*, 7th ed., New York: Cambridge University Press, 1999.
[2] Yariv, A., *Quantum Electronics*, 3rd ed., New York: Wiley, 1989.
[3] Saleh, B. E., and A. M. Teich, *Fundamentals of Photonics*, New York: Wiley, 1991.
[4] Buck, J., *Fundamentals of Optical Fibers*, New York: Wiley, 1995.
[5] Agrawal, G. P., *Fiber Optic Communication Systems*, 3rd ed., New York: Wiley, 2002.
[6] Siegman, A. E., *Lasers*, Mill Valley, CA: University Science Books, 1986.
[7] Chuang, S. L., *Physics of Optoelectronic Devices*, New York: Wiley, 1995.
[8] Proakis, J. G., *Digital Communications*, 3rd ed., New York: McGraw-Hill, 1995.
[9] Shanmugam, K. S., *Digital and Analog Communication Systems*, New York: Wiley, 1979.
[10] Cvijetic, M., *Coherent and Nonlinear Lightwave Communications*, Norwood, MA: Artech House, 1996.
[11] Personic, S. D., *Optical Fiber Transmission Systems*, New York: Plenum, 1981.

[12] Gower, J., *Optical Communication Systems*, 2nd ed., Upper Saddle River, NJ: Prentice Hall, 1993.

[13] Papoulis, A., *Probability, Random Variables and Stochastic Processes*, New York: McGraw-Hill, 1984.

[14] Abramovitz, M., and I. A. Stegun, *Handbook of Mathematical Functions*, New York: Dover, 1970.

[15] Korn, G., and T. Korn, *Mathematical Handbook for Scientists and Engineers*, New York: McGraw-Hill, 1960.

List of Acronyms

ADM	add-drop multiplexer
AM	amplitude modulation
ANSI	American National Standard Institute
APD	avalanche photodiode
AGC	automatic gain control
AOTF	acousto-optic tunable filters
ASE	amplified spontaneous emission
ASK	amplitude shift keying
ATM	Asynchronous Transfer Mode
AWG	array waveguide grating
BER	bit error rate
CATV	cable television
CDMA	code-division multiple access
CSMA/CD	carrier sense medium access/collision detection
CWDM	coarse wavelength division multiplexing
CW	continuous wave
DBR	distributed Bragg reflector
dc	direct current
DCF	dispersion compensating fiber
DCM	dispersion compensation module
DGD	differential group delay
DFB	distributed feedback
DSF	dispersion shifted fiber
DWDM	dense wavelength division multiplex
DXC	digital cross-connect
EA	electroabsorption
EDFA	erbium doped fiber amplifier
E/O	electro-optical
ER	extinction ratio
FBG	fiber Bragg grating
FDDI	fiber distributed data interface
FEC	forward error correction
FM	frequency modulation

FP	Fabry-Perot
FSK	frequency shift keying
FSR	free spectral range
FWHM	full width at half maximum
FWEM	full width at 1/e intensity point
FWM	four-wave mixing
GVD	group velocity dispersion
GRIN	graded index
IEEE	Institute of Electrical and Electronics Engineers
ILA	in-line amplifier
IM	intensity modulation
InP	indium phosphide
IP	Internet Protocol
ISI	intersymbol interference
ITU-T	International Telecommunication Union Telecommunications
IXC	interexchange carrier
LAN	local area network
laser	light amplification of stimulated emission of radiation
LD	laser diode
LEC	local exchange carrier
LED	light emitting diode
LTE	line terminal
LR	long reach
MAC	media access control
MCVD	modified chemical vapor deposition
MEMS	microelectro mechanical switch
MML	multimode lasers
MPLS	multiprotocol label switching
MQW	multiple quantum well
MSR	mode suppression ratio
MZ	Mach-Zehnder
NE	network element
NRZ	nonreturn to zero
NZDSF	nonzero dispersion shifted fiber
OADM	optical add drop multiplexer
OAM	operation administration and maintenance
OAMP	optical amplifier
OCDMA	optical CDMA
OC-n	optical carrier (nth level)
ODMUX	optical demultiplexer
O/E	optoelectrical conversion

O-E-O	opto-electrical-optical conversion
OLT	optical line terminal
OMS	optical multiplex section
OMUX	optical multiplexer
OOK	on-off keying
OSA	optical spectrum analyzer
OSNR	optical signal-to-noise ratio
OTDM	optical time division multiplex
OTS	optical transmission section
OVD	outer vapor deposition
OXC	optical cross-connect
PBS	polarization beam splitter
pdf	probability density function
PDL	polarization-dependent loss
PIN	(P) layer, (I) intrinsic layer, (N) layer photodiode
PLC	planar lightwave circuit
PLL	phase locked loop
PM	phase modulation
PMD	polarization mode dispersion
PoM	polarization modulation
PSK	phase shift keying
PSP	principal states of polarization
PoSK	polarization shift keying
QoS	quality of service
RIN	relative intensity noise
RF	radio frequency
rms	root mean square
RZ	return to zero
SAN	storage area network
SBS	stimulated Brillouin scattering
SCM	subcarrier multiplexing
SDH	synchronous digital hierarchy
SI	System International
SMF	single-mode fiber
SML	single-mode lasers
SNR	signal-to-noise ratio
SOA	semiconductor optical amplifier
SONET	Synchronous Optical Network
SPM	self-phase modulation
SR	short reach
SRS	stimulated Raman scattering

STM-n	Synchronous Transfer Mode (nth level)
TCC	turbo convolutional codes
TDM	time division multiplex
TMN	telecommunications network management
TOF	tunable optical filter
UDWDM	ultra dense wavelength division multiplex
VCSEL	vertical cavity surface emitting lasers
VAD	vapor axial deposition
VOA	variable optical attenuator
VP	virtual path
VoD	video on demand
WAN	wide area network
WDM	wavelength division multiplex
XGM	cross-gain modulation
XPM	cross-phase modulation

About the Author

Milorad Cvijetic received a Ph.D. in electrical engineering from the University of Belgrade in 1984. Dr. Cvijetic has experience in both academia (teaching at the University of Belgrade and Carleton University) and industry (R&D work in the area of high-speed optical transmission systems and optical networks). His research work, related to quasi-single-mode optical fibers, BER evaluation in soliton-based systems, and system performance evaluation in high-speed optical systems with external modulation, has been widely recognized. He currently serves as the chief technology strategist for optical network products with NEC America in Herndon, Virginia. Previously, he has been with Bell Northern Research (later NORTEL Technologies) in Ottawa, Canada, working in the Advanced Technology Laboratory.

Dr. Cvijetic has published more than 40 technical papers and two books, *Digital Optical Communications* and *Coherent and Nonlinear Lightwave Communications* (Artech House, 1996). He has taken part in numerous telecommunications conferences and symposia as a session/conference chairman, technical committee member, or invited speaker. He is a member of the IEEE Communications Society and LEOS.

Index

A

Accumulation effects, 172–75
Acousto-optic tunable filters (AOTF), 58
Amplified spontaneous emission (ASE) noise, 16, 42
    defined, 72
    EDFAs, 223
    power, 173, 198
    total noise current, 83
    See also Noise
Amplitude modulation (AM), 169
Amplitude shift keying (ASK), 230
Arrayed waveguide gratings (AWGs), 59
Attenuation
    curve, 33
    fiber, 86–87
    impact, 128–29
    summary, 171
Avalanche photodiodes (APDs), 17, 45–47
    gain, 162
    illustrated, 46
    parameters, 47
    process, 45
    structure, 46
    See also Photodiodes
Avalanche shot noise, 72

B

Backward pumping, 196
Bandwidth, 6–7
    building blocks, 11
    limited lightwave systems, 179–82
    of multimode optical fibers, 180
    single-mode optical fibers, 181
    system length limitations due to, 181
Bandwidth pipe, 8
Beat noise, 86, 171
Bit error rate (BER), 4, 6
    achieving specified, 153
    false decision and, 151
    functional dependence, 153
    as function of Q-factor, 153
Bit rates, 9
Booster amplifiers, 39–40

C

Carrier Sense Medium Access/Collision Detection (CSMA/CD), 10
Chirped Gaussian pulse
    broadening factor, 100
    illustrated, 99
    spectrum, 97
Chromatic dispersion, 31, 89–103
    cause, 89
    characteristics, 94
    coefficient, 92
    compensation, 205
    components, 93
    of different fiber types, 94
    differential dispersion parameter, 96
    impact, 90, 132–36
    impact analysis, 101
    management, 247–48
    penalty, 134
    PMD impact comparison, 138–39
    power penalty due to, 133, 134, 135
    pulse broadening, 91
    reduction, 93
    summary, 171
    trade-offs, 247–48
Chromatic PMD, 105
Coherent detection schemes, 230–33
Coherent lightwave system, 230, 231
Computer-based modeling, 189–93
    decision circuit modeling, 193
    process illustration, 191
    pulse shaping, 192–93
    random bit pattern generation, 190–91
    signal generation process modeling, 191
    solving wave equations, 191–92
    steps, 190–93
    uses, 193

Connector return loss, 87
Course-WDM (CWDM), 3
Cross-gain modulation (XGM), 240
Cross-phase modulation (XPM), 5, 115–16
  defined, 115
  effect estimation, 115, 116
  effect reduction, 116
  frequency shift, 116
  impact, 145–47
  phase shift due to, 146
  pulse overlapping, 116
  reduction, 146
  summary, 171
  See also Signal impairments
Crosstalk noise, 72, 84–86
  defined, 72
  during photodetection process, 143
  impact, 188
  inband, 84, 85, 86
  out-of-band, 84, 85, 141
  power penalty due to, 167–69
  summary, 171
  See also Noise
Crosstalk ratio, 168
Cyclic-redundancy check (CRC), 189

D

Dark current noise, 72, 79
  defined, 72
  generation, 79
  spectral density, 130
  See also Noise
DCFs
  compensation with, 210–13
  design, 210
  dual-mode, 213
  length of, 210
  nonideal matching, 212
  use of, 211
Dense-WDM (DWDM), 3
Detection schemes, 230–33
  coherent, 230–33
  realization, 233
  SNR, 232
Differential group delay (DGD), 103, 104
Differential PSK (DPSK), 231
Direct optical modulation, 24–25
  efficiency, 24
  extinction ratio, 25
  frequency chirp, 24–25
  initiation, 24

Dispersion compensating modules (DCM), 139, 217
Dispersion compensation
  advanced, 204–20
  chromatic, 205
  defined, 204–5
  FBG, 213–15
  with imperfect slope matching, 212
  in-line, 205, 209–20
  optical fiber pairing, 245
  optical filters, 216–17
  phase conjugation, 215–16
  PMD, 205, 218–20
  postcompensation schemes, 217–18
  postdetection, 205, 209
  predistortion, 205–9
  total, 247
  tunable, 217–18
Dispersion maps
  defined, 213
  periodic, 228
Dispersion shifted fibers (DSF), 93
Dispersion tolerance, 136
Distributed amplification, 227
Distributed Bragg reflector (DBR) lasers, 19, 20
Distributed feedback (DFB) lasers, 19, 20
Double Rayleigh backscattering (DRB) noise, 198
  generation, 199
  power calculation, 199
Duobinary coding, 221

E

Eigenstates, 103
Electroabsorption (EA) modulator, 25, 27–28
  defined, 27
  illustrated, 26
  modulation curve, 27
  operation, 27
  output optical power, 28
  See also Optical modulators
Electromagnetic field specification, 254
Enabling technologies, 195–243
  defined, 195
  detection schemes, 230–33
  dispersion compensation, 204–20
  FEC, 233–37
  modulation schemes, 220–30
  optical amplifiers, 196–204
  wavelength conversion, 237–43
Enterprise Serial Connection (ESCON), 10

Erbium doped fiber amplifiers (EDFAs), 16, 41–42
    ASE noise, 223
    energy diagram, 42
    gain spectrum, 42
    illustrated, 41
    noise figure, 201, 202
    with Raman pumping scheme, 201
    uses, 41
External optical modulation, 25–28
    EA, 25, 27
    MZ, 25–26
Extinction ratio, 67
    defined, 66
    expression, 67
    nonideal, power penalty due to, 163
    power penalty due to, 162–63
    summary, 171
Extrinsic absorption, 34

F

Fabry-Perot (FP) lasers, 18–20
Fabry-Perot interferometers, 54–56
    cascade approaches, 56
    defined, 54–55
    illustrated, 55
    See also Optical filters
Fermi energy level, 265
Fiber attenuation, 86–87
    coefficient, 87
    defined, 86
Fiber-based optical amplifiers, 40–41
Fiber Bragg gratings (FBG), 56–57
    application for dispersion compensation, 214
    dispersion compensation, 213–15
    ripples, 215
    wavelength, 214
Fiber Channel, 10
Fiber distributed data interface (FDDI), 10
Fiber loss coefficient, 128
Fixed multiplexing, 8–9
Forward error correction (FEC), 10, 233–37
    benefits, 233
    codes, 233, 234
    concatenated RS, 236
    decoding process, 235
    defined, 234
    inband, 235
    margin allocation with, 238
    out-of-band, 235
    process, 236

    See also Enabling technologies
Four-wave mixing (FWM), 5, 116–19
    channel spacing in, 143
    degenerate case, 117
    effect, 118
    impact, 141
    nondegenerate case, 117
    power penalty, 142, 143
    process, 117, 118
    reduction, 143
    signal degradation, 118
    summary, 171
    wavelength conversion, 243
    See also Signal impairments
Frequency chirp, 87–89
    adiabatic, 89
    due to SPM, 114
    effect, 88
    summary, 89
    total, 115
    transient, 89
    See also Signal impairments
Frequency division multiplex (FDM), 11
Frequency modulation (FM), 269
Frequency shift keying (FSK), 230
Full-width half maximum (FWHM), 18
    of Lorentzian spectral line, 20
    of sech-function shape, 224

G

Generalized multiprotocol label switching (GMPLS), 11
GRIN rod, 59
Group velocity dispersion (GVD) coefficient, 95, 118, 224

H

Histogram technique, 188

I

Impact
    attenuation, 128–29
    crosstalk noise, 188
    FWM, 141
    modal dispersion, 132–36
    noise, 129–32
    nonlinear effects, 140–47
    PMD, 136–40
    SRS, 141
    timing jitter, 166, 167
    XPM, 145–47

Inband crosstalk, 84, 85, 86
  beat noise component, 86
  illustrated, 85
  occurrence, 85
Indirect bandgap semiconductors, 266
In-line compensation, 205, 209–20
  DCF, 210–13
  defined, 205
  modules, 209
  schemes, 210
  See also Dispersion compensation
In-line optical amplifiers, 40
Insertion losses, 87
Intensity modulation and direct detection (IM/DD), 232
Intensity noise
  defined, 163
  power, 164
  power penalty due to, 163–66
  receiver noise degradation due to, 164
  reflection induced by, 165
  summary, 171
Interchange carriers (IXCs), 2
Intermodal dispersion effect, 90
Intersymbol interference (ISI), 270
  defined, 91
  illustrated, 91
Intramodal dispersion. *See* Chromatic dispersion
Intrinsic absorption, 33–34

J
Johnson noise, 80

L
Langevin forces, 268
Laser intensity noise, 72, 74–75
  back-reflected light, 75
  defined, 72
  estimating, 74
  reflection-induced, 74–75
Laser phase noise, 72, 74–75
  defined, 72
  spectral bandwidth, 75
Laser rate equations, 267–68
  expression, 267
  output radiation phase, 268
Light
  absorption, 260
  emission, 260
  spontaneous emission, 262
  stimulated emission, 260–62, 267
Light emitting diodes (LEDs), 17–18
  defined, 17
  illustrated, 18
  output power, 17
  See also Semiconductor light sources
Light propagation, 254
Lightwave paths
  definition of, 8
  illustrated, 8
  optical component position along, 16
Linear filtering method, 209
Linewidth enhancement, 88
Link power budget, 176
Local area networks (LANs), 2
Local exchange carriers (LECs), 2
Long reach (LR) systems, 4, 176
Lumped amplification, 226

M
Mach-Zehnder (MZ) modulator, 25–26
  defined, 25
  illustrated, 26
  modulation curve, 26, 27
  modulation principle, 26
  See also Optical modulator
Maxwell's vectors equations, 255
Metropolitan area networks (MANs), 2
Micro electromechanical switches (MEMS), 52
Minimum shift keying (MSK), 231
Modal noise, 72, 75–76
  defined, 72
  speckle pattern fluctuations, 76
  See also Noise
Mode partition noise, 72, 73–74
  defined, 72
  MSR, 74
  multimode semiconductor lasers, 73
  See also Noise
Mode suppression ratio (MSR), 74
Modified chemical vapor deposition (MCVD), 36–37
  illustrated, 37
  process, 37
Modulation schemes, 220–30
  duobinary coding, 221
  OOK, 220–22
Monte-Carlo method, 192
Multifiber optical cables, 38
Multimode dispersion, 31
  defined, 31
  single-mode fibers and, 89

Multimode lasers (MMLs), 15
Multimode optical fibers, 29–31
    bandwidth, 180
    defined, 29
    graded-index, 180
    illustrated, 29
    use of, 90
    See also Optical fibers
Multiquantum-well (MQW) laser design, 21
MZ interferometers, 53–54
    defined, 53
    elementary structures, 54
    illustrated, 54
    See also Optical filters

N

Noise, 6, 70–86
    accumulation, 173
    ASE, 42, 72, 83
    avalanche shot, 72
    beat components, 83–84
    crosstalk, 72, 84–86
    dark current, 72, 79
    DRB, 198, 199
    impact, 129–32
    Johnson, 80
    laser intensity, 72, 74–75
    laser phase, 72, 74–75
    modal, 72, 75–76
    mode partition, 72, 73–74
    origination, 71
    parameters, 70–86
    quantum shot, 72, 76–79
    receiving side components, 71
    relative intensity (RIN), 70
    spontaneous emission, 81–83
    thermal, 72, 79–81
    total, 72
    total, power, 129–30
Nonlinear effects, 109, 110
    classification of, 109
    impact, 140–47
    inversely proportional, 111
    SBS, 5, 122–25, 140
    SRS, 119–22
    XPM, 5, 115–16, 145–47
Nonlinear Kerr coefficient, 223, 258, 259
Nonlinear Schrodinger equation (NSE), 228, 259, 260
    defined, 259
    solving, 260

Nonzero dispersion shifted fiber (NZDSF), 33, 94
NRZ coding, 220–21

O

On-off keying (OOK) modulation, 220–22, 269
    formats illustration, 220
    NRZ coding, 220
    with polarization states manipulation, 222
    RZ coding, 221
Operation, administration, and maintenance (OA&M), 10
Optical 3R scheme, 242
Optical add-drop multiplexers (OADMs), 203
Optical amplifier gain, 67–69
    defined, 66
    factor, 68
    as function of output power, 69
Optical amplifiers, 16–17, 39–45
    applications, 39–40
    booster, 39–40
    as enabling technology, 196–204
    erbium doped fiber (EDFA), 16, 41–42, 201–2, 223
    fiber-based, 40–41
    in-line, 40
    output power, 173
    parameter values, 44
    preamplifiers, 40
    pure, 40
    Raman, 16, 196–204
    semiconductor (SOAs), 16, 40, 208, 240–41
Optical cables, 38–39
    multifiber, 38
    outer sheath, 39
    single-fiber, 38
Optical circulators, 51–52
Optical components, 47–61
    active, 48
    circulators, 51–52
    couplers, 48–50
    defined, 17
    demultiplexers, 59–61
    filters, 53–58
    isolators, 50
    multiplexers, 59–61
    passive, 48
    position along lightwave path, 16
    switches, 52–53
    types of, 17
    VOAs, 50–51

Optical couplers, 48–50
  defined, 48
  directional, 48–49
  fused tapered, 48
  illustrated, 49
  N x M, 49–50
  See also Optical components
Optical fiber gratings, 56–57
  defined, 56
  FBG, 56–57
Optical fibers, 6, 16, 28–39
  attenuation, 86–87
  attenuation curve, 33
  bandwidth, 6–7
  bending, 35
  cross-sectional area in, 11, 110
  defined, 16
  dispersion compensation pairing, 245
  drawing process, 36
  fundamental modes, 30
  illustrated, 29
  manufacturing and cabling, 35–39
  material imperfections, 34
  with multilayer cladding, 32
  multimode, 29–31
  single-mode, 29, 31–35
  type selection, 244–45
  types of, 29
Optical filters, 53–58
  for dispersion compensation, 216–17
  fiber gratings, 56–57
  FP interferometer, 54–56
  MZ interferometer, 53–54
  tunable filters, 57–58
  See also Optical components
Optical isolators, 50
  defined, 50
  operational principle, 51
  parameters, 51
Optical modulators, 15–16, 24–28
  direct, 24–25
  external, 25–28
  parameter values, 28
Optical multiplexers, 59–60
  AWGs, 59, 60
  FP filter-based, 61
  illustrated, 60
  MZ filter chain, 60
  thin-film filter design, 61
  types of, 59
Optical networking, 1–3
Optical path length, 249–50

Optical power
  defined, 6
  level, 248–49
Optical preamplifiers, 40
  receiver sensitivity defined by, 158–59
  use of, 40
Optical pulses
  chirped, 97, 99
  unchirped, 99
Optical receivers
  degradation, power penalty and, 161
  degradation due to intensity noise, 164
  sensitivity, 153, 154–60
  transfer function, 269–71
Optical signal parameters, 65–70
  extinction ratio, 66, 67
  optical amplifier gain, 66, 67–69
  origin, 66
  output signal power, 66–67
  photodiode responsivity, 66, 69–70
Optical signals
  equalization, 269–71
  frequency, 260
  modulation of, 269
  quality, 270
  wavelength, 260
Optical signal-to-noise ratio (OSNR), 159–60
  calculated per channel, 174
  defined, 159
  gradually decreasing, 173
  measurement, 159
  Q-factor relationship, 160, 174
  for specific bit rate and Q parameter, 185
  specified, 174
  value, 159–60, 174
Optical spectrum analysis, 187
Optical switches, 52–53
  defined, 52
  electro-optical, 52
  SOA-based, 53
  thermo-optic, 52–53
  types of, 52
  See also Optical components
Optical transmission, 1–3
  enabling technologies, 195–243
  high-speed systems, 182–86
  limitations and penalties, 127–47
  long reach (LR), 4, 176
  network elements, 4
  noise origination, 71
  parameters related to, 5
  quality, 149–54

system definition, 3–11
system performance, 179
systems engineering, 7
systems engineering toolbox, 253–71
ultralong reach (ULR), 4, 177
very-short reach (VSR), 4, 176
viability, 176
Opto-electrical-opto (O-E-O) conversion, 7, 238, 239
Organization, this book, 11–12
Outer vapor deposition (OVD), 36
Out-of-band crosstalk noise, 84, 85
   measurement, 141
   power penalty due to, 169
   See also Crosstalk noise
Output saturation power, 68–69
Output signal power, 66–67
   defined, 66
   total, 67

P

Performance
   evaluation techniques, 189
   in-service monitoring, 189
   monitoring, 187–89
   optical transmission system, 179
Phase conjugation, 215–16
   defined, 215
   for dispersion compensation, 215–16
   for SPM effect compensation, 216
   use of, 215
Phase modulation (PM), 269
Phase shift keying (PSK), 230
Photodiode responsivity, 69–70
   defined, 66
   expression, 69
   increasing, 70
Photodiodes, 45–47
   avalanche (APDs), 17, 45–47
   defined, 17
   noise generation, 80
   noise parameters, 79
   parameters, 47
   PIN, 45–47
   structure, 46
Physical constants, 254
Physical quantities, 253
Pilot tone technique, 187
PIN photodiodes, 45–47
   illustrated, 46
   parameters, 47
   responsivity curves, 70
   structure, 45
   See also Photodiodes
Planar lightwave circuits (PLCs), 217
PMD compensation, 205, 218–20
   application, 219
   defined, 218
   by electrical methods, 219
   equalization, 218
   by optical methods, 219
   See also Dispersion compensation
Polarization-dependent losses (PDL), 5
Polarization mode dispersion (PMD), 5, 36, 103–8
   chromatic, 105
   chromatic dispersion impact comparison, 138–39
   complexity, 108
   defined, 104
   deformations causing, 103
   equalization, 218
   first-order, 104, 105, 106, 107
   first-order, accumulated average, 138
   impact, 136–40
   mitigation, 108
   power penalty due to, 137, 138
   pulse broadening due to, 136
   second-order, 104, 106, 107
   summary, 108, 171
   See also PMD compensation; Signal impairments
Postcompensation schemes, 217–18
Postdetection dispersion compensation, 205, 209
   defined, 205
   linear filtering, 209
   transversal filters, 209
   See also Dispersion compensation
Power budget, 176
   for 200 Mbps transmission, 178
   equation, 182
   expression, 177
   limited lightwave system, 177–79
   for short-reach high-speed transmission, 179
Power level, 248–49
   defined, 248
   enhancing, 248
   per channel, 249
   SNR and, 248
Power margin
   allocation, 175, 179

Power margin (continued)
  allocation for high-speed transmission systems, 184
  allocation illustration, 186
  assignment, 183, 185
  benefit, 184
  defined, 178
  positive, impact, 202
Power penalty, 127
  comparative review, 169–72
  due to chromatic dispersion, 133, 134, 135
  due to extinction ratio, 162–63
  due to FWM, 142, 143
  due to intensity noise, 163–66
  due to nonideal extinction ratio, 163
  due to out-of-band crosstalk, 169
  due to PMD, 137, 138
  due to pulse shape deformation, 137
  due to signal crosstalk, 167–69
  due to SRS, 144, 145
  due to timing jitter, 166–67
  handling, 160–75
  pulse deformations causing, 128
  receiver degradation and, 161
  wavelength conversion, 241
Predistortion dispersion compensation, 205–9
  chirp parameter, 207, 208
  defined, 205
  pulse prechirping, 206–8
  See also Dispersion compensation
Probability density functions, 150
Propagation equation, 257–60
Pulse broadening, 91, 95, 100
  amount of, 101
  determination, 102
  due to PMD, 136
  parameter, 102
  SPM-induced, 115
Pulse dispersion, 6
Pulse envelope
  behavior, 258
  defined, 257
Pulse prechirping, 206–8
Pulse shaping, 192–93

Q

Q-factor, 156
  BER as function of, 153
  defined, 152
  OSNR relationship, 160, 174
  penalties, 188
  SNR relationship, 152

Quality of service (QoS) requirements, 3
Quantum shot noise, 72, 76–79
  defined, 72
  excess noise factor, 78
  factor, 78
  instantaneous current, 77
  photocurrent mean intensity, 77
  spectral density, 78

R

Raman amplifiers, 16, 43–44, 196–204
  advantage, 44
  amplification coefficient, 197
  ASE noise, 198
  backward pumping, 196
  cost-effective design, 43
  deployment, 200, 201
  DRB noise, 198, 199
  gain, 198
  illustrated, 44
  pumping scheme, 204
  schematics, 197
Receiver sensitivity, 153, 154–60
  defined by optical preamplifier, 158–59
  defined by shot noise and thermal noise, 155–58
  degradation and power penalty, 161
  evaluation, 158
  evaluation through Q-factor, 162
  illustrated, 156
  impairments degrading, 153
  OSNR and, 159–60
Refractive index, 112, 113
Relative intensity noise (RIN), 70
RZ coding, 221

S

Scattered Stokes photons, 120, 124
Scattering losses, 34
Self-phase modulation (SPM), 5, 108–15
  frequency chirping due to, 114
  impact on transmission system characteristics, 114
  pulse broadening, 115
  summary, 171
  See also Signal impairments
Semiconductor lasers, 18–22
  DBR, 19, 20
  DFB, 19, 20
  Fabry-Perot (FP), 18–20
  high-power, 43

illustrated, 19
pump, 43
VCSEL, 20
Semiconductor light sources, 17–23
    defined, 15
    LEDs, 17–18
    semiconductor lasers, 18–22
    wavelength selectable lasers, 22–23
Semiconductor optical amplifiers (SOAs), 16
    gain saturation, 208
    in saturation regime, 208
    wavelength conversion, 240, 241
Semiconductors, 262–67
    bias voltage, 265
    conduction capability, 263
    energy bands, 266
    indirect bandgap, 266
    as materials, 262–67
    population inversion, 266, 267
    stimulated emission, 267
Shot noise
    avalanche, 72
    quantum, 72, 76–79
    receiver sensitivity defined by, 155–58
    spectral density, 130
    summary, 171
Signal impairments, 86–125
    chromatic dispersion, 89–103
    fiber attenuation, 86–87
    frequency chirp, 87–89
    FWM, 116–19
    insertion losses, 87
    PMD, 103–8
    SBS, 122–25
    SPM, 108–15
    SRS, 119–22
    XPM, 115–16
Signal-to-noise ratio (SNR), 6, 65
    coherent detection, 232
    degradation, 16
    optical (OSNR), 159–60
    power level and, 248
    Q-factor relationship, 152
    Raman gain and, 202
Single-fiber optical cables, 38
Single-mode lasers (SMLs), 15
Single-mode optical fibers, 31–35
    bandwidth, 181
    characteristics, 112
    chromatic dispersion, 31
    extrinsic absorption, 34
    illustrated, 29

intrinsic absorption, 33–34
multimode dispersion and, 89
propagation equation, 257–60
refractive index profiles, 32
step-index profile, 32
See also Optical fibers
Soliton pulses, 224, 225
    amplification of, 228
    generation, 226
    illustrated, 225
    timing jitter and, 229
    total energy carried by, 225
Soliton regime, 227, 230
SONET/SDH, 10
Spectral efficiency, 245–46
    increasing, 245–46
    multilevel coding and, 246
    system engineering and, 246
    transmission capacity and, 245
Spontaneous emission noise, 81–83
    spectral density, 81
    total power, 82
    See also Noise
Statistical multiplexing, 8, 9
Stimulated Brillouin scattering (SBS), 5,
        122–25, 140
    associated energy levels, 123
    defined, 122
    frequency shift, 124
    for optical signal amplification, 125
    penalty, 140
    process, 123, 124
    summary, 171
    See also Signal impairments
Stimulated emission, 260–62, 267
Stimulated Raman scattering (SRS), 5, 119–22
    bit-pattern dependence, 123
    broadband nature, 144
    defined, 119
    gain spectrum, 121
    illustrated, 122
    impact, 141
    for optical signal amplification, 121, 122
    power penalty, 144, 145
    process, 119
    summary, 171
    threshold power, 121
    See also Signal impairments
Storage area networks (SANs), 10
Synchronous Digital Hierarchy (SDH), 9
Synchronous Optical Network (SONET), 9
Systems engineering, 175–93

Systems engineering (continued)
  of bandwidth-limited systems, 179–82
  computer-based modeling and, 189–93
  of high-speed systems, 182–86
  performance monitoring, 187–89
  of power-budget limited systems, 177–79
  process, 176
  spectral efficiency and, 246
  toolbox, 253–71
  transmission trade-offs, 243–50

T

Thermal equilibrium, 260, 261
Thermal noise, 72, 79–81
  defined, 72
  figure, 81
  receiver sensitivity defined by, 155–58
  reduction, 80
  spectral density, 80, 130
  summary, 171
  See also Noise
Thermo-optic switches, 52–53
Time division multiplexing (TDM), 8, 9
Timing jitter
  defined, 166
  impact of, 166, 167
  power penalty due to, 166–67
  soliton frequency change and, 229
  in soliton transmission, 229
Trade-offs, 243–50
  chromatic dispersion management, 247–48
  optical fiber type, 244–45
  optical path length, 249–50
  optical power level, 248–49
  spectral efficiency, 245–46
Transmission quality
  definition, 149–54
  measurement, 149
Transversal filters, 209
Traveling wave (TW), 40
Tunable dispersion compensation, 217–18
Tunable lasers, 23
Tunable optical filters, 57–58
  acousto-optic, 58
  defined, 57

FP, 57
MZ, 57
See also Optical filters
Turbo convolutional codes (TCC), 236–37
  defined, 236
  illustrated, 237
  modification, 236

U

Ultra-dense WDM (UDWDM), 3
Ultra-long reach (ULR) systems, 4, 177
Units
  physical constants and, 254
  physical quantities and, 253

V

Vapor axial deposition (VAD), 36
Variable optical attenuators (VOAs), 50–51
Vertical cavity surface emitting laser (VCSEL), 20
Very-short reach (VSR) systems, 4, 176

W

Walk-off phenomenon, 122
Wave equation, 257
Waveguide dispersion, 92, 93
Wavelength conversion, 237–43
  application area, 239
  FWM-based, 243
  O-E-O, 7, 238, 239
  power penalty, 241
  signal regeneration and, 238
  SOA-based, 240, 241
  uses, 237–38
Wavelength division multiplex (WDM), 3–4
  course (CWDM), 3
  dense (DWDM), 3
  soliton-based system, 230
  three-band transmission experiment, 204
  ultra-dense (UDWDM), 3
Wavelength selectable lasers, 22–23
  design methods, 22
  versions, 23
Wide area networks (WANs), 2

Recent Titles in the Artech House Optoelectronics Library

Brian Culshaw and Alan Rogers, Series Editors

Chemical and Biochemical Sensing with Optical Fibers and Waveguides, Gilbert Boisdé and Alan Harmer

Coherent and Nonlinear Lightwave Communications, Milorad Cvijetic

Coherent Lightwave Communication Systems, Shiro Ryu

DWDM Fundamentals, Components, and Applications, Jean-Pierre Laude

Frequency Stabilization of Semiconductor Laser Diodes, Tetsuhiko Ikegami, Shoichi Sudo, and Yoshihisa Sakai

Fiber Bragg Gratings: Fundamentals and Applications in Telecommunications and Sensing, Andrea Othonos and Kyriacos Kalli

Handbook of Distributed Feedback Laser Diodes, Geert Morthier and Patrick Vankwikelberge

Helmet-Mounted Displays and Sights, Mordekhai Velger

Introduction to Infrared and Electro-Optical Systems, Ronald G. Driggers, Paul Cox, and Timothy Edwards

Introduction to Lightwave Communication Systems, Rajappa Papannareddy

Introduction to Semiconductor Integrated Optics, Hans P. Zappe

LC3D: Liquid Crystal 3-D Director Simulator, Software and Technology Guide, James E. Anderson, Philip E. Watson, and Philip J. Bos

Liquid Crystal Devices: Physics and Applications, Vladimir G. Chigrinov

Optical Document Security, Second Edition, Rudolf L. van Renesse, editor

Optical FDM Network Technologies, Kiyoshi Nosu

Optical Fiber Amplifiers: Materials, Devices, and Applications,
Shoichi Sudo, editor

Optical Fiber Communication Systems, Leonid Kazovsky,
Sergio Benedetto, and Alan Willner

Optical Fiber Sensors, Volume Three: Components and Subsystems,
John Dakin and Brian Culshaw, editors

Optical Fiber Sensors, Volume Four: Applications, Analysis, and
Future Trends, John Dakin and Brian Culshaw, editors

Optical Measurement Techniques and Applications, Pramod Rastogi

Optical Transmission Systems Engineering, Milorad Cvijetic

*Optoelectronic Techniques for Microwave and Millimeter-Wave
Engineering,* William M. Robertson

Reliability and Degradation of III-V Optical Devices, Osamu Ueda

Smart Structures and Materials, Brian Culshaw

*Surveillance and Reconnaissance Imaging Systems: Modeling and
Performance Prediction,* Jon C. Leachtenauer and
Ronald G. Driggers

Understanding Optical Fiber Communications, Alan Rogers

Wavelength Division Multiple Access Optical Networks,
Andrea Borella, Giovanni Cancellieri, and Franco Chiaraluce

For further information on these and other Artech House titles,
including previously considered out-of-print books now available
through our In-Print-Forever® (IPF®) program, contact:

Artech House	Artech House
685 Canton Street	46 Gillingham Street
Norwood, MA 02062	London SW1V 1AH UK
Phone: 781-769-9750	Phone: +44 (0)20 7596-8750
Fax: 781-769-6334	Fax: +44 (0)20 7630-0166
e-mail: artech@artechhouse.com	e-mail: artech-uk@artechhouse.com

Find us on the World Wide Web at:
www.artechhouse.com